The Cattle of the Sun

The Cattle of the Sun

COWS AND CULTURE IN THE WORLD OF THE ANCIENT GREEKS

Jeremy McInerney

PRINCETON UNIVERSITY PRESS
PRINCETON AND OXFORD

Published by Princeton University Press, 41 William Street,
Princeton, New Jersey 08540
In the United Kingdom: Princeton University Press, 6 Oxford Street,
Woodstock, Oxfordshire OX20 1TW

Library of Congress Cataloging-in-Publication Data

McInerney, Jeremy, 1958–
The cattle of the sun : cows and culture in the world of the ancient Greeks / Jeremy McInerney.
p. cm.
Includes bibliographical references and index.
ISBN 978-0-691-14007-0 (hardcover : alk. paper) 1. Cattle—Greece—History. 2. Pastoral systems—Greece—History. 3. Cattle—Greece—Religious aspects—History. 4. Animal sacrifice—Greece—History. 5. Fasts and feasts—Greece—History. 6. Cattle trade—Greece—History. 7. Greece—History—To 146 B.C 8. Greece—Religious life and customs. 9. Greece—Economic conditions—To 146 B.C. 10. National characteristics, Greek—History. I. Title.
SF196.G8M38 2010
636.20938—dc22 2009039945

British Library Cataloging-in-Publication Data is available

This book has been composed in Sabon

Printed on acid-free paper. ∞

press.princeton.edu

Printed in the United States of America

10 9 8 7 6 5 4 3 2 1

To Maud

Contents

Illustrations

FIGURES

TABLE

Acknowledgments

> "Historians you think," said Miss Tilney, "are not happy in their flights of fancy. They display imagination without raising interest."
>
> —Jane Austen, *Northanger Abbey*

Miss Tilney's words apply, in part, to this study, since it is a product of the author's imagination. Whether it raises interest will be for the reader to decide. The flight of fancy that led to my writing this book began twenty years ago, when I had the great good fortune to be a student at the American School of Classical Studies at Athens. On a trip to Marathon, one of my fellow students (I think it was Kevin Glowacki) observed a single, scrawny cow not far from the ancient battlefield and wryly commented that we had finally seen the famous bull of Marathon. Typically for Kevin and the rest of that cohort of truly exceptional classicists and archaeologists with whom I was privileged to spend a year, what began as a joke soon prompted a more serious discussion. How much meat did the Greeks eat? Where did they run their herds? How did they guarantee steady supplies of cattle for the sacrifices, especially the hecatombs, that figured so prominently in their religious life? Over time these empirical questions stimulated other, more figurative lines of inquiry. Why are so many Greek myths about cows: Zeus appearing to Europa as a bull, or Io punished by sprouting horns and being driven mad by a gadfly, a torment sent by a goddess frequently called by Homer "cow-eyed Hera"? This book is an attempt to answer some of these questions—both the straightforward issues of stock raising and those that arise from what we might call the Greek *imaginaire*, the world of ideas, values, and attitudes that seems to exist separately from the mundane details of daily life. As I hope to make clear, the two zones are not, in fact, separate at all, but are recursively linked.

Before this flight of fancy takes off, however, I should acknowledge the help of a great many people who have commented, posed questions, supplied information, and in many other ways contributed to the way I think about cattle in the Greek world. These include various participants in the Cattle Network discussion group, as well as Carmen Alfaro Giner, Georgios Arsenos, Kim Benston, John Culhane, Denis and Philippe Gerbenne, Philip Harland, Eric Kondratieff, Jeremy Lefkowitz, Brent Shaw, Bronwyn Wikkiser, and Constanze Witt. I owe special thanks to Alex

Perkins and Natalia Bauer, who both served as very capable research assistants. I am fortunate to teach in the department of Classical Studies at the University of Pennsylvania, where my colleagues and students, both past and present, have established a model of scholarly collegiality that is both rare and precious. It is a delight and privilege to acknowledge here my debt to each of these friends. If I single out Tom Tartaron, Cam Grey, Peter Struck, Joe Farrell, Ralph Rosen, and Martin Ostwald, it is only because the entire list would be too long to include.

Most of this book was written while I was on sabbatical. I thank the School of Arts and Sciences at the University of Pennsylvania for generously subventing this precious period of reflection and writing. My sabbatical was spent in rural France, where I passed a good deal of time observing the rhythm of the rural year, especially in relation to the husbandry of Charolais cattle. My wife introduced me to village life here, for which I owe her a debt that a thousand hecatombs could not repay. My children endured a year of school in France in order to make that rural sabbatical possible, and to them too I owe special thanks.

A Note about Spellings and Translations

Spelling conventions in works dealing with the classical world are famously confusing. I choose to be consistently inconsistent, staying close to the Greek where feasible (Herodotos, not Herodotus; Theophrastos, not Theophrastus), but not when this would produce names unfamiliar to Anglophone readers, such as Thoukydides (Thucydides) or Platon (Plato).

All translations are my own unless otherwise noted.

Abbreviations

Abbreviations follow, with a few exceptions, the conventions of Liddell and Scott's *Greek-English Lexicon* (rev. ed.), *The Oxford Classical Dictionary* (3rd ed.), and *L'Année Philologique.*

AA	*Archäologischer Anzeiger*
ABSA	*Annual of the British School at Athens*
AEph	'Αρχαιλογικη Εφημερις
Agora XIX	G. V. Lalonde, M. K. Langdon, and M. B. Walbank, *Inscriptions: Horoi, Poletai, Leases of Public Land* (*The Athenian Agora* XIX). Princeton, 1991.
AJA	*American Journal of Archaeology*
AJPh	*American Journal of Philology*
AK	*Antike Kunst*
Am. J. Hum. Genet.	*American Journal of Human Genetics*
Anecd. Bekk.	I. Bekker, *Anecdota Graeca.* 3 vols. Berlin, 1814–1821.
AntCl	*L'Antiquité classique*
AR	*Archaeological Reports* (supplement to *JHS*)
ArchZootec	*Archivos de Zootecnia*
BAR	*British Archaeological Reports*
BAR-IS	*British Archaeological Reports, International Series*
BCH	*Bulletin de correspondance hellénique*
BE	*Bulletin épigraphique* (published in *RÉG*)
BICS	*Bulletin of the Institute of Classical Studies of the University of London*
BMMA	*Bulletin of the Metropolitan Museum of Art, New York*
BSOAS	*Bulletin of the School of Oriental and African Studies*
CArchJ	*Cambridge Archaeological Journal*
CID	*Corpus des inscriptions de Delphes*, 4 vols. Paris, 1977–.
CJ	*Classical Journal*
ClAnt	*Classical Antiquity*
CMP	A. Furumark, *The Chronology of Mycenaean Pottery.* Stockholm, 1941.

CPCActs	*Acts of the Copenhagen Polis Centre*
CQ	*Classical Quarterly*
CP	*Classical Philology*
CR	*Classical Review*
DK	H. Diels and W. Kranz, *Fragmente der Vorsokratiker*[6], 3 vols. Berlin, 1952.
EA	*Epigraphica Anatolica*
EHR[2]	*The Economic History Review*, 2nd series
FD	*Fouilles de Delphes*, École française d'Athènes
FGrH	F. Jacoby, *Die Fragmente der griechischen Historiker*. Berlin, 1923–30, Leiden, 1940–58.
GHI	P. J. Rhodes and R. Osborne, *Greek Historical Inscriptions, 404–323 BC*. Oxford, 2003.
G&R	*Greece and Rome*
GRBS	*Greek, Roman and Byzantine Studies*
HSCPh	*Harvard Studies in Classical Philology*
HTR	*Harvard Theological Review*
ICret	*Inscriptiones Creticae*, 4 vols. Rome, 1935–50.
IDélos	*Inscriptions de Délos*, 7 vols. Paris, 1926–.
IG	*Inscriptiones Graecae*. Berlin, 1873–.
IGLSyria	*Inscriptions grecques et latines de la Syrie*. Paris, 1929–.
IGSI	*Inscriptiones Graecae Siciliae et infimae Italiae ad ius pertinentes*. Milan, 1925; repr. Rome, 1965.
I.Ilion	P. Frisch, *Die Inschriften von Ilion* (*IGSK* 3). Bonn, 1975.
IMyl	W. Blümel, *Die Inschriften von Mylasa* (*IGSK* 34–35). Bonn, 1987–88.
I.Oropos	B. C. Petrakos, Οἱ ἐπιγραφὲς τοῦ Ὠρωποῦ. Athens, 1997.
IPArk	G. Thür and H. Tauber, *Prozessrechtliche Inschriften der griechischen Poleis: Arkadien (IPArk)*. Vienna, 1994.
Iscr. di Cos	M. Segre, *Iscrizioni di Cos*. Rome, 1994.
JAOS	*Journal of the American Oriental Society*
JBL	*Journal of Biblical Literature*
JDAI	*Jahrbuch des Deutschen Archäologischen Instituts*
JFA	*Journal of Field Archaeology*
JHS	*Journal of Hellenic Studies*
JRS	*Journal of Roman Studies*
KA	R. Kassel and C. Austin, eds. *Poetae comici graeci* 1. Berlin, 1983.
LGPN	P. M. Fraser and E. Matthews, *A Lexicon of Greek Personal Names*, 4 vols. Oxford, 1987–2005.

LSAM	F. Sokolowski, *Lois sacrées de l'Asie Mineure*. Paris, 1955.
LSCG	F. Sokolowski, *Lois sacrées des cités grecques*. Paris, 1969.
LSCG Suppl.	F. Sokolowski, *Lois sacrées des cités grecques, Supplément*. Paris, 1962.
MAL	*Memorie. Atti della Accademia nazionale dei Lincei, Classe di Scienze morali, storiche e filologiche*
MH	*Museum Helveticum*
ML	*R. Meiggs and D. Lewis, A Selection of Greek Historical Inscriptions to the End of the Fifth Century BC.* Oxford, 1969.
NEA	*Near Eastern Archaeology*
NFE	C. Austin, ed. *Nova fragmenta Euripidea: In papyris reperta*. Berlin, 1968.
NGSL	Lupu, Eran. *Greek Sacred Law: A Collection of New Documents* (Religions in the Graeco-Roman World 152). Leiden, 2005.
OAth	*Opuscula Atheniensia*
OJA	*Oxford Journal of Archaeology*
PCPhS	*Proceedings of the Cambridge Philological Society*
P&P	*Past and Present: A Journal of Historical Studies*
PP	*La parola del passato*
Proc. R. Soc. B	*Proceedings of the Royal Society B*
Prott-Ziehen	H.T.A. von Prott and L. Ziehen, *Leges Graecorum sacrae e titulis collectae*. 2 vols. Leipzig, 1896–1906.
RA	*Revue archéologique*
RDAC	*Report of the Department of Antiquities, Cyprus*
RÉG	*Revue des études grecques*
RhM	*Rheinisches Museum*
RIDA	*Revue internationale des droits de l'Antiquité*
RivFil	*Rivista di filosofia*
RSA	*Rivista storica dell'antichità*
SEG	*Supplementum Epigraphicum Graecum*
SIFC	*Studi italiani di filologia classica*
SNR	*Schweizerische Numismatische Rundschau*
Syll.[3]	W. Dittenberger, *Sylloge Inscriptionum Graecarum*, 3rd ed. Leipzig, 1917; repr. Hildesheim, 1960.
TAPA	*Transactions and Proceedings of the American Philological Association*
ZPE	*Zeitschrift für Papyrologie und Epigraphik*
ZVS	*Zeitschrift für vergleichende Sprachforschung*

The Cattle of the Sun

CHAPTER 1

Cattle Habits

> The greatest and most moral homage we can pay to certain animals on certain occasions is to kill them. . . .
> —Ortega y Gasset, *Meditations on Hunting*

Animal Rights

The epigraph above is an example of what Michael Pollan has recently called "hunter porn," an overblown style of writing that assumes "that the hunt represents some sort of primordial encounter between two kinds of animals, one of which is [the writer]."[1] Men face danger, men kill, men provide meat, men rule. Actual hunter-gatherer societies are more likely to survive on the staple supply of grains provided by women, but the symbolic capital vested in the hunt is not based on the scientific measurement of where calories come from in the diet. Understandably, then, with the rise of both feminist critiques and the articulation of animal rights, we are now witnessing the emergence of many more critical approaches to the issue of meat production.[2] Animal rights activists assert that animals have the same inalienable rights as humans, or, following a strand in Jeremy Bentham's utilitarianism, that because they are sentient, can feel pain and pleasure, they have interests that deserve to be taken into consideration, investing them with rights.[3] The growth of the deep-ecology movement and dismay over how animals are raised in modern factory-farm conditions have also fueled debates about how we treat our animals, and whether humane butchery is possible, or a hideous contradiction in terms. A widely seen exhibition sponsored by PETA juxtaposing images of the Holocaust with the abuse of factory-farm animals is the latest and most graphic illustration of the assault on the older, Cartesian notion that animals are mere mechanisms to be used for our pleasure and consumption.

It is unlikely that any such debate ever took place in the Greek world.[4] Certainly, Porphyry argued against eating animals and Empedokles asserted that it was a universal law not to kill living things, but Aristotle was closer to the mood of ordinary people when he asserted that Nature had made animals for mankind, "both for his service and his food."[5] For Aristotle, the separation between humans and animals was simply too great for animal behavior even to be judged by human standards. Discussing vice and virtue, he notes,

> There is no such thing as virtue in the case of a god, any more than there is vice or virtue in the case of a beast: divine goodness is something more exalted than virtue, and bestial badness is different in kind from vice. (Aristotle, *EN* 7.1.2 1145a 15, trans. Rackham)[6]

The extremes for humans are a type of goodness that cannot match the goodness of the gods, and a type of badness that rarely reaches the bestiality of animals. In fact, for Aristotle the goal of a life lived properly is to rise above our animal natures, which are equated with our appetites, and to exercise our moral and intellectual abilities so as to be better humans. In such a scheme, animals are subservient to the needs of humans, their inability to communicate clearly with us rendering moot the question of their needs or wants.[7] It was also not a difficult step to see uncivilized people as brutes and to equate them with animals, as Strabo does in commenting on the behavior of Corsican mountaineers brought to Rome: looking at them, he says, you could "see and marvel at the degree to which the nature of wild beasts and grazing cattle is manifested in them."[8]

The Bovine Idiom

Strabo and Aristotle reflect a common paradox in the human interaction with other large mammals: the tendency to insist on the utter difference between us and them, and an equally powerful tendency to see a deep affinity between our species. The former impulse makes possible the scientific study of animals based on empirical observation. Aristotle, for example, was well informed about cows, commenting on the relationship between pasture and milk production, preferred types of feed, and techniques for increasing animal size. He correctly notes that cattle suffer from both ticks and lice, and observes that they are susceptible to diseases of the hoof and lungs. He describes the techniques for castrating calves, and was familiar with both Epirote cattle and Paionian bison, which he describes in close detail.[9] Aristotle's empirical observations represent one approach to investigating the position of cattle in the Greek world. One may ask, like Aristotle, where the cattle were raised, under what conditions, and how the market in meat operated. In answering these questions one can learn a great deal about changes in society. Between 1965 and 1993, for example, cattle numbers in Greece plummeted from well over 1,100,000 head to 608,000.[10] Similarly, the growth of cattle markets may indicate enormous social change. Between 1867 and 1868, for example, Abilene, Texas, went from a smattering of log huts to a railhead capable of handling 1,000 railcars of cattle per month—transformed by the railway and, in the process, transforming the regional geographies of the United States as profoundly as the Civil War.[11]

There is, however, another way of exploring the cattle system of ancient Greece. Empirical matters of the sort explored by Aristotle were of little interest to Homer, for whom cattle were a fixture of heroic society. Instead, cattle in epic function as measures of wealth and status, to be fought over, raided, paid as dowry, and perhaps most importantly, sacrificed to earn the favor of the gods. We glimpse the profound complexity of the web of values and associations surrounding cattle in the Greek imagination in the episode from which this study takes its title: the story of the cattle of the Sun. Immediately before the slaughter of the suitors, Odysseus is interviewed by Penelope and tells her, in his guise as Aetion the Cretan, of her husband's long travails. Explaining Odysseus' absence he says,

> . . . As he was sailing out
> From the island of Thrinacia, Zeus and Helios
> Hit him hard because his companions had killed
> The cattle of the Sun. His men went under,
> But he rode his ship's keel until the waves
> Washed him ashore in the land of the Phaeacians,
> Whose race is closely akin to the gods.
>
> (Homer, *Od.* 19.302–7, trans. Lombardo)[12]

The long wanderings of Odysseus, then, are the result of divine punishment, the explanation for which takes us straight into a world of gods and heroes who, somewhat prosaically, act very much like humans. They sail the seas and tend their herds. This is hardly surprising. Greek gods feel jealousy, lust, and rage as passionately as the humans whose lives they dominate, if not more so. Their attachment to cattle, however, is more complex than a mere extrapolation of human activity into the realm of the divine. Hermes may rustle the cattle of Apollo like a common thief, but humans cannot transform themselves at will, like Zeus, into a fine white bull as a way of seducing Europa. Only gods can inflict a transformation into the form of a cow, as Hera does to Io (Hera herself having some totemic relationship with cattle, as her epithet "cow-eyed" suggests). In the Greek imagination, cattle hold a special place. They can be the objects of veneration, as in the bull cult celebrated on Crete, and their sacrifice, particularly a hecatomb, constitutes the greatest and most sumptuous offering humans may make to the gods. Zeus is well-disposed toward the Trojans precisely because his altars have never lacked for sacrificial offerings, and Poseidon is indignant that the Achaians should build a wall at Troy without offering a hecatomb.[13] Cattle are the preferred medium for exchange between us and the divine.[14]

The relationship between the empirical and the symbolic resists easy analysis, in part because each is constantly in flux and carries the past with it. Take, for example, the particulars of stockraising. These underwent constant change from the Neolithic to the age of Pericles. What began with the domestication of the auroch would culminate in the Panathenaic procession as hundreds of head of cattle were led to the Acropolis, slaughtered to make an Athenian holiday. In the course of these changes cattle assumed a central position in the imaginations of people in the eastern Mediterranean. In part this is because herding is a livelihood unlike any other, yet it is also because cattle lead us down so many interesting paths. They facilitate trade by giving us valuables to barter; they incline us to sacrifice, so as to render their murder more palatable; they foster social stratification by giving us a commodity to control, exchange, or share. If hunting favored the growth of the hominid brain, then herding favored the growth of human culture.[15] Edward Evans-Pritchard coined the term "the bovine idiom" to suggest that under certain conditions the phenomenon of herding might serve as a controlling metaphor for the way a society understood itself.[16] He employed the expression to suggest that the relationship between the Maasai and their cattle was so intimate that it shaped profoundly the Maasai understanding of the world. More recently Bruce Lincoln has noted that among the Nuer, "just as the social idiom is an idiom of cattle, the religious idiom is one of cattle too."[17] The Maasai lived by herding and lived off the products of their cattle—meat, milk, and blood—and their very cosmology was shaped by the cattle experience: the first men, for example, entered the universe having slid down from heaven on a bull's pizzle.[18]

While Classical Greece was not a society of transhumant pastoralists, neither did it ever entirely abandon the herder's habits of mind. Perhaps if not a bovine idiom, then what the Greeks retained was a bovine register. It is important to stress this since we are going to explore not only how and under what conditions stock breeding was practiced, but also the place of cattle in the Greek *imaginaire*. There, thanks to the operations of metaphor and metonymy, cattle took the blue ribbon. I am not proposing that the cow is simply a symbol of something else (Hera, wealth, docility, the object of desire) but that the accumulation of experience between humans and cattle—hunted, tamed, bred, nurtured, yoked, milked, killed, eaten, worshipped—fixes them firmly within the *habitus* of the Greeks. Since this is not a matter of deliberate intent on the part of the Greeks, I should explain exactly what I mean by this. In a postface to the 1967 edition of Erwin Panofsky's *Gothic Architecture and Scholasticism*, Pierre Bourdieu wrote as follows:

> . . . culture is not just a common code, or even a common repetoire of answers to common problems, or a set of particular and particularized forms of thought, but rather a whole set of particular and particularized forms of thought, a whole body of fundamental schemes, assimilated beforehand, that generate, according to an art of invention similar to that of musical writing, an infinite number of particular schemes, directly applied to particular situations. This *habitus* could be defined, by an analogy with Noam Chomsky's 'generative grammar,' as a system of internalized schemes that have the capacity to generate all thoughts, perceptions, and actions characteristic of a culture, and nothing else.[19]

This approach to culture reserves a space for human agency within a recursive system, whereby humans are equipped by their society's values and prevailing epistemology to act, and in acting so themselves become full participants within their culture, agents of change within their society, in ways that make sense to them. This is a theoretical approach, then, that allows space for historical contingency, that is, for the values, ideas, decisions, and actions that occur within a cultural matrix that informs an individual's conscious and unconscious choices.

The importance of this for our evaluation of cattle systems is twofold. First, herding foregrounds particular practices and experiences that end up dominating entire cultural fields: institutions from marriage to war, concepts of prestige and value, modes of social interaction, negotiation of social hierarchies, all end up being refracted through the prism of herding. Second, because *habitus* is neither fixed nor inflexible, it can continue to reflect notions, values, and experiences that inform the individual's perceptions and the culture's shared grammar of symbols and ideas long after the empirical circumstances that gave rise to any part of it are changed or lost. Put differently, it is possible to speak of the Greeks as both a pastoral and post-pastoral society, practicing farming, manufacturing, and trade, yet still wedded to cattle because of their rich accumulation of significance. Such incongruities are not uncommon. Referring to the Gogo people of southern Africa, whom he describes as semi-pastoral, Peter Rigby notes that "the basis of subsistence in Ugogo is primarily agriculture."[20] Yet the society he goes on to describe is one in which cattle underpin cosmology, residence patterns, property, inheritance, marriage, clan structure—everything, in fact, that falls under that difficult term, culture.[21] Similarly, describing the importance of cattle to the pastoral Fulani of Nigeria, Akanmu Adebayo has recently observed, "It is difficult for a non-pastoralist to understand what cattle mean to the Fulani. Everything begins and ends with cattle. The life of their men and women revolve around cattle. All activities, all conversations, and all thoughts center on cattle."[22] The bovine idiom is not as intense in Greek culture as in the case of these African societies, but it remains ubiquitous. The comparison with contemporary pastoral and semi-pastoral societies may help to show how the Greeks continued to live in the shadow of the bull, and so I pursue some such comparisons in the next chapter.

Another feature of Bourdieu's notion of *habitus* that makes it a useful interpretive tool is that it opens up a space for the operation of memory. Through memory we participate in reformulating our *habitus* individually and collectively in a covert collusion with a past both imagined and real. Is there any past that is not, at some level, imagined? Like notions of chivalry or courtly love, aspects of our imagined past are woven into the values and imaginings of the present. In fact, the conscious version of this sensation, the feeling that some part of what we were is lost, nostalgia, can be one of the most powerful elements of our *habitus*. The Greeks of the Archaic and Classical periods were especially prone to experiencing the world this way, not merely because their Bronze Age past was physically manifest around them in Mycenaean citadels or tholos tombs, but also because the unifying cultural product of the Classical world—epic—was an imagining of the heroic past. That heroic past has a stratigraphy as rich and distinctive as the layers of the tel at Hissarlik, and one of those layers, buried deep in the Greek *imaginaire*, was the cattle stratum.

Closely related to this feature of *habitus* is the problem facing any diachronic study of the Greeks, namely, the risk of anachronism. One might fairly object that Homer's poems are no guide to fifth-century polis religion. Some feel they aren't even a guide to Greece in the eighth or seventh centuries. But the charge of anachronism is too easily used as an excuse not to look for the threads of culture that bind the practices of one age to another, and to impose sterile boundaries between different times. Certainly Homer's Achaians are not identical to Pericles' Athenians, but neither are they unrelated. In a recent essay on meaning in history, Eelco Runia has called attention to the difficulties created for historians by such elements of the human experience as memory, *lieux de mémoire*, and trauma, none of which fit neatly with the concern for narrative and emplotment that have dominated history (or at least discussions of the philosophy of history) since the publication of Hayden White's *Metahistory* in 1973. Runia argues that a way forward is to recognize the existence of "presence" in history, which he defines as "the unrepresented way the past is present in the present."[23] Runia's suggestion may help us to understand the continuing hold of the cattle complex on the imagination of the Greeks. Quite simply, the pastoral experience was ever present, even if pastoralism was only one, very specific and limited aspect of the economic life of the Greeks. Consider, for example, the simple fact that cattle need a good deal of water. It was therefore desirable to pasture them near marshland, a practice that helps to explain the artistic convention of showing cattle with wading birds such as egrets.[24] The same observation has recently helped Thomas Tartaron to identify a Mycenaean industry of processing animal hides on the shores of Glykys Limin in Epiros. Here the large quantities of flint flakes found by the shore are consistent with scraping hides, while the landscape, characterized by what Tartaron calls "flat, swampy terrain," was well suited to grazing cattle and satisfying their enormous need for water.[25] Homer, too, often associates cattle with coastal areas. On the Shield of Achilles, cattle are being released from their byres and graze on reed beds. Similarly, when Telemachos arrives at Pylos he witnesses Nestor's sacrifice of nine times nine black bulls on the shore. Later, Pausanias would note that only cows were allowed to graze by the magical waters of the Milichos River, near Patrai, since the waters caused any animals grazing there to bear only male offspring. Similarly, Aristotle comments on the pernicious effects of the waters of the Sybaris River on cattle, while Pausanias tells the story of the bull of Kerkyra, which wandered close to the seashore trying to alert the locals to the great schools of fish nearby.[26] It is not the veracity of any of the separate stories that matters, so much as the perpetual repetition of the association of cattle and water in folklore and myth. Accordingly, when set against these persistent associations, Zeus' appearance in the form of a bull emerging from the sea as Europa plays by the shore becomes, in one sense, unremarkable.

Yet even if there was a perennial cattle presence in Greek culture, it is the tension between the actual conditions of pastoralism and the symbolic importance of cattle that offers us a fuller understanding of the Greeks. That is to say, the concern for cattle, the desire for meat, the need to sacrifice, the religious valuation of cattle—these were not passive because they were ever present. These are active elements in Greek culture that are subject to change, yet also shape change. Accordingly, the chapters that follow will attempt to flesh out our understanding of the importance of cattle to the Greeks in a variety of settings and at different times, where the bovine idiom worked actively in different ways. In chapter 2 I deal in the broadest possible terms with the changing relationship between humans and cattle that arise from the symbiotic process of domestication. This involves examining the role of large animals in shaping the emergence of human consciousness in prehistoric times from its previous embeddedness within the natural world. This process left a complex legacy in which aurochs were both potentially dangerous yet uniquely important to us as sources of protein and of draft power (after domestication), and symbolically as the focus of newly developing religious and political systems. I employ ethnographic parallels in order to show that pastoralism produces a highly distinctive set of social practices that can be seen to operate in the Greek world. This may strike some readers as either simplistic or so general as to be banal, but I am not trying to argue that all pastoral societies are the same, nor that every aspect of Greek culture arises simply from their taste for red meat. Rather, I wish to show how deep-seated patterns of thought, feeling, and symbolism that arise from the interaction between humans and cattle continue to crop up across the spectrum of cultural production, in plastic arts, in myth, and in performance, as well as in specific institutions from marriage to imperial pageants. Cattle raiding, bride price, and the close association of cattle with sacrifice and feasting are all signs of Greece's status as a cattle culture, if we take this to mean a culture in which the relationship between humans and cattle remains a defining feature, even long after the society may have ceased to be nomadic or pastoral.

In chapter 3 I look at the Bronze Age through the lens of cattle culture. Following the dominant trends in Aegean Bronze Age archaeology, I focus on the palatial system on Crete, both before and after the arrival of the Mycenaeans from the mainland. What emerges when we concentrate on the rich evidence for the bull in Bronze Age cultures is a reading of the Minoan and Mycenaean societies in which the prestige of cattle made them an item of unique value in the reciprocal economy. Both the control of breeding and the distribution of cattle, either for sacrifice or as working animals, reinforced the position of palaces at the heart of Bronze Age culture, a position also strengthened by the ceremonial use of cattle in cult and especially bull-leaping ritual. Cattle became the most potent embodiment of power and social rank. Subsequently I deal with the symbolic significance of the bull, a more complete survival, in certain respects, of the Bronze Age than any other institution of the time, and one that would have more lasting effect on the Greek society that developed from the ninth to the seventh century. The associations of cattle with power and status were critical to the emergence of a distinctive notion of the hero. This can be traced in the *Iliad* and the *Odyssey*, the subjects of chapter 4, but also in the many stories that involve Herakles' cattle adventures, which are treated separately in chapter 5. In both chapters the study of myth cycles and epic stories focused on cattle demonstrate that the cattle idiom allowed the Greeks to do cultural work—that is, to tell stories that actively made sense of the changing world of Archaic Greece. This was especially true of their encounter with non-Greeks and the indigenous peoples of the western Mediterranean. The continuing presence of cattle in the imagination and experience of the Greeks supplied them with the means of expressing power, authority, and status in a way that was meaningful and historically rooted in the world of heroes, which is to say, the past. I develop the notion of cattle breeding as a distinctive practice that inflects an entire range of values and beliefs in chapter 6, which deals with the emergence of the Greek pantheon. Here I argue that, although there is evidence for a variety of responses to the emergence of the gods familiar in Homer and Hesiod, Archaic Greece witnessed the development of a standardized form of religious practice centered on sacrifice and communal feasting. This Panhellenic system was laid over a more disparate scheme of local heroes and deities. I examine the cattle associations of major gods of the pantheon, such as Zeus, Hera, and Poseidon, in order to show that the universal (for the Greeks) symbolic significance of cattle—prestige, power, status—bolstered the emergence of a Panhellenic, "Olympian" system. I then further investigate the development of a Panhellenic system through the articulation of a distinctive Panhellenic space expressed in the Homeric Hymns. In particular these hymns explore the themes of the gods' birth, wanderings, and arrival at his or her sanctuary as tropes connected to the assertion of control of sanctuary space by the Olympian gods, and in the creation of a coherent, identifiably Greek space. Throughout these stories the gods interact with each other and with humans through cattle: stealing them, eating them, sacrificing them in endless variations, all of which reflect and reinforce the centrality of cattle to the imaginative life of the Greeks.

Throughout these chapters we find patterns relating both to the use of cattle in the economic life of the Greeks and in the *imaginaire* as well. It is, however, a mistake to equate origin with explanation. Societies change, the economics of husbandry change, and the significance of a symbol changes, so our examination of cattle in Greek history is an attempt not to reduce cattle to a single, fixed place, but to understand the ways in which cattle came to have so peculiar and so distinctive a place in the world of the Greeks. It would be foolish and reductive to assert that cattle rearing dominated either the economic or religious experience of the Greeks. Indeed, I am not arguing that Greece was primarily a pastoral society or that livestock were the principal commodity around which the Greek rural economy was based, as has recently been argued, for example, by Hans Derks.[27] Nevertheless, the bovine idiom, to borrow Evans-Pritchard's phrase, is an undeniably important aspect of Greek society and curiously understudied, especially in the sphere of economic studies. Rather than review the entire span of a century's scholarship on the ancient economy, let me note some recent works to illustrate my point. Thomas Gallant's 1991 study, *Risk and Survival in Ancient Greece*, is a stimulating exploration of the rural domestic economy of ancient Greece. Gallant explores the strategies used by subsistence farmers to minimize risk both seasonally and over the course of the life cycle of the household.[28] In the course of his painstaking analysis of the ancient data his comments on the position of cattle touch on livestock prices, the preferability of plow oxen to hoeing, the ability of larger holdings to sustain greater numbers of livestock, and the use of strontium/zinc ratios in calculating the amount of meat in the diet of classical Athenians. Gallant also uses modern data to tease out his picture of the precarious subsistence farmer, but at every step his data, his models, and his conclusions are firmly focused on the farmer as cultivator. Herding is pictured as either literally peripheral, pushed to the extremes of cultivated territory as more land is cultivated, or confined to those holdings large enough to support fallow land and pasture. This is not to say that Gallant is wrong. His analysis is compelling, but one wonders: where and how did this world of small farmers produce the vast quantities of meat needed to supply Delphi, Olympia, and Isthmia, or Thebes and Athens?

The question is not posed to be disingenuous, but to suggest, rather, that our reconstructions of the Greek countryside tend to be skewed, concentrating on agriculture and ignoring the impact of a large-scale cattle industry. Consider the case of Jason of Pherai. In 370 BC he intended to celebrate the Pythian Games with special magnificence and sent orders to Thessaly for cattle and other sacrificial animals to be collected. Xenophon reports that more than one thousand head of cattle and ten thousand other livestock were assembled from Jason's allies throughout Thessaly, with a gold crown awarded to the city that raised the finest bull to lead the herd to Delphi.[29] Are we to imagine that herds of livestock of this size could be brought to Delphi without there being an apparatus for collecting and driving these animals well over one hundred kilometers? In early English law the term *agistment* was used to describe the arrangements made for pasturing animals on someone else's land, a practice still common in Australia both in times of drought, when the cattle may be moved to less vulnerable areas, or immediately prior to market, when animals are moved to rich lands to be fattened. Is this what is meant by *epinomia*, a term frequently used in regulations promulgated by sanctuaries in relation to the use of sacred lands? This question leads us to the second half of this study. Having examined the movement of cattle breeding into the religious sphere in the myths and cult practice of the Archaic period, I focus in chapter 7 on sanctuaries as central places in the emergence of a sacred economy. The breeding and supply of cattle to market were largely regulated by sanctuaries. In terms of supply and demand, then, sanctuaries were the force that drove the cattle industry. In practice this was structured in a variety of ways, from direct breeding and control of sacred lands to leasing. By examining various arrangements, especially from Delphi, Delos, and Eleusis, we see that no single system existed, but that the sacralization of the cattle economy was also a response to a more mundane development: the bringing of increasing amounts of land under direct cultivation. Sanctuaries mediated more than elite competition. They also reconciled competing pressures on how to exploit land. Since cities occupied the land in a very different way from the great Panhellenic and international sanctuaries, their handling of meat from production to consumption was somewhat different, and in chapter 8 I focus on how a commercial meat industry developed in an urban setting. The distinction between religion and economics becomes even more blurred here, and results in unusual developments in both spheres.

An unintended yet profound consequence of the formation of this cattle nexus, founded on sanctuaries and the sacred economy, was that sanctuaries were also driven to promulgate laws and regulations for the administration of the complex cult activities that they sponsored. Since they controlled rich land and often land that lay in contentious regions, sanctuaries were implicated in the business of legislation: who could use sacred land, who could visit sanctuaries, how sacrifices were to be conducted and when. An entire body of sacred law evolved that had nothing to do with doctrine but everything to do with ritual. These sacred laws regulated actions within the sanctuaries of the gods and in wider society, defining taboos, imposing restrictions, and guiding activities within the sphere of the sacred. The impact of this was profound because, as Kurt Latte recognized eighty years ago, the emergence of a body of law, even though procedural rather than statute law, supported by the authority of the gods provided a model for the evolution of law in a secular setting. Accordingly, the status of sanctuaries, reinforced by their role at the heart of the cattle system, in turn made them engines of change in areas well beyond the sphere of religion. The distinctive shape of Greek legal culture was shaped by the sanctuary/cattle nexus. This is the subject of chapter 9. In the final chapter I develop the argument that the authority of sanctuaries also provided the basis for the development of a fiduciary economy. As cattle and cattle production increasingly came within the orbit of the sacred business of sanctuaries, as the supreme commodity of value for exchange, dedication or consumption, they embodied value and wealth. The fiduciary economy was made possible, as Aristotle understood, by a monetized economy that in turn depended on a highly developed sense of value, measured, as we shall see, by the Greeks in cattle.

From the Bronze Age, and even earlier, the Greeks were heirs to a tradition in which cattle were much more than objects, whether of veneration, trade, or consumption. So complex and so embedded was the relationship between human and bovine societies that cattle unwittingly became the means of doing cultural work for the Greeks. How to understand slavery? Analogize the slave to the beast of burden. How to understand insemination? The woman is the field plowed by the male. How to understand demagoguery? Imagine a glutton stuffing his stomach with sausage to the point of bursting. Communities are bound by many ties, and for the Greeks the consumption of meat was one of the most fundamental. The cattle of the Sun were mythical beasts located on the periphery of the

Greek world, and yet also symbols of what was present at the heart of every Greek sacrifice: the flesh whose consumption defined our relationship to the gods and each other.

Taste and Gluttony

The symbolic value of cattle helped shape some of the most distinctive practices and habits of the Greeks, reinforcing the importance of sanctuaries and framing notions of the divine, of heroic action, of elite performance, even making possible the emergence of a monetized economy. These associations were leitmotifs running through the mentality of the Greeks and were instantiated in daily life in the nexus of sacrifice, feast, and distribution that was a recurrent feature of Greek life. This is reflected in official documents, such as these early-fourth-century regulations for the cult of Asklepios:

> Decision of the People. Athenodoros proposed: Concerning the things that the priest of Asklepios Euthydemos says, the People vote thus: In order that the preliminary sacrifices that Euthydemos, the priest of Asklepios, has specified may be carried out and in order that the other sacrifice may occur on behalf of the Athenian people, the People decree that the Commissioners of Asklepios are to conduct the preliminary sacrifice that Euthydemos has specified using the money from the Surgical Revenues . . . and they are to deposit the remaining funds inside the treasury within the temple. And so that the Athenians may be able to distribute as much meat as possible, the sacrificial officers who are in office are to see to the holding of a public festival. They are to distribute the meat from the leading ox to the *prytaneis* and the nine archons, and to the sacrificial officers and those who participate in the procession, while they are to apportion the rest of the meat to the Athenians . . . (*IG* II^2 47.23–35)

Meat is a medium for a complex set of exchanges: between the human community and its gods, between the community and its officials, and between the community and the individuals who constitute it. The circulation of the meat, first on the hoof, then on the altar, then in butchered pieces, facilitates the expression of a social order. Beginning with the procession and continuing with sacrifice and distribution, the entire system is strictly controlled by an apparatus that fuses civil and religious authorities into a single, coherent expression of power. The players, the relationships, and circulation of the commodity conform neatly to Bourdieu's definition of social capital: "the ensemble of actual or potential resources that are tied to the possession of a durable web of relationships, more or less institutionalized, of inter-acquaintance and inter-acknowledgement."[30] The degree to which this "cattle complex" was embedded in Greek culture is signified by the fact that no political revolution ever questioned it or subverted it. No change of constitution at Athens, no choice of oligarchy over democracy, no lurch toward tyranny ever undermined the constant repetition of religious performances based on sacrifice. Quite the opposite, in fact. An institution associated in the Homeric world with kings and heroes can become a central institution of the Athenian democracy. As Peter Rhodes has shown, the *boule* received a report on the status of public sacrifices, and Folkert van Straten's iconographic study of sacrifice demonstrates that the Greeks never tired of seeing representations of themselves leading animals to the altar and roasting the god's portion.[31] Iconographically, what Bronze Age object resembles its Classical descendant as closely as the Ayia Triada sarcophagus, whose depiction of a sacrificial procession anticipates the Panathenaic frieze? Nor did the utopias imagined by Plato and Aristotle ever advocate an alternative to the model of the city as a community bound by sacrificial obligations. With the possible exception of the Pythagorean community, not even the most fantastic ideal city, eugenically engineered and elite dominated, could be imagined by a Greek thinker without utterly conventional performances of piety, expressed through sacrifice. In this respect, the Asklepios regulations cited above may be more "typical" evidence of Greek culture than any passage from Sophocles, Plato, or Thucydides, since it represents neither a poet's nor a philosopher's reading of the culture but is a direct expression of the Athenian community.

Despite the fact that the system of sacrifice, consumption, and distribution was at the heart of the polis-sanctuary complex, there are signs of resistance to its viselike hold on the imagination of the community, or at least signs that subversive voices were to be heard on occasion. One sign of disquiet is the persistent association of meat eating with gluttony, often focused on the figure of Herakles, and the other was through the expression of religious and philosophical schemes that rejected meat. The equating of meat eating with gluttony, in particular, seems to have arisen in the milieu of popular attitudes.[32] It is reflected, for example, in popular expressions such as "He is carrying an ox in his jaws," applied, according to the *Suda*, to gluttons.[33] Aristophanes takes this association one step further as part of his satire on contemporary politics. He uses gluttony as a metaphor for the assembly's unquenchable appetite for endless speeches, and plays on the state's role as sponsor of sacrifices to make the Athenians laugh at themselves. The chorus in *Knights*, for example, criticizes Demos (the People) for facing every clever politician with a gaping maw, feeding on their speeches, to which Demos replies, picking up the metaphor, that he likes to fatten politicians only to lead them to the slaughter, thereby keeping control of politics and satisfying his own gluttony.[34] If Demos is eager to consume politicians and their speeches, the most demogogic politician is guilty of the same flaw: Kleon is described as a "yawning gulf, a devouring Charybdis," by the chorus. In a similar vein, the Sausage Seller, an everyman figure who challenges the voracity of Kleon, boasts,

> "And I, when I have bolted the tripe of an ox together with a sow's belly and swallowed the broth as well, I am fit, though slobbering with grease, to bellow louder than all the orators and to terrify Nicias." (Aristophanes, *Eq.* 356–58)[35]

This sets the stage for an exchange of insults of epic proportions between the antagonists. Like two barkers in the marketplace they abuse each other in increasingly graphic language, threatening to gut, stuff, slice, and otherwise butcher each other.[36] The poet employs the violence and the voracity implicit in episodes of sacrifice, butchery, and feasting to tap into the more brutal side of consumption hidden behind the official treatment of sacrifice as a sacred matter. Our meat eating is evidence of our carnal selves. The blood running off the altar may be holy, but the grease dribbling down your chin is not.

The Greeks are constantly reminding themselves of the prosaic side of meat eating by channeling it into myth, in particular into the figure of Herakles the glutton. Karl Galinsky has warned, correctly, that Herakles' roles are so varied that no "all-embracing exegesis" can sum him up, but tales of his gargantuan appetite were very popular.[37] This aspect of Herakles is attested as early as Pindar and a host of playwrights whom Athenaios quotes on the subject of Herakles' gluttony, including Astydamas, Epicharmos, and Ion of Chios.[38] He participated in an eating contest with Lepreus, a story incorporated into a panegyric to Herakles composed by the brother of Theopompos the historian.[39] The theme of his gluttony becomes a mainstay of Attic comedy, but as Nicole Loraux has observed, the identification of Herakles with his belly works in ambiguous ways.[40] His appetite is at once hypermasculine, and equates his eating with a voracious sexual appetite, while at the same time his belly is like a woman's, especially that of a pregnant woman. The hero is even given a *peplos*, a woman's garment, by Athena. Some of Herakles' enormous popularity resides in the fact that this model of masculinity also has something faintly ludicrous about him. He is closer to his appetites than the Olympian gods, retaining part of his humanity in his excesses, his appetites, and his silliness.[41]

If the picture of Herakles devouring an entire animal plays with the deep identification of masculine power with the bull, there is second aspect of the age-old symbolic identification that is also mildly subversive:

the predictable metonymy of the bull's castration and cuckoldry. The symbolic and metonymic layers here are wonderfully rich. Like the castrated bull, the cuckold has been emasculated. In the case of the bull the emasculation is physical, in the case of the cuckold, metaphorical. But the reference to the cuckold as a man with horns derives from something else. A steer still has horns. Viewed head on, the animal still gives the appearance of virility. Thus the horns represent an ambiguity surrounding the cuckold's state. To outward appearances he appears the same man, just as the steer has the horns of the bull. On the other hand, the cuckold has horns but (metaphorically) no testicles. This play of symbols can traced back to at least as early as the fourth century, in the riddle of Euboulides entitled "The Horned Man": "What you have not lost you still have. But you have not lost your horns. So you still have horns."[42] Just as the ox still has its horns (but not its testicles), so too the cuckold has the appearance but not the substance of virility.

Herakles casts a long shadow. For example, Theophrastos describes the ancient version of a bodybuilder as follows: "If he is invited to a shrine of Herakles somewhere, you can be sure he will throw off his cloak and try to lift the bull to twist its neck."[43] Even Herakles' feat of eating an entire ox, the draft animal he took from Theiodamas, was emulated by strongmen, like Milo of Kroton, who replicated the feat at Olympia after carrying the beast up and down the stadium on his shoulders.[44] Pankratiasts loved to model themselves on Herakles, but sometimes emerged looking as foolish as the hero. Polydamas of Skotoussa, for example, was remembered, according the *Suda*, for launching himself "into a herd of cattle and seizing the biggest and fiercest bull by one of its hind legs. He held fast the hoof and did not let go; the bull escaped, and Polydamas was left holding the hoof."[45] Euripides' dismissive comments regarding athletes as incapable of learning to live well because they were slaves to their jaws and appetite also reflects the identification of great, beef-eating jocks such as Theagenes of Thasos, who ate an ox on his own, with the comic figure of Herakles.[46] A darker side of this comic vision also existed: how better to suggest the utter despair of Ajax at being robbed of the *aristeia* for the finest fighter at Troy than to have him slaughter all the Greeks' cattle and the cowherds too. Such profligate waste and such a violation of the strict code of slaughter![47]

The inverse of the meat-eating glutton was the vegetarian ascetic. The followers of Orpheus, for example, were famed for avoiding eating anything once animate.[48] Aside from this religious group, some individuals gained notoriety for their vegetarian habits. Atheniaos devotes a number of pages to men who existed on liquid diets, such as Philinos, who survived on milk alone for his entire life, and a host of others such as Anchimolos and Moschos, a pair of sophists from Elis who were known as "water drinkers" and who ate only figs. They were, notes Athenaios with some surprise, as healthy and vigorous as anyone else, although their body odor was so offensive that everyone avoided them at the baths.[49] Many of the people who cultivated a meat-free diet are labeled philosophers or sophists and the tradition of avoiding the consumption of meat was unambiguously linked to the common philosophical posture of setting oneself at odds with the norms of the community. This meatless diet could raise the ire of more conventional folk, as when Theseus berates Hippolytos, characterizing him, in the words of Albert Henrichs, as "a radical fanatic who practices vegetarianism":

> Go, boast that you eat no meat, that you have Orpheus
> For your king. Read until you are demented
> Your great thick books whose substance is as smoke.
>
> (Euripides, *Hipp.* 952–54, trans. Grene)[50]

The supreme example of this philosophy was, of course, Pythagoras, whose followers renounced all meat as part of an ascetic regime, the so-called *bios Pythagoreios*, that incorporated strict vegetarianism with drinking only water, avoiding humor, wearing a single garment, and cultivating alarming habits of personal hygiene.[51] The list of ancient philosophers who either advocated or at least gave serious consideration to vegetarianism is impressive and sometimes overlooked, but it would be wrong to conclude that the renunciation of meat was either widespread or frequent. In fact, it is its very marginality that allows vegetarianism to serve as a marker of the philosopher's disdain for convention.[52] It could hardly fail to be so since eating meat was so conventional. In fact, in Plato's discussion of evolutionary change in Book 6 of the *Laws*, cannibalism and vegetarianism are both cited as examples of earlier dietary regimes that still existed among some people, but which were clearly not common practice among the Greeks.

If vegetarians renounced meat on the grounds that a suffering animal might have the reincarnated soul of a human, this metaphysical position challenged the orthodox views of life and death that generally held sway. For most, the souls of the dead were in the Underworld, not in the dog whose yelp convinced Pythagoras it was the voice of his dead friend.[53] If their renunciation was based on an appreciation of an animal's capacity to feel, and the assertion that it possessed a soul, extraordinarily modern-sounding arguments that are nevertheless clearly expressed in both Plutarch's *On the Eating of Flesh* and Porphyry's *De abstinentia*, such observations were drowned out by the drums, trumpets, and songs that marked the ritual killing of animals at sacrifice. The very same Plutarch who could ask indignantly, "Who were the first people to claim that we owe no justice to dumb animals?" was equally capable of waxing lyrical about the ritual of sacrifice. Of his time as a priest of Apollo serving at Delphi he writes: "You know I've served Apollo for many years, but you won't hear me say, 'Plutarch, you have had enough sacrifices, enough processions, enough choirs.' "[54]

A recognition that animals suffered for what Plutarch calls our "better fare," meaning gourmet cooking rather than staples, was part of a more general disquiet regarding luxury and indulgence, but the religious associations of meat trumped such anxieties. A similar tension can be seen in the story of Empedokles: he is said to have sacrificed an ox at Olympia after making it from spices, an act that was taken as proof that he approved the teachings of Pythagoras—but the story also shows that he was not prepared to abandon conventional piety.[55] Pythagoras resolved the same problem by sacrificing at the altar of Apollo Patroos at Delos, because, as Diogenes Laertius reports, "wheat and barley, and cheesecakes, are the only offerings laid upon it, as it is not dressed by fire; and no victim is ever slain there."[56] Some may have advocated the meat-free diet as healthier, but even in antiquity it was noted that Homer's heroes are only ever seen consuming grilled and roasted meat, so that eating meat already had heroic status for Greeks in the Archaic period.[57] Renouncing meat as unhealthy defied conventional wisdom. When the Hippokratic writers investigated questions of diet, they imagined that in the earliest times humans, like cattle, had grazed on raw cereals, but this is seen as too harsh a diet, rather than beneficial. In the treatise *On Ancient Medicine* Hippokrates writes,

> I hold that not even the mode of living and nourishment enjoyed at the present time by men in health would have been discovered, had a man been satisfied with the same food and drink as satisfy an ox, a horse, and every animal save man . . . For many and terrible were the sufferings of men from strong and brutish living when they partook of crude foods, uncompounded and possessing great powers—the same in fact as men would suffer at the present day, falling into violent pains and diseases quickly followed by death. (Hippokrates, *On Ancient Medicine* 3, trans. Jones)[58]

The critical change, then, was when humans learned to prepare food, particularly by cooking it. In fact Hippokrates has little to say about the consumption of meat, suggesting that it was not a major feature of the common diet under investigation. He discusses the theory that it was beneficial to give raw meat to those with a weak constitution (an idea he dismisses), but he was clearly more concerned with the properties of raw versus cooked rather than one food group compared to another. The Hippokratic ideal was a moderate diet that avoided food in its most brutal state, and it is this notion of moderation as the characteristic feature of the civilized Greek male that made *omophagia*, the ritual destruction of animals and the devouring of their raw flesh practiced by the female followers of Dionysos, so potent a symbol of the destructive power of the gods. Sacrificial meat was thus a medium for expressing piety within proper bounds.[59] To avoid it altogether was to turn one's back on the community, while to eat meat raw was to surrender to divine madness that reduced us to savagery.[60] As groups emerged wanting to resist assimilation into the mainstream, avoiding sacrificial meat was an easy, dramatic, and symbolically powerful way of advertising their desire to stand apart. Accordingly, St. Paul offers this casuistical advice to his followers:

> If any of them that believe not bid you to a feast, and ye be disposed to go, whatsoever is set before you, eat, asking no question for conscience sake. But if any man say unto you, This is offered in sacrifice unto idols, eat not for his sake that showed it, and for conscience sake. (1 Corinthians 10:27–28)

Presence

It is clear that cattle remained a powerful presence in Greek culture, affecting institutions, practices, and values in every corner of Greek life. It is also apparent that there were sufficient disruptions between the Bronze Age and the Iron Age to make continuity of most social practices and institutions impossible. So one cannot simply say that the raising of cattle for sacrifice was important because it had always been so. On the other hand, the breakdown once imagined to be so complete in Greek history, exaggerated by our reliance on studying history in discrete epochs, is now a matter treated with more nuance. It is the subtle continuity of some ideas, some values and some practices, serving new ends and representing new social realities that constitute the strange presence of the past in the present. Put another way, the lords of Knossos may have banqueted on sacrificial cattle, as did the princes in Homer, and the chiefs at Lefkandi, as did the people of Classical Athens. No group is the same as the other, nor are they entirely different.

The present carries the past within it, either hidden or in plain view, changing established patterns to serve new ends, yet also replicating those patterns when no change is forced upon it. This can be illustrated with phenomena from two very different data sets. One is linguistic: the term *boukoloi*. Not only does it mean "cowboys" (and is the title, in fact, of a comedy by Kratinos), but in the Hellenistic age it was also the word for bandits, while in Attic Greek the related verb *boukoleo* was synonymous with theft.[61] Behind its use in that sense stands the figure of Hermes the cattle rustler. Yet at the same time, the *boukolos* and *archiboukolos* were also priests of Dionysos, and the *boukoleion* was the residence of the Archon Basileus at Athens.[62] Here the religious association is pronounced, but not the connotation of theft. A different god looms. The second example is coroplastic: the figure of the bull in Cypriot art from the Early Bronze Age to the Archaic period. For much of this time bull imagery was ubiquitous in Cypriot art: as protomes added to bowls, as figurines, as masks, and in two remarkable instances as parts of sanctuary models depicting the performance of religious rituals.[63] Yet this bull imagery oscillates between two poles, one of continuous presence and the other of continuously changing meaning, depending upon contingent circumstances: a dedication at a sanctuary differs from a votive deposit in a grave in every way, from its creation to its function.

In one way, then, the Greeks were as much a people of cattle as the Maasai or the Dafla, in that their religion and mythology employed the human relationship with cattle as a touchstone to which they constantly returned. Accordingly, when Demosthenes wanted to stress how Athens' response to the rise of Philip exemplified the city's righteousness, he measured this in the city's public proclamation of sacrifices in time of crisis.[64] Similarly, when Theokritos composes *Idyll* 17 in honor of Ptolemy II, it is the king's sacrifices that serve to mark his legitimacy and piety.[65] And yet at the same time, the Greeks were not people of cattle in the same way as the Maasai, since they also farmed, traded, sailed, colonized, served as mercenaries, and so forth. (Some would say that, as people of cattle, not even the Maasai are Maasai, since some of their constituent tribes also farm, or do not speak Maasai, or do not employ Maasai institutions such as age classes.[66]) It is in this paradox, in this exploration of the slip between what is said and what is done, the difference between what the Greeks were once and what they became, that we see how the past continues to be a part of the present. The bovine orientation of sanctuaries both drives the formulation of a critical set of institutions—sacrifice, dedication, feasting—and is driven by those same institutions. In some cases the development takes place primarily at a sanctuary within the polis, such as in the case of the Acropolis at Athens; at other times and places, it takes place primarily in a sanctuary outside the polis, such as at Olympia, and on occasion the sanctuary is outside but tied to the polis, as in the case of Eleusis.

These multiple trajectories might lead one to abandon any attempt to categorize and classify sanctuaries, yet there are common features that unite them and arise from broadly shared features of Greek culture. The religious authority located at sanctuaries was a critical agent, we shall see, in at least three areas of Greek life. It counterbalanced the weakness of the early state, it provided the basis for the articulation of law, and it made possible the emergence of a monetized economy. In fact, it was only through the recursive relationship that existed between polis and sanctuary that the *poleis* of Classical Greece were truly able to achieve the complexity of developed states. In this process, cattle existed as mediums for exchange, measurements of value, and conveyors of status. They were constant companions, providers of traction, milk, meat and hides, and their symbolic value was as crucial to the formation of Greek culture as was their material importance. Remove the ox, and one contemplates a society with nothing to put on its coins and no reason to have coins. The story of the Greeks and their cattle began with hunting and domestication, that is to say, before there were Greeks or Greece. It is perhaps fitting that millennia after the Greeks stopped sacrificing cattle to the Olympian gods, who have fled or are not listening, that we should recognize the symbiosis between humans and cattle that was a constant presence in Greek culture.

CHAPTER 2

The Paradoxes of Pastoralism

> Bestiality, and its principle of uncertainty, must be killed in animals.
>
> —Jean Baudrillard[1]

Domestication

The various pastoral regimes practiced by the Greeks were made possible by the much earlier domestication of cattle. This resulted not only in a steady supply of protein for the pastoralist but also helped give pastoral societies a distinctive shape: the custodial care of the herd, the ritualized treatment of butchery, the elaborate use of cattle as measures of wealth, and the central importance of cattle in the rich imagination of pastoral people are all evidence of the profound impact of domestication on human societies. In order to understand the significance of these practices to the Greeks it is worth examining the phenomenon of domestication in greater detail. Far from being a simple matter of human domination of the wild, domestication, we shall see, was a complex set of processes that changed human societies as profoundly as it changed the species that underwent domestication.

The earliest domesticated species was the dog and the domestication of cattle first occurred around the middle of the seventh millennium BC in the Near East, although recent genetic studies have shown that there were probably at least three separate domestication events in Africa, Eurasia, and Asia.[2] The earliest reliably dated evidence comes from Ras Shamra, in northern Syria, and from Gritille and Hayaz in Turkey. At Ras Shamra, for example, the deposits of large auroch bones (*bos primigenius*) are gradually replaced by those of smaller domesticated cattle (*bos taurus*). Herding caught on quickly, and the practice reached Greece by no later than 6300 BC, as humans and their cattle moved westward along the Mediterranean coast.[3] At this time the bones of domesticated sheep, goats, and cattle all make their appearance in the archaeological record at Argissa Magoula in Thessaly.[4] The domestication of cattle was itself only one in the series of complex changes known collectively and somewhat misleadingly as "the domestication of plants and animals" or "the Neolithic Revolution." Both phrases elide a number of separate but related processes that unfolded over thousands of years. There was considerable variation in the impetus behind each and in the effects of these domestications, and it is important to understand the distinctive features of cattle domestication since the symbiosis between humans and cattle established patterns and practices that remained profoundly influential in Greek culture.

Earlier definitions of domestication emphasized the morphological differences between domesticated and wild breeds. According to this approach, sometimes referred to as the osteological definition, animals undergoing domestication displayed a physical divergence from their wild ancestors that could be identified in the archaeological record. Domesticated pigs, for example, can be distinguished by a smaller lower wisdom tooth (M3) than that of wild boar. Another significant marker is bone size: the earliest domesticated sheep, goats, and cattle are all smaller than their wild ancestors. The Neolithic auroch, ancestor of domesticated cattle, often stood 1.55 m at the withers, but its domesticated descendants in the late Neolithic and Bronze Ages rarely exceeded 1.22 and 1.23 m, respectively. Empirically, the identification of domesticated species by osteological analysis remains a valuable tool in charting the shift from hunter-gatherer societies to pastoral or mixed pastoral-farming communities, but the phenomenon of domestication is now recognized as more complex than simple morphological changes in the species concerned. Domestication is signaled by such changes, but not defined by them.[5]

Beginning in the 1960s, studies of domestication shifted to cultural definitions, emphasizing human involvement in the process, with particular attention to breeding and exploitation.[6] As researchers have come to appreciate the complexity of domestication, definitions have grown more complicated. For example, one researcher proposed the following criteria for domestication:

1. Humans must have a recognized need or desire that can be satisfied by controlling and breeding the animals.
2. Humans must control the population of the animals in question to the extent needed to satisfy this need or desire.
3. Humans must nurture and protect the animals concerned.
4. Humans must be involved in directing the breeding of the animals concerned through selection.[7]

The domestication of cattle fits these conditions well. Cattle provide 2,360 cal/kg, and so from a nutritional point of view are highly desirable.[8] There has been speculation that animals were first domesticated for their milk products and that meat production came later, but the evidence for this is at best inconclusive.[9] In any case, optimality models derived from the study of omnivorous primates and hominids suggest that our dietary preferences had already evolved toward animal matter.[10] Prior to domestication, hunting was the only way to satisfy the human inclination to carnivory, a high-yield food practice to which early hominids may have been genetically inclined, if, as has been suggested, meat-adaptive genes were linked to longer life spans and brain development.[11] A single Thomson's gazelle yields the equivalent of 300 figs worth of energy, 2,000 figs of protein, and 666 figs of fat.[12] But carnivory based on hunting is unreliable: figs are easier to catch than a gazelle. Accordingly, the domestication of animals for food went along with the domestication of plants and the beginnings of agriculture. As the shift to sedentary agriculture raised the proportion of carbohydrates in the Neolithic diet, finding a steady source of animal protein became even more desirable, since the shift to a cereal-based diet brings with it a host of health problems, including an increase in osteomalacia and porotic hyperostosis. A diet too high in cereals can lead to deficiencies in lysine and tyrosine as well as riboflavine and niacine, and has been linked to increases in child mortality rates.[13] Breeding cattle, as opposed to hunting, therefore brought two benefits: access to a steady supply of animal protein and a corrective to the imbalance of a cereal-based diet. From an alimentary point of view, then, pastoralism fits the first condition listed above, fulfilling a human need for a reliable source of protein. Similarly, the other criteria listed describe our relationship with domesticated cattle well: it is easier to protect, breed, and cull a herd than to hunt successfully; and we are capable of forming attachments to large mammals and their offspring.

Definitions that emphasize exploitation, breeding, and nurturing are also appealing because they fit with our recent experience of animal husbandry. Since the eighteenth century European husbandry has been devoted to highly controlled breeding. Robert Bakewell (1725–95) pioneered the bioengineering of specific traits in cattle. He accomplished this by ruthlessly inbreeding the herd on his Leicestershire farm, culling any animals that did not contribute to his scheme, thanks to which the long-boned, lean cattle of his day gave way to shorter, thicker, fatter, and meatier animals. Such was his impact that between 1710 and 1795 the average weight of beef cattle at the Smithfield market rose from 310 pounds to 800 pounds. The cattle husbandry familiar today is a system essentially created by Bakewell.[14] Without him the modern industrial farm, with batteries of hens and hundreds of pigs confined to narrow pens, would be unimaginable. After Bakewell the growth of railroad networks contributed to the development of markets and increased the pressure on pastoralists to produce animals that would bring a greater return at market.[15] Such programs did not exist in the Greek world, of course, although the marked increase in the size of domesticated animals in the provinces of the Roman empire has led to the suggestion that animals were being deliberately bred up.[16]

The breeding that occurred in the early stages of domestication, however, was of a very different sort. The 20 percent diminution of cattle size in relation to wild aurochs that took place between 6500 and 1500 BC shows that Neolithic herders were hardly trying to increase the size of their animals.[17] Of greater concern to them was that their cattle should be tractable and docile. It is a pleasant conceit to imagine Neolithic herders breeding for temperament in the way that modern breeders try to improve milk yield, but the two approaches are quite different. The modern breeder actively chooses which animals will reproduce on the basis of desirable qualities that they exhibit (hardiness, leanness, fertility, and so forth). But traditional herders, like the Lapps who live by herding reindeer, do the opposite. They are not intentionally breeding for docility, but protecting the herd from aggression, which they limit by castrating most of the males of the herd.[18] At first glance the results seem the same: human intervention alters the gene pool in favor of desired characteristics, either selecting for desired qualities or selecting against undesirable ones. But, in fact, selecting against aggression goes hand in hand with a response to domestication from the animals themselves: domesticated species tend to display neoteny, the characteristics typical of the young and adolescent of the species.[19] This may include smaller overall size, and larger head-to-body ratio. It was certainly easier for humans to develop feelings of attachment toward animals that appealed to the human capacity to nurture the young, but it is unimaginable that these crucial mutations were intentionally planned by Neolithic herders. In any case, the phenomenon is not confined to large animals domesticated for food, such as sheep, goats, and cattle. Cats, dogs, rabbits, and pigs all exhibit this inclination toward neoteny, as if in giving up their wildness they became not only tame but infantile. Compared to their wild relations, domesticated animals are less wary, less flighty, less aggressive. Since this phenomenon is so widespread among domesticated species, it is highly unlikely that it is the result of intentional breeding in each case. Instead, it points once again to the symbiotic nature of domestication. Both humans and animals contributed to the domestication of the species.[20]

Definitions of domestication, then, are apt to reflect unwittingly modern systems of husbandry, viewing the relationship as entirely one-sided: humans are solely in charge and animals are commodities to be manipulated.[21] Consider, for example, Achilles Gautier's 1990 definition of domestication: "the process whereby wild animals through various forms of cultural control acquire domestic traits which allow humans to exploit them."[22] Here "cultural control" is another way of saying that humans set the conditions, animals respond, and humans then exploit them. As in all these definitions and reconstructions, it is the animals that change. The constant element is the agent of these changes: *homo domesticans*. By focusing on exploitation such definitions fail to take into account the symbiotic nature of domestication, and run the risk of underestimating the impact of domestication on humans. It is a relationship in which humans do not exert complete power.

A more nuanced approach would see humans as participants in a broader process, both effecting and affected by changes. Instead of assuming an inevitable progression from hunting to agriculture, one might explore domestication in terms of the impact of sedentarism on Neolithic humans. A number of French scholars have suggested that the move to settled farming set in train a shift in the relationship between humans and nature. No longer wholly integrated into the natural world, the people of the Neolithic, according to Jacques Cauvin and Daniel Helmer, experienced a degree of disengagement from nature. Since "nature" does not exist except insofar as we perceive it, this disengagement consisted of reimagining the place of humans in the natural world.[23] This resulted in a fascination with the human form, which was now invested with a special force. In the Neolithic, humans become aware that they are distinct from the natural world around them; they come to appreciate their "personal phylum." Divorced from nature, they appear on a higher plain, which permits them to develop a new conception of their place vis-à-vis animals. In this interpretation of the Neolithic, man is no longer the constant, unchanging agent, inexorably extending his sway over the beasts. Instead, domestication can only come about after the Neolithic has witnessed a dramatic shift in human consciousness. According to Cauvin, domestication depended upon humans forming a conception of their place in nature. Cauvin refers to the changes that occurred just before domestication as "the revolution of symbols," and recent work has demonstrated that there was an explosion in animal symbolism in the ancient Near East immediately *prior* to the domestication of animals.[24] Furthermore, although many of these figural representations of animals, such as foxes and vultures, disappear after 7500 BC, the one pair that persists is that of the human female and the bull, suggesting that the two were integrated into a complex symbolic system of life and death that set the stage for domestication. "Neolithicization" includes not only empirical changes, but affective changes as humans became separate from (and eventually superior to) the rest of the natural world.

A similar approach has recently been taken by Ian Hodder, who argues that domestication occasioned a split between two realms: the *domus*, the world of the household and the accumulation of wealth based on the mastery of nature, and the *agrios*, the world of the wild, evoked by hunting and feasting.[25] Hodder's study concentrates on Europe and the archaeological record of pottery, burials, and figurines, but his system can be more broadly applied. One could compare the hunters of Lascaux, who were in awe of the ferocity of the wild animals they encountered, and who were entirely embedded in nature, with a later world inhabited by keepers of sheep, like Abel, and tillers of the ground, like Cain, who could imagine that they had been given dominion over every living thing that moveth upon the earth (Gen. 1.28 and 4.3). No longer in nature but above it, they are witnesses to the full break with the natural world that comes with domestication.

Herds and Humans

By the beginning of the seventh millennium BC, only dogs had been habituated to human company, as attested at Natoufian sites such as Hatoula and Mureybet.[26] At around the same time, circa 12,000 BP, burials at two sites in Israel point to the beginning of our special relationship with dogs: at Mallaha, a puppy was buried under the right arm of a human skeleton, while at Hayonim two dogs were buried between two humans with a third human placed on top.[27] By no later than 6000 BC, however, first sheep and goats and then cattle had begun the transformation of human societies, making it possible for the people of the Neolithic to practice a mixture of farming and herding, and in some places to rely almost exclusively on herding.

How were humans equipped to handle the challenge of becoming herders? Our nearest dealings with large animals had been as hunters, and it is tempting to see herding as an extension of hunting. To kill a large animal is also to dominate the animal, while the herder exercises control over an entire herd and will eventually kill at least some of them. Two aspects of hunting remained central to the relationship between humans and cattle after domestication. From at least as early as Paleolithic times humans connected animals with death and the afterlife. The hunt was the earliest human activity to acquire symbolic force. From the paintings of Lascaux, which depict hunters killing and being killed by bison, to the cult chambers of Çatal Höyük, adorned with plastered bucrania, prehistoric people dwelt on the profound significance of killing large animals. The exact meaning of these depictions is irrecoverable, but their symbolic power is undeniable. Through the hunt, elk, deer, bison, and auroch became our companions in confrontations with death, theirs or ours. This is especially notable in Neolithic Sudan and Nubia, areas of dynamic change where many human communities clustered around the temporary lakes west of the Upper Nile, increasing the proximity of humans and cattle and favoring domestication.[28] In many cases bucrania were placed strategically in burials, while in other cases hundreds of bucrania were used to mark the approaches to elite burials.[29]

The second legacy of hunting is more troubling: a proclivity for violence. The millennium of great domestications is bracketed by a notable increase in the levels of violence within human society. There are earlier possible cases of murder, such as at Krapina in Croatia, but these remain ambiguous.[30] By 10,000 BC, however, clear evidence emerges of episodes of violence beyond anything previously witnessed. At Djebel Sahaba in Sudan, fifty-nine subjects were massacred and buried.[31] Another massacre took place around 5000 BC in Tallheim, Germany, involving thirty-four individuals. Their skeletons reveal they had been beaten and often wounded, with trauma to limbs and head, and in most cases their craniums had been smashed in a coup de grace.[32] This increase in violence toward other humans first arose among hunter groups, but will have been fueled by the surpluses that first appeared among Neolithic cultivators and pastoralists. Both had stores of food, making them an attractive target for raids. At the same time, the differentiation into different types of social groups—farmers, herders, and those who continued as hunter-gatherers—meant that sedentary cultivators were tied to their fields and vulnerable to attack. Like prey around their water holes, they invited the attention of human predators.

So profound was the differentiation of human societies into pastoral and non-pastoral types that it established a recurring pattern of conflict between the two. Those cultures, like that of early Israel, that identified with the pastoral ethos favored Abel, the shepherd, over Cain, the cultivator. Why else is Abel's offering pleasing to the Lord and Cain's not? Even as Israel evolved into a complex state it retained a strong identification with pastoralism as normative: the patriarchs are, above all, herders. In the desert they are safe, while Sodom, Gomorrah, and Babel—cities are the culmination of sedentarism—are *loci* of evil.[33] The hold of pastoralism is equally strong in Zoroastrianism. In the *Avesta*, for example, Ahura Mazda states that his second name is "Herd-Giver," while the faithful pray "to Mithra of wide cattle pastures, whose word is true."[34] On the other hand, in communities that chose to identify primarily as cultivators, the cultivated field stood for order, perpetually under threat from raiders unattached to the land. The earliest collection of Japanese myths, the *Nihongi*, for example, tells of an impetuous raid by Susano-o, the storm *kami*, who defecates in the well-tended rice paddies of his sister Amaterasu before hurling a flayed pony into her weaving hall.[35] Weaving and cultivation here stand for productive order, while uncontrolled aggression is a nomad on horseback. The Romans, whose empire spanned the Mediterranean but was identified with a city, looked at the Corsicans who lived as shepherds in the mountains and saw only brigands and savages, "in whom is manifested the nature of wild beasts."[36] Another imperial voice echoes the same sentiments. Edmund Spenser explained the Englishman's superiority to the Irish simply in these terms: the latter were pastoralists. Spenser observed, "Loke into all Countries that live in suche sorte by kepinge of Cattle and you shall finde that they are bothe verie Barbarians and uncivill, and allsoe greatlye given to warr."[37] The revenge of the pastoralists is to embrace their reputation for lawlessness. Among the Koutsovlachs of northern Greece being herders and being *kleftes* (thieves) go hand in hand.[38] Perhaps the first thinker to try to reconcile nomadism and sedentary civilization was Ibn Khaldun. He characterized the pastoral Arabs as rough, impatient of authority, and hostile to sedentary civilization: "As a result of their mode of living, their very presence is inimical to the existence of buildings, which are the very foundation of civilization."[39] But he also regarded cities as lacking the quality of social cohesion, *asabiyah*, that was found in tribal societies. In the Arab conquest of the Mahgreb the two would merge. As Ernest Gellner puts it, for Ibn Khaldun, "economically, the tribe needs the city: politically, the city needs the tribe."[40]

The roots of this division of human societies into different types lie in the Neolithic, and the domestication of cattle is only one piece in the rich mosaic of Neolithic life, yet it was a crucial one. While hunting had established some patterns for the ways humans would look upon large animals (as, for example, sources of protein, spirit guides, or potential killers), in this period the emergence of herding had a special role in shaping key ideas and values that remained fundamental to human societies throughout the eastern Mediterranean. Quite simply, herds require management. An examination of two societies that are focused on cattle-breeding reveals surprising similarities between them and the culture of classical Greece, similarities that help to explain features of the Classical world that are frequently overlooked.

Pastoral Societies

The present-day Dafla (or Nishi) people of eastern Bhutan and Arunachal Pradesh are a society consisting of autonomous longhouses loosely organized into villages, each longhouse consisting of up to twenty relatives. Until recently they lived primarily as herders.[41] In traditional Dafla society there were few permanent possessions, no concept of individually owned land, and family groups changed their houses every two or three years. Dafla culture revolved entirely around raising cattle, both zebu (*bos indicus*) and the even more valuable mithan (*bos frontalis*), a long-horned animal with a pronounced dorsal ridge that was preferred for sacrifice.[42] The domestic mithan is believed to be descended from the wild gaur

Figure 2.1 Indian gaur (*bos gaurus*), wild cousin of the mithan. Photo credit: Elephas Creations—Nature & Wildlife Photography Solutions.

(Fig. 2.1).[43] The horns of the mithan served no practical purpose in Dafla society but were used for display; hung on the posts of the longhouse, they were the visible record of the achievement and wealth of the leading men. One nineteenth-century observer claimed that the mithan was the cornerstone of Dafla life.[44] Mithan herds were the final measure of wealth, but it was a wealth that circulated in elaborate ways through Dafla culture. Mithan were also used as a way of carrying on trade with the Apa Tani, a sedentary people with whom the Dafla had extensive trade ties. The Dafla exchanged their mithan for prestige items such as Tibetan bells.

The mithan was also at the heart of a central institution in Dafla life: raiding. These raids were often aimed at the same Apa Tani with whom the Dafla traded. Raiders went in search of mithan to carry off, or human captives whom they held for ransom. These raids were often settled by arbitration, with, once again, the cattle at the heart of the system. In one case fifteen mithan were traded for seven men and women. In addition to serving as both the object and means of violent and semiviolent exchange, the mithan was also the mechanism for cementing peaceful ties. They were given as bride price by the groom, and returned as marriage gift by the father-in-law. They were the objects of elaborate gift exchanges, called *dapos*, designed to increase prestige and the ties between leading individuals. In the *dapo* ceremony the exchange of mithan established a pact of nonaggression between families. A similar institution among the neighbouring Aka people consisted of a highly ritualized set of guest visits, called *thumona*, that recall the conventions of *xenia*. The mithan was the preferred gift exchanged to cement the alliance.

Mithan sacrifices long remained central to the religious life of the traditional societies of the eastern Himalayas, despite the presence of Hinduism, Buddhism, and Christianity, religions all more or less hostile to the value placed on the mithan. At sacrifice time, special care is taken to trick the gods so that they will not be offended by the sacrifice of a mithan. Sometimes the ritual of sacrifice is prolonged over five days, with the animal being harrassed, attacked, wounded, and tortured so that its spirit will be too exhausted to seek vengeance on its attackers.[45] The cosmology in operation here resonates for anyone familiar with the story of Prometheus, who stole fire from heaven and attempted to placate Zeus with a trick. Sacrificing a bull, he burned the animal's long thighbone wrapped in sweet-smelling fat, but kept the meat for humankind. In both episodes, Greek and Dafla, humans are the nutritional winners, and in both systems cattle have a special place mediating between the human and the divine. In both the act of sacrifice arises from a particular milieu, the raising of cattle. In fact, the consumption of the meat after sacrifice is the culmination of the entire scheme.

The reason for recounting these details should be clear to any reader who recognizes in the Dafla a community resembling Odysseus' Ithaca: a society of households that lived by raiding, practiced the conspicuous consumption of meat, and exchanged cattle as a part of a broader trade system in which gifts, honor, and women were also commodities. Homer's world shows all the hallmarks of a stock-raising society. We shall return to this theme in the following chapter, but the similarity of Dafla life to the heroic world of the Greeks suggests that the Greek conception of wealth was heavily influenced by the distinctive ways that wealth is construed in a herding culture. In the case of cattle wealth the most desirable commodity is itself active and alive; as the herd grows, wealth produces itself. It cannot be passively stored, like money in a bank, because paradoxically it must diminish even as it grows. Cattle get sick, and even the most healthy must eventually die. So there is an imperative to use this wealth, to put it into circulation before it is lost or consumed. It follows, therefore, that cattle wealth favors a range of complementary human institutions: bride price, gift exchange, and raiding are all integral parts of a cattle culture. They are social institutions that are, so to speak, cattle driven. Cattle are the ultimate measure of wealth in a society that lives exclusively on cattle, but the association is so deeply rooted that the habits of the herder do not pass away after cattle cease to be the major source of wealth. The experience of herding lurks in the recesses of cultural memory, continuing to inform the *habitus* of societies as they change.

This is not to say that there is a universal type of pastoral society, nor would I want to shoehorn the Greeks into this category. Cattle-raising societies vary, but it is the communalities they share and the continued influence of cattle in their values and institutions that makes them a valuable guide to Greek society.[46] Consider, for example, the Bahima of southern Uganda, whose cattle habits also point toward a cattle idiom shared by early Greek culture. Like the Dafla, the Bahima exploit cattle not only for their products, but as the medium for significant social exchanges, both peaceful (gift giving) and violent (raiding). Rites of passage among the Bahima are marked with cattle-focused ceremonies. At the age of four months, for example, a boy is placed on a cow's back with a bow, arrow, and rope. He is said to be "put down on a cow" (*okuteekwa aha nte*). Symbolically, his life as a man has begun as he takes up the weapons of the hunter and the tools of the herder. Later he will marry a girl at a ceremony in which the couple spit milk at each other. The instances of cattle-based activities, rituals, and institutions goes on indefinitely: recitations of cattle names while praising them (*okugambente*), offering cattle as gifts (*empaano*) in order to establish reciprocal relations (*irembo*), giving cattle as compensation (*empongano*), and a whole class of words describing the various procedures associated with marriage, beginning with the bride price paid by the groom (*enjugano*).[47] Similarly, among the Maasai, marriage and the distribution of cattle are understood to go hand in hand. "Marriage," as one recent commentator notes, "dictates the ownership of cattle."[48]

One particularly revealing feature of east African cattle cultures is that language often reflects a distinctive epistemology characteristic of pastoral societies: the language of the Bahima, Runyankore, shows that the pastoralist experiences the world through cattle. Cattle are categorized in terms of their breeding status. Thus, for example, herds are made up individual cattle such as *ekyoota*, a heifer approaching maturity; *ekibanga*, a heifer mature but not yet ready for mating; *enshumba*, a cow that is being mated; and *ejigija*, a cow that has calved once. These are clearly good statuses. On the other hand, an *emberera* is a barren heifer, and an *ekicura* is a cow too old to bear calves. In this vocabulary we encounter an entire life cycle conceptualized in terms of the animal's ability to bear offspring. Of course hunters, too, were capable of seeing whether they were hunting a male or female animal, and whether it was young or old, but in herding the herder does not merely encounter an animal, but must manipulate an entire herd according to these categories. It is this close engagement with the animal, necessarily a part of the pastoralist's control of the herd, that distinguishes the cattle complex—specific, referential, transferable from animal herd to human society—from the more general lesson of classification that the hunter learns by observation of the animal world. In an essay on animal totemism Claude Lévi-Strauss famously labeled animals "goods to think with" and claimed that the observation of their differences offered "conceptual support for social differentiation."[49] Pastoralism took this process one step further, creating in the domesticated herd a model of one type of human society. It is hard to imagine that pastoralists could employ categories such as *entsigasi*, a virile young bull, and *engundu*, a mature bull who can mount cows, solely with respect to their herd and not apply them, at some level, to human society as well.

Pastoralism favors a view of the world in which a good deal of human activity is focused on the animal. The Bahima, for example, take exquisite care to note the color of each animal, as well as any impurities or distinguishing features, and the animal's ancestry.[50] These are expressed in the animals' names and woven into elaborate bovine genealogies, the subject of heroic song recounting the beauty of the cattle (*ebirahiro*). Another genre, *ebyevugo*, comprises accounts of cattle raids in which descriptions of the cattle again figure prominently.[51] Taboos based on these color-coded genealogies also dictate who can feast on which animals. At every stage cattle shape Bahima culture.

Even for philosophers the domesticated animal stimulates reflections on human society. To get the best hunting dogs or fighting cocks, asks Socrates, do you breed them indiscriminately or from the best? And similarly, to get the best horses and other animals, don't you use animals in their prime? In the ideal republic, he concludes, we should apply the same principles to breeding humans.[52] Eugenics is the bastard child of domestication. Even more than Plato, Aristotle uses our contact with domesticated animals as a way to understand human relations: "The use which is made of the slave differs but little from the use made of tame animals; both he and they supply their owner with bodily help in meeting his daily requirements."[53] In the same passage, the infamous justification of slavery, Aristotle applies the same teleology to domesticated animals and slaves. Both are ruled by free men, and thereby become useful to society. As ever, Euripides can be relied on to subvert convention. Women would be better off as cattle, he has Medea bitterly remark. At least cattle don't pay for their servitude, whereas women must have a dowry as "a down payment on marriage," after which they are owned by their husbands.[54]

Which came first? Is the domesticated herd modeled on human society or is it the model for human society? Among the Bhutanese Ao people, humans and their cattle are believed to share parts of multiple souls. One of these souls is the "celestial mithan." Each man's fate dwells in the sky, and it comprises three souls. One of these is incarnated in a particular mithan living here on earth. Accordingly, human beings are really made up of eight parts, half earthly, half celestial, joined by the cattle that are the most tangible link between one realm and the other. Among the Dinka, cattle are especially esteemed because it is believed that humans imitate them.[55] The same notion, at least expressed metaphorically, can be glimpsed in Philostratos' condemnation of Kritias for making an alliance with Sparta and helping "the monstrous design of Sparta to make Attica look like a mere pasture for sheep by emptying her of her human herd."[56] An early Dutch anthropologist in Aceh noted that local people designated a "captain" in each herd, an animal distinguished by its color or size, and believed to safeguard the vital spirit of all the animals in the herd, just as the chief was seen as the soul of the village.[57] Similar ideas were expressed among the Thonga: "The chief is the land . . . he is the bull. . . . Without him the cows remain infertile."[58] The pairing of bull and chief works particularly well in epic settings, as is illustrated by an episode in the *Táin*:

> "That is a fine heifer going by," said Noisiu. Derdriu replied, "Indeed it is, for the heifers grow big where there are no bulls." Knowing she must be the famed Derdriu and therefore Conchobor's property, he warily retorted, "Sure you have the bull of the province to all yourself, the King of Ulster." But she faced him square on and said " A hoary old animal! I'd rather have a game young bull like you." (*Táin Bó Cúailnge* 11, trans. Kinsella)[59]

The same attitude is reflected in the Greek term *agelarchês*, meaning either the leader of a herd, or the captain of a company.[60] Herding creates an imaginative space in which *homo sapiens* can see a society taking shape under human control. The herder maps ideas about social organization onto pliant subjects. The herder is no hunter; he does not stalk the cow in order to fall upon it suddenly and kill it. Instead he watches over *ekyanya*, all the young calves still suckling, and he entices the cow to be milked (*okubarabaisa*), he stimulates the cow to help it become pregnant (*okuhagirana*), sometimes having to travel long distances for water for his cattle (*okuhanganga*), taking them to pasture (*okuseetura*), and making sure that each one gets a turn at the watering hole (*okuhinda*). The Runyankore glossary is replete with labels applied to the cattle, and verbs describing the actions of humans with respect to those cattle. Even Bahima counting revolves around the experience of herding: their decimal system uses *engundi* for ten and *eigana* for one hundred, from a typical herd of one hundred cows, but *engundi ikumi* (ten bulls) means one thousand because ten bulls will each have a herd of one hundred cows. Pastoralism favors categorical thinking and demands a range of specialized activities that all arise from our mutual dependence but are directed by the herder. It is also an experience that allows the pastoralist to distinguish, *modo grosso*, between the tame and the wild. In the Book of Enoch the descendants of Adam, such as Abraham and Isaac, are bulls, while the gentile descendants of Noah are, David Halperin and Gordon Newby note, "wild beasts and birds of prey."[61]

Sacrifice

Cattle, then, are important not only as a very particular type of wealth, but as the focal point for a complex set of human behaviors and values arising from the experience of stock raising. Domesticated cattle elicit a specific set of responses from their human herders, and this forces the herders to adapt to the new relationship by becoming nurturers and managers. In doing so humans manipulate the herd, exerting a control over it that helps both humans and cattle to thrive.[62] Since cows are capable of repaying the herder's attention by giving milk year after year, the close relationship between humans and cattle often expresses itself in a deep resistance to killing the animals in any setting other than sacrifice. An African example, once again, makes the point neatly. In the 1850s the Xhosa people were struck by two calamities: military defeat at the hands of the British and an outbreak of lung sickness among their cattle. These blows fed a millenarian movement that centered on a young Xhosa girl who prophesied that the dead were about to arise and that they were bringing with them glorious new herds of cattle. In the meantime, she claimed, the Xhosa were required to slaughter whatever animals they owned. In the space of two years, over 400,000 head of cattle were slaughtered, but many Xhosa resisted, appalled at the notion of killing their animals. When one man was urged to do so he replied that he would rather kill his father than kill his cattle.[63] At the farthest extreme is the elevation of the herd animals to the level of friends and confidants. When the Cyclops, who milks his sheep and eats only human flesh, has been blinded by Odysseus, he addresses the lead ram as follows:

> My poor ram, why are you leaving the cave
> Last of all? . . . Are you sad
> About your master's eye? A bad man blinded me, . . .
> Noman—but he's not out of trouble yet!
> If only you understood and could talk,
> You could tell me where he's hiding. . . .
>
> (Homer, *Od.* 9. 445–455, trans. Lombardo)

When cattle are raised primarily for meat consumption, the very ties that bind pastoralists to their animals become a hindrance. Furthermore, slaughtering cattle for meat reduces the overall size of the herd that the herder has worked hard to increase, so that our desire for meat, in nutritional terms an efficient way of getting protein, can only be satisfied by diminishing the herd. In modern cattle systems, these contradictory pressures are balanced by routinely slaughtering year-old bull calves, since the herd does not need an equal number of bulls and cows to reproduce successfully. This probably explains the preference for slaughtering bulls in ancient cultures, as well: it was a mechanism for the orderly culling of the herd, applying the same concept of herd management to the killing of animals as to their breeding. The Greeks, however, like many other ancient societies, did not simply butcher animals in response to utilitarian forces of supply and demand. They sacrificed their animals. Butchery was therefore conditioned by a particular kind of sacralized violence: it was conducted according to prescribed formulae in ritualized settings. It was bound up with cosmological meaning, serving as a moment of rapport with the divine, and had very specific social outcomes, usually manifested as a feast.[64] In all these respects sacrifice constitutes a highly idiosyncratic human institution, and the connection between this and pastoralism remains a contested issue. At one extreme are those who have argued that domestication took place precisely as a means of guaranteeing a supply of animals suitable for sacrifice.[65] At the other end of the spectrum are those who have gone so far as to predict the death of sacrifice as a conceptual category, complaining that it is a grab bag of "elements taken from here and there in the social fabric of societies."[66] This is going too far: the very ubiquity of the phenomenon demands that it be conceptualized in the broadest possible terms.

Attempts to understand sacrifice as a general phenomenon fall into various schools of thought, but two are of special significance with respect to the question of domestication. Walter Burkert argues that the origins of sacrificial ritual lie in the activities of bands of hunters, and accentuates the ambivalence inherent in the hunt. The hunters will slaughter a fellow creature, and will take nourishment from its death. Poorly equipped by nature, with minimal strength, pathetic claws, and no fangs, we are not natural predators. We have made ourselves into killers, and we profit enormously from this traumatic act by increasing our supply of animal protein. To cope with this paradox we have evolved an elaborate sacrificial system whereby the violence inherent in the act is sacralized and the benefits are shared in a feast.[67] A second interpretation postulates a fundamental difference between hunting and sacrifice. While sacrifice resembles hunting in that it involves the killing of an animal, sacrifice is conceived as an offering. Hunting is predatory, and the animal involved is stolen from nature, not produced. Accordingly it cannot be offered. Furthermore, hunting is opportunistic and immediate, and the killing of a hunted animal does not involve the temporal dimension implicit in sacrifice: we offer a gift to the gods in the expectation of a future return of either bumper crops or increasing herds.[68] Sacrifice thus involves us in a different kind of contact and contract with the gods than does hunting. It is not uncommon for discussions of the hunting origins of sacrifice to overlook these distinctions. Nanno Marinatos, for example, in an otherwise insightful essay contrasting Minoan and Greek sacrificial iconography, insists that the human-animal aggression depicted on Minoan seals and frescoes parallels the aggression played out in sacrifice. Yet the scenes depicting hunting that Marinatos adduces from Kato Syme, Vapheio, and elsewhere mainly show goats, boars, and deer, while the Ayia Triada sarcophagus and Malia sealing depicting sacrificial scenes clearly show cattle trussed on the altar. Iconographically, hunting and sacrifice are distinctly different.[69]

The domestication of large comestible animals, we have seen, is a symbiosis. It alters the composition and nature of both species and herd, on one side, and elicits a new set of behaviors and values on the other. The cattle breeder is as artificially selected as is the cattle breed. If the cattle have become smaller and more docile, the herder has had to become a nurturer, fond of his cows, singing to them, searching for the lost sheep, protecting the herd from wolves, and taking on the authority of the pack leader. All these performances work to both sides' advantage: the cattle are stabled in the cold weather, while the herd's impulse to reproduce is skillfully managed so that more infants survive. The herder's patience and attentiveness are rewarded with increase. In no part of this relationship does the predatory nature of the hunt intrude, until, as usually happens, we decide to eat the animals we have raised. In some respects, eating meat violates all the tendencies that make domestication possible and successful. In the domesticated relationship, if animals have become our young, we have become their parents, and this demands a change in the business of killing animals, since killing a domesticated animal is tinged with betrayal in ways that would never trouble a hunter. Little wonder that in Greek, two of the words used to describe grazing animals could also mean deception.[70] There is only the faintest echo of the hunt as the sacrificial animal is asked to give its consent with a nod, recalling the hunter's prayers that the animal will agree to give up its spirit, yet even this may be a no more than a coincidence. The domesticated animal is just as alive as its wild cousin, and the act of taking a life is as fraught for the herder as the hunter, a fact recognized in Plutarch's famous observation: "They considered the sacrifice of living animals a very serious matter, and even now people are very wary of killing an animal before a drink-offering is poured over him and he shakes his head in assent."[71] In every other respect, however, the sacrificial kill is totally unlike the hunt. The mischances and failures, punctuated by occasional success that characterize the hunt give way to an awful certainty: the domesticated animal is sure to end up dead and consumed. The hunter's success depends upon luck or the cooperation of the gods. In the case of some prey, such as a boar or wild bull, the hunted animal may turn the tables on the hunter. The animal deserves respect for its strength. But the herder faces no such reversals or upheavals. He has a power that far excels that of the hunter, the power to decide exactly when the animal is to die. Just as he chooses to nurse the sick lamb, just as he drags the exhausted cow out of the mud, just as he stays awake and drives off the wolves, just as he determines when the bull will mount the cow, so too he chooses when the sheep or goat or bull dies. He betrays his creatures and becomes a god. It is not a coincidence that patriarchal pantheons arise in stock-breeding cultures. These gods mirror the profound power of the herder who has acquired a more complete power to take life than even the most skillful hunter. Understandably, then, Plato defines piety as the human nurturing of the gods in the same way that cattle farming is the nurturing of cattle. Nor is it a coincidence that the fantasy of the animal who offers itself for sacrifice should be so widespread.[72] Such miracles absolve us and hide the truth, namely, that cattle, gods, and humans are involved in a bloody ménage à trois.[73] Similarly, as Aristotle understood, the capacity of the animal to move hints at its having a soul that directs that movement, a complication that will always make eating meat qualitatively different from eating cabbage.[74]

The killing of domesticated animals, therefore, favors a sacralized treatment that will conceal the betrayal that lies at the heart of their death. The victims are washed, garlanded, and feted on the way to the altar, as if being tricked, as Iphigeneia is tricked into believing she is about to marry Achilles when in fact her imminent death is the result of treachery. The knife used to cut the animal's throat is concealed in a basket, and the animal is showered with grain so as to nod, and thereby assent to the sacrifice. The act of eating meat is not itself the cause of much anxiety, as the celebration following sacrifice makes clear. It is the way in which the meat gets to the table that is disquieting. At a profound level, killing the animals we have raised is an abuse of our power. Sacrifice transforms our brutality and duplicity, displacing it into the realm of the holy where power renders ethical concerns irrelevant. In sacrificial systems it is the gods' abuse of power that we try to defend ourselves against, but it is our abuse of power that underlies this very conception of divinity. Similarly, sacrifice is often performed as if to expiate some guilt or to purge impurity, whereas the actual guilt and impurity lie in the very sacrifice itself, doubly transformed into the animal and into the sanctification of the act of killing it.[75] The profound tension between trust and betrayal deeply embedded within pastoralism is thereby channeled into an elaborate performance hedged around with rituals. The frequent connection in Greek cult between the act of sacrifice and the flight from the altar of an ax-wielding character hints at the ambiguity at the heart of sacrificial business.[76] This most characteristic of Greek institutions carries in it traces of the deep ambivalence that was the Greeks' pastoral inheritance.

Becoming Human

The dilemma of learning to kill the very animals we raised was increased by the fact that domestication took place during the Neolithic period, when the separation of human from Nature had only just begun. Animals were still close; humans were not yet wholly separate from them, nor superior. This is reflected in the archeology of Neolithic Europe, particularly in bone assemblages and wall paintings. At Fontbregoua, a French site occupied around 5000 BC, excavators found human and animal bones showing similar signs of butchery, suggesting that the meat had been scraped off both human and animal carcasses by flint knives. The long bones of both had been smashed to allow the extraction of rich marrow, after which the bones were flung indiscriminately together into garbage pits. The excavators of the site have no doubt that this is clear evidence of cannibalism, but some scholars have preferred to explore the mentality operating here: were humans being reduced to the level of animals, or were animals being elevated to the level of humans?[77] Either way, there is no clear evidence of humans asserting a notion of distinctness, much less of human superiority. Nor is this case unique. Similar deposits have been reported at Zauschwitz, where the long bones and skulls of six humans, smashed to allow the extraction of brains and marrow, were found in a heap of animal bones and household refuse; five individuals were similarly treated at Fronhofen in southern Germany; thirty-eight such partial skeletons were found at Hohlestein in Baden-Württemburg. The list goes on. These instances have forced a reappraisal of the often romantic picture of Neolithic life and suggest that Neolithic societies in Europe experienced remarkable degrees of violence well before the rise of the state-based civilizations of the Bronze Age. In Greece, too, the Neolithic has been the subject of recent reevaluation, leading some scholars to see signs of intercommunal warfare, perhaps fueled by competition over grazing land.[78] Some have taken the deposits in Northern Europe as evidence of alimentary cannibalism, but of equal significance and greater certainty is the evidence they offer for the relationship between humans and large animals. It is so close that one cannot distinguish between the treatment of one species and another. There is no hierarchy. They are equivalent.

In fact, the relationship between human and animal we can infer from Paleolithic and Neolithic art would seem to be one in which humans expressed awe in their encounters with the great animals they hunted. Hunters adorned sacred sites such as the Lascaux and Chauvet caves with depictions of bison hunts, and it is sometimes the human who lies dead at the feet of the huge animal. They created realistic reliefs of these animals in clay and stone, such as the examples from Le Tuc d'Audoubert. In many cases the human and the animal merge, as in the example of the famous "sorcerer" disguised as a stag from the Trois-Frères cave in southern France; such images surely represent attempts by humans to harness the power of the animal.[79] The recent discovery in Germany of the so-called *Lowenmensch*, an ivory figurine half human, half lion, dating to the Aurignacian period (33,000 BP), supports the view that shamanism existed as early as the Paleolithic.[80] In the Neolithic this continued. This art not only blends human and animal worlds, but often does so in a manner suggesting a highly ritualized context for the meeting of the two. Even in instances that are not explicitly shamanistic, important activities take place in the presence of animals. At the Grotta di Romito in central Italy, two humans, one an adolescent suffering from chondrodystrophic dwarfism, were carefully buried underneath three finely engraved aurochs.[81]

In the Neolithic period, then, we witness the beginning of a transformation in the relationship between human and animal that will result in complete rupture. Up to this time the relationship between humans and large animals had been close. Shamanistic experiences elided the barrier between species. Humans prayed to the animals and engaged in rituals in which humans imitated animals and took on their powers, often no doubt so that they could hunt them. Such was the animals' importance that when Neolithic people depicted these hunts the humans were antlike in stature compared to the gigantic stags, bulls, and boar around which the humans swarmed. In the paintings from Çatal Höyük, for example, the hunters are insignificant in comparison to their prey.

Yet, from the seventh millennium on, human societies began to adapt quickly to the domestication of the very beasts they had formerly known only through the hunt and through religious experience. In one sense, pastoralists became even closer to the same species of large animals as they raised their pigs, goats, sheep, and cattle. But they did so by insinuating themselves into the wild animal's life, taking control of it and creating a hybrid family of human and animal, the domestic herd, in which the majestic power of the auroch was diminished into the meek dependence of the cow. In manufacturing this odd creature the first pastoralists were also creating a conundrum: we may hunt animals and tend crops, but only large, domesticated animals require us to combine these operations. They are, as it were, both cultivated like vegetables and slaughtered like wild animals. The sacralization of violence that occurs in sacrifice is an attempt to separate the nurturing features of herding, which make a feast possible, from the killing that is also necessary if we are to enjoy the fruits of our labors. Sacrifice, it has been noted, is an essentially communal act, but the fact that societies unfamiliar with breeding do not practice sacrifice alerts us to its deeper significance: sacrifice resolves the paradox of stock raising.[82] For hunter-gatherers, there is no such dilemma, and hence there is no need to make killing into a sacred act. The religious systems of indigenous Australians, for example, display extraordinary complexity and subtlety, but sacrifice simply has no place in any Aboriginal religious system, either conceptually or operationally.[83] As hunter-gatherers, but not pastoralists, they have no need for the practice or the concept.

The Shadow of the Bull

A further complication in the already convoluted relationship between humans and cattle arises from the fact that, of all comestible domesticates, none except perhaps the pig has a wild cousin that can so powerfully threaten human life. Wild sheep and goats are bigger and more aggressive than their domesticated cousins, no doubt, but only the wild bull has the size and ferocity to match predators such as lions. Accordingly, for pastoral societies, there looms close behind the domesticated cow a wild bull. If the docile cow incarnates order, sustaining us and evoking a nurturing response from us, as we help her to breed and to suckle, the wild bull is her opposite, the rampaging beast that threatens to bring chaos and undo order. Cattle therefore embody a tension between ferocious and docile, wild and domesticated, natural and civilized. As complex, stratified societies emerged in the Bronze Age, this symbolic ambiguity was especially useful, since the development of agriculture had brought about a bifurcation in religious systems: the earth as mother and sustainer of life was complemented by the masculine power embodied in the bull.[84] At the same time, the social structure conforming to this cosmology was centered on kingship, often sacral, of which the bull was a potent symbol. Yet as a symbol of royal power, it resonates in ways that are quite different from other royal animal symbols, such as lions and griffins. These beasts may be majestic and magical, but the bull is different: it is not a predator, its aggression toward humans is usually in defense of its herd, and it is clearly related to a species that has been domesticated. Its wildness is therefore not restricted to the animal world of the wilderness; the bull stands at the intersection of the wild with the domesticated, hinting at the incompleteness of that process. These resonances are especially useful as complex states take shape at the end of the Neolithic because they aid in the articulation of kingly ideology, by suggesting that the king as bull possesses a power as yet untamed and beyond the control of ordinary people.

A benign leader may figure himself as the good shepherd, but early kings preferred a more intimidating profile.

The evidence for the bull's role in conveying messages about kingship is generally fragmentary and disjointed, yet so commonplace are images relating to the bull throughout the world of the ancient Near East that it is clear that the bull was critical to the way the rulers of the first complex, hierarchical states projected their power. For example, around 2250 BC Naram-Sin erected a stele celebrating his victories. At the apex he is shown wearing a horned helmet as he towers over his defeated enemies. Since the horned helmet was commonly associated with deities, as in the case of Shamash giving his blessing to Hammurabi's law code or Ishtar's multihorned war helmet, Naram-Sin's message is clear: part bull and part divine, he is on a level above men. The clearest example we have of such a bull-king is Gilgamesh. His epic has been read as an investigation of how Gilgamesh is socialized, but a more accurate reading would be how the bull-king can be tamed.[85] Right from the outset, Gilgamesh surpasses "all other kings, heroic in stature." He is the "brave scion of Uruk, wild bull on the rampage." This is not entirely a rhetorical flourish, since Gilgamesh's genealogy makes him the son of a goddess who is herself a wild cow. He is called

> Wild bull of Lugalbanda, Gilgamesh, perfect in strength,
> suckling of the august Wild-Cow, the goddess Ninsun.
> (*Epic of Gilgamesh*, SB 1.35–36, trans. George)[86]

As well as being the child of Ninsun, the wild cow, he is also descended from Lugalbanda, whom the god Zangara calls "the wild bull of the mountains."[87] Gilgamesh's bullishness fits his stature, a power so enormous that it threatens the very city he rules. Gilgamesh has internalized the destructiveness of the bull, which makes him not only supremely powerful, but also reckless and essentially unconcerned with anyone else. There can be no social contract with such a king:

> Gilgamesh lets no son go [free] to his father.
> Day and night he behaves with fierce arrogance,
> [*King*] Gilgamesh, [who guides the numerous people,]
> he who is shepherd of Uruk-the Sheepfold!
> Gilgamesh lets [no] girl go free to [her bridegroom.]
> The warrior's daughter, the [young man's bride,]
> The goddesses were listening to their complaint.
> (*Epic of Gilgamesh*, 1. 85–93, trans. George)

The twinning of Gigamesh and Uruk to shepherd and sheepfold is revealing. The first city to exist in literature is conceptually mapped onto the shelter constructed by the shepherd, another sign of the importance of the pastoralist experience to the imagining of city life. These references also point to the tension at play in the story: from the point of view of the citizens of Uruk, Gilgamesh should be a shepherd, caring for them, his sheep, but actually he "lords it over the men like a wild bull" (1.212, trans. George), the animal whose might shatters order. It is therefore not enough to locate the epic's anxiety in the socialization of Gilgamesh, as though he were a wild child needing to be taught how to hold a knife and fork. Gilgamesh is the king, but so destructive that he must be tamed. The epic is thus informed by competing and contradictory tendencies: in the orthodoxy of the story the king is the semidivine hero whose mighty deeds are praised in song. Heterodoxically, however, the king is the problem. He is a wild animal that must be made to submit, his power harnessed to some more socially productive behavior. Both views of Gilgamesh employ the figure of the wild bull, favorably to suggest his divinity, but also pejoratively to suggest his intractability. It is a literary device exploiting the ambiguity in our dealings with cattle, domestic and wild, to express an ambivalence about power and kingship.

The first stage in the taming of Gilgamesh is projected onto the complementary figure, Enkidu. If Gilgamesh is the wild man in civilization, then Enkidu is a man from the wilderness still in his natural state.[88] He eats with the gazelles and breaks hunters' traps, another performance that combines literal and metaphorical resistance to the social world of humans. Like Gilgamesh he is a threat to ordinary humans, but he is from the outside and it is to the inside that the citizens turn, to their own king. His first response, when they appeal to Gilgamesh for help, is to send a temple prostitute to seduce Enkidu, who starts wearing clothes borrowed from the woman, drinks beer, and scares away the wild animals who were his earlier companions. He has lost his innocence. He has been domesticated and emasculated through sex. Once he arrives in the city he wrestles with Gilgamesh; the two are said to grapple like bulls. (2.6.218) The fight ends with Enkidu's submission. How could it not? The domesticated has ever only a shadow of its former strength, while the wild bull retains its power. But civilization itself cannot survive the unchanneled ferocity of the bull. If the external has not been victorious upon being brought into Uruk, then the internal must be sent out: Gilgamesh and Enkidu leave on a quest.

Adventure is from the outset a political narrative. It allows the hero, the imaginative projection of the king, not only to perform deeds advertising his strength and cunning, but to do it in a location far away. In prehistory distance is prestige, and so traveling constitutes part of the adventure and part of the message about the hero's status. Gilgamesh and Enkidu's first adventure is to destroy the demon, Humbaba. He is a guardian of the cedar forest, possibly a bear deity, and the story may well spring from some actual expedition to acquire trees from Lebanon.[89] The reasons for the expedition are not clearly laid out, but Gilgamesh brings back a single cedar from which he fashions a door for the temple of Enlil. The heroes are successful and as a reward Gilgamesh is offered a place in Ishtar's bed. He refuses, she is incensed, and implores Anu to release the bull of heaven. Each time the bull snorts the earth cracks open and hundreds of the people of Uruk fall into hell. Finally Enkidu gets hold of the bull:

> So Gilgamesh, like a bull dancer,
> svelte and mighty then,
> plunged his sword into the throat held fast by Enkidu.
> They butchered and bled the bull and then cut out its heart
> to offer as sacrifice before Shamash.
> Then Gilgamesh and Enkidu retreated
> from the altar itself and stood afar
> in deep respect as they did pray.
> (*Epic of Gilgamesh*, 1. 85–93, trans. Jackson)[90]

The entire episode reads as a charter myth for the practice of sacrifice. The animal must be subdued, the killers must be brutal yet pious, and when they have stabbed the animal they proceed to bleed it, cut it into pieces, and offer a portion to the gods. Enkidu flings a portion at Ishtar who, with her attendants, goes into ritual mourning for the beast. The poem also refers to hanging up the intestines of the slaughtered bull, polishing its horns as a memento of the heroes' deed, and holding a celebration, elements anticipating later features typical of sacrifice: garlands, bucrania, and the feast. Indeed a detailed study has demonstrated the overwhelming similarities between the Bull of Heaven chapter in the epic of Gilgamesh and the sacrifice of water buffalo carried on throughout southern India to this day.[91] The scene is shown on Assyrian cylinder seals (Fig. 2.2) and may have been already established in Akkadian oral and graphic traditions before it had even been written down.[92]

Although Gilgamesh is sometimes referred to as a "culture hero," the expression tends to be employed loosely.[93] He does not introduce agriculture or stock breeding; both are imagined as already existing in the world of the poem. He does not bring fire, brewing, wine, or writing to Uruk, but he is a genuine culture hero, introducing sacrifice. The implications of this, since he is also the bull of Lugulbanda and the son of the wild cow Ninsun, are worth considering more deeply. Moreover, the epic is emphatic concerning the identification of the king with the divine bull: just as Gilgamesh gave the inhabitants of the city no peace with his rampaging at the beginning of the first tablet, so too the Bull of Heaven, even though it was sent to punish Gilgamesh, afflicts the entire community,

Figure 2.2 Gilgamesh slaughtering the Bull of Heaven. Porphyry cylinder seal. Ca. 2300 BC. Paris. Musée du Louvre. Photo credit: Réunion des Musées Nationaux/Art Resource, NY.

smashing dykes and plowing up the fields. Seals depicting the killing of the bull also reflect an awareness that the king and the bull are, at some level, twins: the bull is shown with the face of a man and Gilgamesh with the horns of a bull. What, then, are we to make of the sacrifice? The sacrifice of the bull is in some ways the sacrifice of the king as well. Since he is the agent who performs the sacrifice he is sacrificing himself. One interpretation would emphasize the substitution of bull for man. By emphasizing the similarities between the two the story invites the sacrificial community (those who participate in and profit by sacrifice) to read their own institution as the reenactment of a cosmological narrative with clear consequences for social order. The king's power is asserted in deed, but the ultimate sacrifice that would serve the community and confirm his fitness to rule is displaced onto a victim sent from heaven. It is heaven's gift to Uruk, which Uruk both accepts and returns. The king's authority is thus implicitly confirmed since the king is the pivot on which the exchange between heaven and earth turns.

Recent discussions of Gilgamesh have correctly identified the hero's story as reflecting the emergence of human consciousness from what Roy

Willis and Patrick Curry call "its embeddedness in Nature through the Paleolithic and early Neolithic periods."[94] Such interpretations run the risk of missing a corollary of this process: the emergence of a social consciousness. Gilgamesh's Uruk is the prototypical stratified society. It fashions a cosmology populated by gods and demons in which sacrifice serves as the means of situating humans, kings, and gods in meaningful relationship to each other. In the epic it is the king and culture hero who first kills the bull of heaven and processes a divine affliction into a blessing. The repetitions of this, like any other such sacred narrative, took place in storytelling and in performance, both the domain of professionals, bards and priests. Furthermore, since the contest between Gilgamesh and the Bull of Heaven was seen in the constellations of Orion and Taurus, the myth was part of a larger system of astrological knowledge also under the control of a priestly class. Sacrifice as a social institution regulated by priests gave them the opportunity to turn their knowledge into performance. The Bull of Heaven episode is not just a stage in the process of an arrogant individual's personal enlightenment. It is a story exploring the king's proper role in society that hints at the interconnectedness of human and divine (where the divine is anthropomorphic rather than Nature), and that authorizes sacrifice as the proper medium for exchange between human and divine, thereby affirming a hierarchical social order in which kings and priests dominate because of their knowledge and their actions (reading the heavens, telling stories, going on quests, sacrificing animals, communicating with the gods). At its core, the epic of Gilgamesh is much more than a story of "historically novel individualism."[95] It is a profoundly ideological document, emerging from a complex social order grappling with the tensions that surround kingly authority.

The epic of Gilgamesh ends by asserting the hero's humanity, which is conveyed by his increasing self-awareness as he experiences death and bereavement. This fascination with the human condition, which occurred in response to the rise of ranked, state-based societies in various parts of the Bronze Age Near East, is accompanied by the evolution of ever more complex pantheons. Gods fall into benevolent and malevolent categories, either as anthropomorphic deities accompanied by animals that mirror or suggest the deities' relation to humans, or as demons. A recent study tabulates this as follows:[96]

Benevolent Deities	Destructive Divinities
Anthropomorphism	Animal Gods, monsters
Domesticated species emblematic of deities: bull, calf, bird, cow	Undomesticated species emblematic of monsters: snake, serpent

While gods and monsters do fall into opposite categories, this scheme is flawed since it mistakenly consigns all references to bull gods to the benevolent sphere, on the assumption that the conceptual category "lies within the orbit of cultural domestication."[97] Baal and El, for example, are frequently shown as bulls or referred to as bulls, but it is doubtful they carried this designation because they were identified as divine draft animals. It is precisely because the bull evokes a power outside of domestication that these gods are figured as bulls. Numbers 24:8 refers to an image of the divine having "the horns of the wild ox," while Ugaritic texts refer to "the fierce young bull of the storm god, Adad."[98] What made the bull a suitable symbol and sometimes an avatar of a god (and king) was precisely the explosive power that was barely contained. This is the fundamental experience of the divine that underlies the religious systems of the eastern Mediterranean in the Bronze Age. This divine power is mirrored in royal power; each reinforces the other, and both draw upon the contradictory associations with cattle as ways of configuring this power as potentially overwhelming and thus having to be appeased. Eventually the entire community can be figured a bull or cow. In a late-third-millennium lament from Uruk, as the city falls to attackers it is likened to "a great wild bull which has been wounded with an arrow" and "a wild cow which has been pierced with a spear," while the invaders "let the blood of the people flow like that of a sacrificial cow."[99] The imagery here is especially interesting in that it draws on both hunting and sacrificial motifs.

Both Egyptian pharaohs and Hittite kings alike would identify themselves with the bull in his divine form and thereby exploit the ambiguity of his might: it must be respected, and it was never entirely controlled. This identification of the bull with the king can be attested in the earliest Egyptian records, such as the Narmer Palette from Hierakonpolis (c. 3100 BC), which shows a triumphant pharaoh smiting his enemies under the gaze of bull-headed deities, while on the obverse a bull gores a victim in a scene that serves as a double of the king's victory.[100] Just as the bull was the king's twin, so too it incarnated the divine. The Apis bull was regarded as the servant of Ptah, whom the Greeks referred to as Epaphos.[101] Plutarch reports that the Egyptians regarded the Apis bull as the living image of Osiris, while Diodorus claimed that Osiris manifested himself to successive ages of men through the Apis bull.[102] The bull was central to a nexus of divine and royal power not just in the cult of the Apis bull but also in cults of the Mnevis bull at Heliopolis and the sacred bull of Hermointhis.[103] These bulls were living embodiments of divine power. In the Pyramid Texts the Mnevis bull cries: "I am the wild bull of the grassland, the great-faced bull which came out of On. I have come for you a wild bull of the grassland, for I am he who always fashioned you and will continue to fashion you."[104] Power, virility, divinity are located in these living symbols of godly and royal authority. They also configure that power in a specific way: as wild bulls they can break down irrigation ditches and trample crops, so it is a power that must be managed. Priests and hieratic kings are managers, harnessing the power of the wild bull in a cosmic struggle patterned on the experience of the pastoralist. The messages encoded in Egyptian bull cults can therefore be seen as an extrapolation of the ambiguity of our relationship with cattle: domestication is our triumph, but it is a victory only ever half-won. The wild always threatens to irrupt into society.

So potent was the identification of hieratic royal power with the bull that the symbolism was adopted from Egypt by the Hittite kings. Around 1250 BC the Hittites took the motif of pharaoh and bull, found on Seti I's temple at Abydos and used in the reliefs cut into the rock at Yazilikaya where Teshub, the sky god, and Hebat, his spouse, are depicted standing before a calf, a symbolic representation of their son Sharruma.[105] Similarly, a relief from Alaca Höyük shows a king and queen worshipping before the statue of a bull while a seal of Murshili III shows the king standing behind a chariot yoked to two bulls and driven by the Storm God.[106] The bull and bull-man, in fact, are recurring symbols in Hittite art, deployed to fashion a coherent royal iconography that asserted the unity of the royal and the divine.[107] A similar package of symbols would be employed in the Bronze Age Aegean, where, as we shall see, early Greek culture inherited the same paradox of domestication.

CHAPTER 3

Cattle Systems in Bronze Age Greece

> . . . Pasiphae promoted breeding cattle
> To make the Cretans bloodier in battle.
>
> —Byron, *Don Juan* 2.155

Cattle and the Redistributive Economy

Domesticated cattle (*bos taurus*) are attested at Neolithic sites in Greece such as Argissa Magoula. Descendants of these early cattle can be seen today in one of the two indigenous breeds surviving in northern Greece: the Greek steppe (or Sphakia type) cattle. Steppe cattle stand between 1.10 and 1.25 meters at the shoulder, and weigh up to 250 or 300 kilograms (cow and bull, respectively).[1] The animal has a long face and distinctive long horns in the shape of a lyre, which were prized as drinking vessels.[2] The steppe breed came to Greece from Anatolia, and depictions of long-horned cattle in Minoan art resemble steppe cattle in shape, if not coloring. The second indigenous breed, the shorthorn, is also familiar from art. Bronzes dedicated at Greek sanctuaries from as early as the Archaic period often show these animals. They tend to be a little shorter than the long-horned steppe breed but reach similar weights, around 200 and 300 kilograms (cow and bull). Short-horned cattle breeds are more common in northern Europe and there have been suggestions that the first Greek shorthorns come either from Illyria or from Greek contact with the Celtic world.[3] Neither assertion has been demonstrated conclusively, but divergent lineages in the mitochondrial DNA of European, African, and Asian cattle breeds suggest two separate episodes of the entry of domesticated cattle into Europe. The earlier and faster wave was across the Mediterranean along the coastline, while a second movement entered northern Europe more slowly via the Danube basin.[4] The two indigenous Greek breeds conform to this scenario. By the fourth century, however, such distinctions were no longer evident, although Aristotle could identify different breeds of cattle, distinguishing between the smaller cows of Phasis (intriguingly, the capitol of Aeetes, son of Helios, and hence a place already connected with cattle) and the larger northern cattle in Epiros.

Neolithic herds were small, and the limited size of Neolithic settlements meant that there was no competition between cattle and crops for land.

The wide availability of forest and brush in the mountains guaranteed plenty of forage and permitted low-intensity cattle raising at little effort, but with a good return on labor expended. Cattle were also invaluable as a source of traction power, their importance attested by models of early plows.[5] If Neolithic husbandry was practiced on a small scale, the same cannot be said of the Bronze Age ranching practiced on Crete. Here palatial society arose even before 1700 BC. Cattle, especially bulls, played an important role in Cretan society, both in economic and symbolic terms. We begin with the cattle system of Minoan Crete and will then consider the religious significance of the bull in Cretan cult.

The palaces of Minoan Crete were multifunctional, serving as the center for complex economic, political, and religious systems, and were a microcosm of the Minoan world.[6] They were also not static. Toward the end of the Bronze Age, they appear to have undergone a transformation in which their ceremonial functions came to dominate. Critical to this transformation was a new emphasis on the slaughter of bulls at sacrifice and two correlated practices: feasting and the distribution of meat. Although sacrifice occurred prior to the LM II period, as is suggested by the association of bucrania with double axes and other sacrificial paraphernalia, Paul Rehak and John Younger have observed that in the Final Palatial period the killing of bulls becomes the culmination of bull-centered rituals. They speculate that this reflects Mycenaean influence. They also point to the appearance of the Minotaur on seals at this time. Their tentative suggestion is that these Minotaurs may represent men in masks "appropriating the power of the bull for symbolic, even shamanistic purposes."[7] A parallel shift in religious practice centered on the bull appears to have occurred in Cyprus around the same time. Here, as metallurgy developed, votive dedications included bull figurines and the so-called Ingot God. Sophocles Hadjisavvas describes the phenomenon this way: "Whereas at the beginning of the Late Bronze Age artists favored a female human being and male cattle, toward the end of the period they invented another deity combining the male and the bull. The new image of the deity in the form of a god standing on an ingot and wearing a horned helmet appears only at sites where copper production was predominant, such as Enkomi, while the bull preserved its importance at all other settlements."[8] Toward the end of the Bronze Age, then, on both Crete and Cyprus, a fusion occurred of bull, god, and man, attesting to the enormous symbolic significance of cattle.

The increase in bull sacrifice in the LM III period raises a question: how was an intense demand for sacrificial cattle accommodated? It is well established that the redistributive economy of the palaces controlled enormous resources. Knossos, with a population of between 12,000 and 17,000 people, possessed flocks numbering by some estimates up to

100,000 sheep.[9] These numbers mean that the wool industry at Knossos would have required at least 100,000 hectares of grazing land, implying direct control of much of the Mesara plain from Knossos. Some estimates go as high as 200,000–300,000 hectares, which would mean that Knossos controlled as much as a third of the island.[10] And just as Knossos was the center of a wool and textile industry, so too there seem to have been other regional specializations similarly centered on palaces: barley at Ayia Triada, wheat at Khania, and wine at Zakro.[11] Palatial domination of the hinterland was reinforced by smaller satellite palaces or villas, such as those at Amnisos and Tylissos. In a study of the villas at Ayia Triada, outside Phaistos, Vance Watrous has been able to show that the palatial system was replicated on a smaller scale. Scribes, smiths, potters, carpenters, weavers, cooks, and farmworkers are all attested working on estates, paid in rations of wheat, olive oil, barley, figs, wine spices, wool, cloth, and copper.[12]

In this highly centralized system the position of cattle has received less attention, despite the fact that at least a third of the Linear B records from Knossos deal with livestock. This oversight is due to the predominance of sheep breeding and the much smaller numbers of animals recorded in the cattle-related tablets.[13] Cowherding as a distinct occupation is not mentioned in the Knossos tablets and may have been performed by other categories of workers, further diminishing its visibility. Tablet PY Ae 04 from Pylos, for example, reports that *Ke-ro-wo* the shepherd at *A-si-ja-ti-ja* was assigned the job of watching over the cattle of Thalamatas. In some respects, however, the cattle system was very different from the sheep industry. Unlike sheep and goats, for example, oxen were given names, which bestowed on them an individuality not shared by the smaller animals. In the Linear B tablets from Knossos we find oxen called *ai-wo-ro, ke-ra-no*, and *ko-so-u-to,* corresponding to later Aiwolos (Nimble), Kelainos (Blackie), and Xouthos (Tawny).[14] Cattle are also unlike other commodities in the Linear B records in other important ways: sheep, grain, and oil were handled in huge quantities, and the records of these pertain to quotas coming into the palaces. Cattle, on the other hand, appear in small numbers and are usually listed in records of goods being dispersed from the palace. Take, for example, this tablet from Knossos (Ch 902):[15]

> *Mi-ru-ro* at *Si-pe-we*: one ox, twelve young calves
> The *ko-re-te* of Odrus: one ox, twelve young calves
> The *ko-re-te* of Wa-to: one ox, twelve young calves. Da-nu-wo of Wa-to: one ox, twelve calves
> The *ko-re-te* of Si-ra-ro: one ox, twelve young calves
> The *E-ra-ne* at *Pa-ko-we*: one ox, twelve young calves

The *U-wo-qe-we* at *Odrus*: one ox, twelve young calves
The *ko-re-te* of *Ri-jo-no*: one ox, twelve young calves
The *ru-ki-ti-jo*: one ox, twelve young calves
The *ko-re-te* of *A-pa-ta-wa*: one ox, twelve young calves
The *ko-re-te* of Ku-ta-i-to: one ox, twelve young calves
The rowers of *Re-na*: one ox, twelve young calves
The *we-re-we* . . . one ox, twelve young calves

It is unlikely that these oxen were being dispatched to pull plows. Draft animals are referred to as "working oxen" (*we-ka-ta*). When the palace distributes them to dependent communities they are recorded as being sent "to the village" (*da-mo*), and are listed in even numbers (for example 2, 6, and 50), since plowing was done by yoked pairs.[16] Since the oxen on Ch 902 are listed singly, are not described as working animals and are sent to recipients of various sorts including officials, it is unlikely that they were intended as plow-oxen. Rather, they were probably being distributed for sacrifice, a fate that seems not to have been shared by rams even though Knossos bred many more of them than cattle.[17] The oxen were the means by which a network of obligation was fashioned between Knossos and peripheral localities and subordinate groups. In the case of the calves, it is reasonable to infer that these were animals from a central herd raised near the palace and dispatched to smaller communities further afield. The sex of the calves is not specified. If, as is likely, each dozen comprised a mixture of male and female, then a variety of purposes can be imagined: heifers would be kept for breeding purposes, while bull calves would either be castrated and raised as plow oxen, or allowed to reach maturity intact, at which time they might be put to stud, used in the bull games, or sacrificed for meat, their by-products being processed for leather, tallow, and glue.

The tablet is valuable for the light it sheds on the size and nature of the network of exchange centered on Knossos. Although not all locations mentioned can be identified with assurance, some are well established. *Ru-ki-ti-jo* refers to the people of Lyktos, located 25 kilometers southeast of Knossos, close to the western edge of the Lasithi Plain, while *A-pa-ta-wa* corresponds to Aptera, situated well over 100 kilometers west of Knossos and only 10 kilometers from Khania. Another tablet, C59, mentions fifty working oxen "at Kydonia" in western Crete, and it has been suggested that cattle breeding was concentrated in that region.[18] The list is especially valuable as evidence for the use of cattle as an item of exchange between different groups and individuals within the administrative hierarchy of LM III Knossos. Six of the dispersements are made to an official called the *ko-re-te*. The etymology of the word is uncertain, but Michael Ventris and John Chadwick note a possible connection to the verb *koren-*

numi ("to satiate").[19] If this is correct, the *ko-re-te* may have been envisaged as the official who "satisfied" the needs of the local population by providing sacrificial animals, and by extension, the meat consumed in communal gatherings (cf. Latin *curator* and *procurator*). Another group to whom the standard distribution is made is referred to as the *pa*$_3$*-ko-we e-ra-ne*. The first of these terms is a locative, translated as "at Pa-ko-we," but the identity of the recipients is opaque. Ventris and Chadwick, somewhat mischievously, asked, "Surely not Hellanes?," but a more likely explanation is that the term refers to the members of an *eranos*, a commensal group. The other recipients, the *U-wo-qe-we*, are generally thought to be officials perhaps serving as overseers, while the *re-na-jo e-re-ta* are probably rowers, another important group serving the interests of the palace.[20]

What sets the ox tablets apart from other transactions involving livestock and agricultural produce is that they demonstrate that the Late Minoan economy depended not only on the redistribution of staple goods, but on reciprocity and the giving of high-value goods in exchange for staples. Since outlying territories supplied Knossos with wool and staples such as oil, grain, and wine, there was little incentive for suppliers to cooperate with the palace center if the only reward was to receive back a percentage of the very goods they themselves had produced. Cattle served as a reciprocal gift, a luxury item of enormous value. Oxen were a powerful source of traction, a rich source of protein, and an abundant source of leather. So valuable were they, in fact, that when metal ingots were traded they were shaped like hides, as if the shape symbolized their value.[21] The pattern of exchange here resembles that of more recent empires, in which raw goods from the colonies were exchanged for manufactured goods from the imperial center, although a better analogy, certainly closer in time, may be the exchange of grain for sheep recorded in the Persepolis Foundation Tablets from late-sixth-century Persia. Describing that system, Wouter Henkelman has recently argued, "In the western sector of the region under its purview, the administration had the possibility of exchanging grain or wine and thus providing animals for sacrifices it wanted performed."[22]

Although the bureaucratic records of Linear A and B create the impression of total control exerted from the palatial centers, recent studies have emphasized that the redistributive economy was never the sole system operating on Crete. It was complemented by decentralized economic activity and private farming as well.[23] It is unlikely, however, that cattle herding took place outside the palace-dominated areas, since pastoralism on a scale producing hundreds of calves every year requires access to very considerable water supplies and the ability to move across the landscape freely. Large-scale pastoralism is difficult to accomplish alongside small-scale farming.[24] It is more likely, in fact, that the disproportionate value of cattle encouraged a tight control over the cattle system by the palace. Thanks to a monopoly on cattle, Knossos, and no doubt the other palaces, affirmed networks of regional cooperation that included within their parameters some districts under very direct control, administered by palace-appointed overseers, as well as more loosely affiliated areas retaining considerable degrees of autonomy. The redistributive economy relied upon a degree of cooperation from sectors beyond direct control. Crete in the Late Bronze Age was not a monolithic temple estate administered from Knossos so much as a network of regions, towns and estates each operating in a different way within the palace's sphere of influence.

According to this scheme cattle wealth was used in practical and symbolic ways to assert palatial authority. Religious rituals that focused on the bull were ideologically charged, not simply because they drew on the same metonymy of bull strength and royal power as in the Near East, but also because they complemented the very real importance of cattle in Minoan society as the embodiment of palatial largesse. For this reason the Minoan interest in bull and cattle-related imagery is not limited to the religious performance of bull leaping. Cows and bulls stand for prosperity, and capturing them is an assertion of power.[25] A good illustration of this is to be seen well before the end of the Bronze Age on the so-called Naval Expedition fresco from Thera.[26] One panel of the fresco shows a boat pulling in to shore before a town whose inhabitants can be seen clustered around a building. Behind them looms a hill. Figures of drowned men in the water show that the landing has been resisted forcefully but warriors have made their way to shore. They are armed with boar's-tusk helmets and full-length shields made of cow's hide and can be seen marching uphill. Above them cattle of various sorts—rams, cows, and possibly deer—are being driven toward the hills. The warriors and the cattle are composed on three planes, zigzagging their way up and away from the town and into the mountains behind. There have been attempts to identify the scene with specific locations such Libya or Crete, and to assign ethnic identities to the sailors, soldiers, and defenders, but a more important point is often overlooked: the fresco does not show the sack of the city, and the attention of the raiders is focused on chasing the herds that are being driven to safety in the mountains. In the next scene, to the right of the one shown above, the sea battle continues while on the shore wait two bulls, suggesting that the raid has been successful.[27]

Sarah Morris has demonstrated that the West House frescoes reflect many of the same motifs as epic poetry, prominent among which is the topos of the raid. She notes that Homer refers to the raids of Achilles in the Troad (*Il.* 6.421–24, 20.89–93), and to these references one could add the many stories told by Odysseus of his attacks on the Kikones, his theft of the Cyclops' sheep, his reconnoitering of the land of the Laistrygonians rich in sheep and cattle, and the most disastrous raid of all, the theft of the cattle of the Sun.[28] Yet the presence of such formulaic elements in both visual and oral narrative in no way detracts from the value of the frescoes as indicators of activities familiar to Bronze Age viewers. The Minoan world was no stranger to raiding, and cattle were an especially desirable commodity.[29] The difference between the Bronze Age practice shown on the frescoes and the recollection of it in epic poetry is that by Homer's day, in the eighth century, cattle were no longer a critical part of the intricate exchange networks that connected different districts. Odysseus' men slaughtered the cattle of the Sun for six days and gorged themselves stupid on the meat, but the warriors on the Theran frescoes would return to the orderly, tranquil world depicted on the south frieze. At home they would have a variety of uses for their captured livestock.

Raiding was one way of beefing up the supply of cattle; another was to capture wild animals. A clay larnax from Armenoi shows hunters driving an assortment of animals into an enclosed area. They spear the large animals and use nets to trap their calves.[30] Similarly, the two famous cups from Vapheio show contrasting methods of trapping bulls. One way was to use a sexually receptive cow as a decoy, the other was to wrestle the animal into submission. On the "violent" Vapheio cup, after bursting through the net the bull is shown charging over one would-be captor and tossing another high into the air in a scene that appears to parody bull leaping. The first of these two cups is Minoan but the second is a Mycenaean production based on Cretan models.[31] The very high degree of imitation achieved by the Mycenaean artist is itself a testimony to the pervasive identification of the bull with Minoan culture.

Bull Leaping

In Cretan culture, the bull is ubiquitous.[32] Palaces such as Knossos, Mallia, and Phaistos were prominently adorned with horns of consecration. Great ceremonial axes of bronze, suggesting the sacrificial slaughter of bulls, stood on display in the great courts of the palaces. Objects such as the Ayia Triada sarcophagus show cattle trussed in preparation for sacrifice.[33] In every medium imaginable, from rings to terracotta figurines, from stone seals to frescoes in relief, the image of the bull permeates the Minoan world. Furthermore, depictions of bulls and bull leaping figure prominently in the pictorial decoration of Neopalatial Knossos, where the major entrances leading to the center of the palace complex, on the west, north, and southeast sides, were adorned with wall paintings of bulls and bull leaping.[34] Notable among these depictions are those found on rhytons. The Cretan rhytons were probably derived from Syrian antecedents, but the Cretan vessels evolved into objects of exceptional size and beauty.[35] The bull's-head rhyton from Knossos is carved out of steatite and decorated with rock crystal and gold.[36]

Vessels like this one and a similar vessel found at Zakro were too heavy to be used in anything but ritual settings; the boxer rhyton weighs 1 kilogram empty and 3 kilograms when full.[37] They are perforated at the bottom and could only hold liquid as long as they were plugged. These, then, are ceremonial containers. One possibility is that they dispensed wine and were used in a banquet setting. Nanno Marinatos, however, has persuasively argued that they were used to hold blood from sacrificial animals, which was collected and then poured onto the ground as a libation.[38] One side of the Ayia Triada sarcophagus shows a bull trussed for sacrifice while on the other side women pour what may be blood from the sacrificial animal into a large krater.[39] A religious function is also suggested by the portraitlike quality of the rhytons. In contrast to the stylized depiction of priestesses, princes, and even deities, the bull's-head rhytons are remarkable for their vividness and their individuality. They are portraits that would have been the centerpiece of any gathering at which they were used. If such a gathering were the feast following a sacrifice at which the bull was consumed or its meat distributed, a formal libation from a vessel imitating the bull's head would have constituted a ritual reenactment of the bloodletting that began the sacrifice. The savagery of the animal's slaughter was replaced with the formal dignity of libation. The disposal of the rhyton after the ceremony amounted to a second killing.[40]

A deep-rooted tension between the wildness of the bull and the need to master it also appears to underlie the most famous Minoan institution involving bulls: bull leaping. The details of this practice are much contested. At one extreme are those who deny such performances ever took place, while at the other are those who offer a detailed breakdown of how the ritual was performed. The latter base their conclusions on the various depictions of bull leaping shown on seals, rings, ivory figurines, and frescoes. Based on these artistic representations, Sir Arthur Evans believed that the process consisted of four clearly defined phases: the leaper approached, grasped the bull by the horns, vaulted over onto the animal's back, and then sprang onto the ground. Subsequent studies by Agnes Sakellariou and John Younger assembled evidence for variations on Evans' schema, until a range of styles was identified and associated with specific periods in Late Minoan culture. These changes in style make little sense divorced from an actual practice; they are therefore taken as evidence that a "sport" of bull leaping did exist. The conclusion of Younger's study sums up the connection between artistic representations and the performances they recall: "In conclusion, bull-leaping begins to appear in artistic representations toward the beginning of the Late Bronze Age in Crete and on the Mainland. The main system of performance probably followed that of the Diving Leaper Schema. When bull-leaping itself was discontinued, perhaps toward the close of the LB IIIA or the beginning of the LB IIIB period, later representations depicted the leaper in the floating pose (Type III), a pose not copied directly from the sport."[41] The ceremonial courts characteristic of Minoan palaces are usually taken to be the locations for this activity, serving as Minoan bull rings.[42]

Despite the assurance of Evans' reconstruction, or perhaps in reaction to it, there are those who seriously doubt whether it is at all possible to leap a charging bull, particularly when, as in most of the scenes depicted in Minoan art, the bull has its head up, with its horns vertical. Alexander MacGillivray, for example, asserts that no person ever jumped over a bull's back on Crete or anywhere else. He suggests instead that the artistic depictions of bull leaping are representations of a celestial drama. "Orion confronts Taurus, composed of the Hyades and Pleiades (the seven sisters), while Perseus somersaults with both arms extended over the bull's back to rescue Andromeda."[43] Yet a cosmological reading need not preclude a ritual performance, and certainly does not disprove its existence altogether.[44] Moreover, the lack of any Minoan liturgical text comparable to, for example, the epic of Gilgamesh, makes it hazardous to speculate about Minoan cosmology in anything but the broadest terms. We can read the Assyrian seals with some degree of confidence because we have a textual narrative that explains the scene depicted, but we do not have anything comparable for Crete, unless we rely on later stories regarding Minos, Pasiphae, and the Minotaur, stories that were not written until after Crete had come under Greek control. If there were Bronze Age myths accompanying the stories of Pasiphae and Minos, they certainly were contaminated in the process of transmission. Here we come up against an abiding difficulty in recreating the particulars of Bronze Age religion: interpretation without textual support must remain tentative.[45] Since we do not have the Cretans' stories about the stars, we cannot assume that they identified Orion, Taurus, Andromeda, and the rest in the same way as the later Greeks or their own contemporaries in the Near East. Even if we were to suppose, reasonably, that the astrological traditions of the Near East were familiar to the Cretans, their existence alone would not be enough to dismiss bull leaping as modern invention, or to read the scenes of bull leaping as purely symbolic.

The variety and specificity of the scenes, in fact, point toward actual performances. Why else would assistants be shown holding the horns? Why in some cases would two animals be shown? In most depictions the bull is charging, but in at least one instance the great beast is calmly seated on the ground with its legs folded neatly beneath it. In another notable case the bull has begun to mount what looks like an altar, its forelegs raised above the ground and awkwardly hooked onto the stone blocks.[46] These variations make no sense if all the artistic renderings refer to the same single cosmic drama. If, on the other hand, they commemorate specific performances, the differences between various depictions are intelligible. As with the realistic rendering of individual animals on bull's-head rhytons, the scenes of bull leaping suggest an interest in commemorating actual events rather than representing a cosmological story.

The only compelling objection to the existence of real bull leaping, in fact, is the assertion that it is physically impossible, a claim often made in discussions of contemporary bull sports.[47] Yet in the southwest of France, a version of bull jumping is still practiced regularly in a form that is an almost exact parallel of the Minoan version: the *course landaise*. The animals employed are not attacked, stabbed, or slaughtered by the participants.[48] In one event the leaper vaults over the body of the charging cow. In one variation of this, the *saut de l'ange*, the jumper leaps straight along the same axis as the charging cow (Fig. 3.1). The technique looks exactly like the flying leap depicted in Minoan glyptics. Other leaps include a somersault over the cow's back, the *saut pieds joints*, in which the leaper thrusts his legs forward while leaping vertically, and the most difficult of all, the *saut vrillé*, in which the sauteur performs a pike while leaping.

Comparisons with the *course landaise* raise some intriguing possibilities for our understanding of Minoan ritual. For example, depictions of Minoan bull leaping that involve the killing of animals are rare, and it is possible that the point of the exercise was not to kill the animal but to demonstrate superior skill.[49] The organization of the *course landaise* also raises questions regarding bull leaping as an institution. Senta German has recently proposed that bull leaping should be seen as a performance carried out by young men of high status.[50] This is a plausible interpretation, but the organization of the *course landaise* also raises the possibility of professional performers. In the modern version of the sport, the performance is only one part of a larger system of both cattle production and specialized training, consisting of sixteen *ganaderias*, which function as breeding farms for the 1,200 *vaches landaises*, the breed specifically raised for the sport, and as *écoles taurines*, training schools for the *toreros*. The same may have applied in Crete, so that rather than being high-status youths engaged in acts that functioned to advertise and reinforce social hierarchy across Minoan society, these Cretan bull jumpers were specialists from the place that was the home of bull leaping: Knossos. Commenting on the distribution of bull imagery in Minoan art, Rehak noted, "So prominent is this bull imagery that one suspects that its dissemination reflects an institutionalized source, perhaps the élite who controlled the

Figure 3.1 Sauteur performing the *saut de l'ange. Course landaise.*

Knossos palace."[51] Depictions of bull leaping, though popular at Knossos, are not attested at the other Minoan palaces, although actual performances surely took place in the great courts of the other palaces. Bull leaping elevated the palaces as ceremonial centers, since only palaces were designed for the spectacle, but the institution required more than priests and hieratic processions. It depended on skilled athletes and suitable animals. By monopolizing the stock—we have already seen evidence for the distribution of cattle from Knossos (Ch 902)—and the specialized personnel on which the institution was based, Knossos asserted its cultural dominance of Crete.

Bull leaping was thus a centerpiece of Minoan life. It may have been viewed as a sacred reenactment of a cosmic drama with roots going back to earlier Near Eastern religious systems. Such connections with Egyptian and Near Eastern cosmology helped make the Cretan bull cult exportable: excavations within the past twenty years in Egypt at the Hyksos capital of Avaris (Tell el-Daba'a) have brought to light wall paintings of the sixteenth century BC that depict scenes of bull leaping, suggesting that the practice was already known outside of Crete well before the end of the Bronze Age.[52] Similarly, Canaanite seals reflect an awareness of the practice, as do seals found at Alalakh, in Syria, dating to the seventeenth century BC.[53] The exact details of the cult, especially the narrative that complemented the ritual, are probably irrecoverable, but the existence of such a dangerous performance shows that Cretan religion reflected the same legacy of pastoralism witnessed throughout the stratified states of the Bronze Age eastern Mediterranean: the compulsion to demonstrate in myth and ritual man's control of nature through mastery of the bull.

As in Egyptian and the Near Eastern bull cults, Minoan bull leaping gave expression to a tension that underlies man's somewhat tenuous mastery of nature, reaffirmed each time human triumphs over animal. Not coincidentally, such cults flourish in societies becoming increasingly stratified, as the affirmation of human prowess serves by analogy as an affirmation of social order. For this reason bull leaping was a public performance, displaying the theatrical quality that is a distinctive feature of Late Minoan society. Recent studies of Knossos have emphasized this aspect of Minoan culture, noting that there are few signs of habitation in the last phases of the palace at Knossos. Neopalatial Knossos may have been as much a ceremonial as an administrative center, which would help to explain the prominence of theatral areas here and at the other palaces.[54] The carefully designed entryways leading to these performance areas were decorated with reliefs that repeated certain topoi such as processions and bull leaping, suggesting the very rituals that one would witness upon entering. If the frescoes throughout the complex can be read as guides to the performances that occurred here, then there is every reason to believe that the palaces were the focal points for an increasingly hieratic society, in which religious celebrations such as harvest festivals were the major social events.[55] There is also evidence for the use of complex proportional schemes such as the Fibonacci sequence in the articulation of units of ashlar masonry in the west façades at Knossos, Mallia, and Phaistos during the Second Palatial period, suggesting that the design of the public space, like the decoration of the palaces, was meant to evoke the importance of the activities that occurred here.[56] Like any rituals, these performances were ephemeral, but rings produced from steatite molds depicting bull leaping commemorated the events. These rings were popular among members of the Minoan elite, who favored scenes of bull leaping as personal markers. We can infer the high status of the individuals who used these rings from the fact that identical sealings from the rings have come to light at sites all over Crete, from Khania to Zakro. Fifty-three sealings from ten "Knossos replica rings" have been found so far, and of these ten, six depict bull leaping.[57] The ideological significance of the ritual was paramount.[58]

Feasting

It is clear, then, that in Late Minoan culture cattle enjoyed a unique status both as symbols of power and as a unique commodity in an elaborate network of exchange, despite the fact that the palatial economy was not exclusively or even primarily focused on cattle breeding. Bull leaping was a nodal point in this cattle system; another was the banquet. Formal commensality of this sort serves a double purpose: it unites those groups who share the feast and advertises the underlying power relations that exist among the participants and within the broader community. Feasting can also convey a powerfully conservative message. Later regulations concerning sacred associations and their banquets frequently contrast the proper, formalized performance of the feast with its opposites: savagery, unruly conduct, and gluttony.[59] Minoan banquets are likely to have emphasized orderliness in much the same way. A series of social performances is played out: the powerful provide, the dependent receive, the insignificant are excluded. These messages can be reinforced by the quality and quantity of the food and drink, or even by the value of the vessels used. In Bronze Age feasting the lavishness of the banquet paraphernalia is especially noticeable. Inventories from both Knossos and Pylos refer to thrones, footstools, tripods, and goblets, as well as rhytons and ewers in the shape of a bull's head (*qo-u-ka-ra*).[60]

Changes in feasting practices match ideological change. There is evidence to suggest that ritual feasting assumed a high level of importance in Cretan society in the LM III period. Feasting had been a feature of Cretan social practice in earlier periods, in a variety of settings: by tombs, as funerary banquets, at caves and peak sanctuaries, as religious rites, and in the banquet halls of the palaces. In LM IIIB and C, however, competitive feasting began to emerge as a means of reinforcing the exclusive status of the elite. The process can be seen at Phaistos. Here, on the basis of very different types of pottery desposits, Elisabetta Borgna suggests that two forms of feasting were being conducted: public banquets, or celebratory feasts, which took place on the Acropoli Mediana and were open to heterogeneous social groups, and more restricted competitive feasts that were held at the Casa a ovest del Piazzale I.[61] Borgna notes the use of elaborately decorated pottery at the elite banquet, which she also sees as a sign of competitive display. Elite self-definition, social stratification, and regional relations were all modulated through palace-sponsored feasting.

Taken with the Linear B evidence for the control of cattle production by the palaces, and coinciding with the decline in bull leaping suggested by changes in glyptic representations, the signs of palace-centered banqueting underscore the ideological significance of cattle wealth in Late

Minoan times. This is the period in which Crete is firmly under Mycenaean control, and it is surely under Mycenaean influence that bull leaping passes from practice into memory. Soon the realistic bull of the ceremonial rhyton will give way to the fanciful Minotaur (whose only taurine feature is exactly that which the rhyton depicts: the head). It is with the Mycenaean control of Crete that the symbolic significance of the bull takes second place to the more tangible benefits to be had in controlling access to herds for breeding, plowing, and eating. In other words, the distinctive features of the LM III cattle system, in terms of production, distribution, and consumption, must be understood in the context of the Mycenaean domination of Crete.[62] Like most imperial conquests in antiquity, this was maintained not by massive military occupation but by the cooption of local elites. Mycenaean Crete is not heavily fortified. Instead, the Mycenaean control of Crete was accomplished by winning over the Minoan elite. Those who administered the palaces' estates and resources found under the Mycenaeans fresh opportunities to assert their status through the particular emphasis given to cattle wealth in the Mycenaean world. Cattle were no longer destined for elaborate bull games that made sense within a Minoan theology, but were now used as a means of rewarding those satellites that cooperated with central authority, while palace-sponsored feasting on a scale not previously practiced on Crete allowed the elite both to sponsor banquets for others and to engage in ritualized commensality among themselves. It was from Mycenae that Late Minoan society acquired its taste for banqueting, and it is to the Mycenaean mainland we now turn.

The study of feasting in Mycenaean society has advanced considerably in recent years and a number of studies give a more detailed picture of commensality in the Mycenaean world.[63] The most striking evidence comes from Pylos. The Ta series of Linear B tablets found here list eleven tables, six thrones, and sixteen stools, which along with other pieces of furniture, vessels such as tripods, and accessories such as axes and knives were destined for use in sacrificial feasts. Frescoes from the palace show seated banqueters facing each other and raising their cups as if to toast each other, from which we can infer that those participating at banquets sat in pairs.[64] This interpretation fits well with the twenty two miniature *kylikes* found in the same room. We can imagine eleven pairs of high ranking men and women seated at eleven tables on their thrones and stools, each with a ceremonial cup used for pouring libations. This is clearly a restricted event, and it may even be possible to speculate on the identity of the high-status participants. The six thrones would serve the *wanax* and *lawagetas*, the lord and war chief respectively, as well as three *telestai*, officials representing landholders, and one representative of those who worked the land, the *worgioneion ka-ma*. The sixteen stools will then have been for the sixteen *koreteres*, or governors of the sixteen principal localities in the Pylian kingdom.[65]

What makes the Linear B evidence particularly interesting is that the Ta series tablets come from room 7 inside the palace, the same location as a rich deposit of animal bones that shed further light on feasting practices at Pylos. The bones were found in six depositional units, each of which consists of the remains of between five and eleven head of cattle. The bones showed signs of dismembering and decarnation consistent with the removal of flesh before the bones were burned. At the same time, the bones found in the deposits were nearly all mandibles, humeri, and femurs, the large bones of the upper forelegs and rear legs.[66] Taken together, these details point to sacrifices of between five and eleven animals at a time, any one of which would have produced vastly more meat than could have been eaten by a gathering of only twenty-two worthies. At a conservative estimate the small bovines raised in the Bronze Age could provide 100 kilograms of meat. One of the larger sacrifices could have resulted in over 1,000 kilograms of meat, enough to feed hundreds of people. Clearly the bone deposits from the palace imply the consumption of vastly more meat than could be consumed at an elite banquet.

The two most recent studies of the bone deposits in room 7 have assumed that the animals were sacrificed and then consumed in communal feasts. This is possible, but not the only necessary conclusion. Either the elite banquet was matched by a more public banquet held elsewhere, along the lines of the Phaistos model proposed by Borgna, or else most of the meat was butchered and distributed for private consumption. This seems to have been the practice at Tsoungiza, the Mycenaean site close to the classical site of Nemea, where important evidence regarding feasting has come to light.[67] The bone deposits here indicate that deliberate butchery took place on site, but the bones are not a random sample from all parts of the skeletons. Instead, bones from the head and feet dominate the record, suggesting that the animals were dismembered and that most of the flesh-carrying joints were taken away and consumed elsewhere. This helps to explain the discrepancy between the enormous quantities of meat made available by these sacrifices and the very small numbers of banqueting objects—thrones, tables, vessels, tripods, and knives—recorded in the palace inventories. In general, only members of the elite participated in the feast, while most of the meat was given away. A recent study of the newly discovered Linear B tablets from Thebes shows that it was commonplace for wheat and flour also to be widely distributed.[68]

The feast, then, should be seen as one component within a system of food distribution that also included sacrifice and dedication. There is, however, an interesting discrepancy in the material culture. The faunal evidence from Ayios Konstantinos on the Methana peninsula reveals a marked preference for feasting on pigs that does not correlate with the hundreds of figurines, primarily bovine, deposited in the sanctuary.[69] Together, feasting and meat distribution comprised a central social institution in the Mycenaean world for establishing the social hierarchy. Meat eating in any agrarian society is already an act of conspicuous consumption. The feast teased out the significance of this and tied that consumption (and that conspicuousness) to an assertion of status. The privilege of participation demarcates the elite, while the distribution of meat and grain to those excluded from the feast reaffirms their status as subordinates and dependents. Symbolically, the giving of meat becomes a way of acquiring prestige. Through the feast, cattle wealth is transformed into social capital.[70]

Nestor's Cattle

While feasts were a focal point for Mycenaean social ritual, they were only made possible by an extensive system of husbandry. The evidence from Pylos is the most complete available to us from a Mycenaean kingdom. Here, as at Knossos, sheep were raised in great numbers, but unlike Knossos, the palace at Pylos also took an active interest in all aspects of cattle breeding throughout the kingdom. Herding was carried out at three levels: under direct palatial control, by the subordinate communities scattered across the kingdom, and by wealthy individuals. Palatial involvement is reflected in the palace records: although the Pylos tablets mention only five or six shepherds the same tablets refer to 280 cowherders (*qu-o-ko-ro*).[71] The palatial herds may have been extensive, although if "cowherder" is the equivalent of "plowman" then we might only have to double the numbers of herders to reach a plausible number of cattle. This would make sense of a number of tablets from Knossos that preserve the name of a herder and a single pair of oxen, such as *Ta-za-ro* and his oxen, *Ai-wo-ro* and *Ke-ran-no*.[72] One tablet from Pylos (PY An 20) notes the disappearance of ninety cowherders, a number as mysterious as the circumstances surrounding their disappearance. Does this point to hundreds of cattle abandoned, or to a less traumatic episode involving local men failing to fulfill an obligation to plow the land of the *wanax* or some other high official? Sixty herders mentioned on PY An 830 are tied to land belonging to the *ko-re-te* and are described as *o-pi-da-mi-jo* (cf. *epidemioi*, "locals"). Perhaps the missing cowherders had fled from a particularly odious official. It is also possible that *qu-o-ko-ro* designates an individual's status with respect to his obligations to the central administration, but does not translate into a description of his full-time occupation.

Whatever the exact meaning of "cowherder," there is no doubt that the Mycenaean palaces were able to organize the movement of animals on an impressive scale. The *o-pa* tablets from Thebes demonstrate that pigs, goats, sheep, and cattle were brought from as far away as Karystos and Amarynthos on the island of Euboia.[73] One thinks of the speeches of Eumaios and Philoitios in the *Odyssey* in which they lament the herds of Odysseus sent to graze on the mainland and on Kephallonia.[74] This might be a description of Dark Age pastoralism, but it could equally preserve a memory of the more organized practices of the Late Bronze Age. If cattle were regularly driven to palatial centers over considerable distances, and conveyed over water too, this will have alleviated pressure on the satellite towns within the Mycenaean kingdoms to turn over fertile farmland to grazing. This would help to explain why a region like Nichoria, well watered and ideally suited to grazing, actually supplies evidence of less cattle production in the LM III period than before or after. The percentage of cattle among all animals reported from Nichoria by period drops from 20 percent in MH to 15 percent in LH, before rising dramatically to 35 percent in the Early Iron Age.[75]

Aside from herds in the possession of the palace and its officials there were also herds owned by communities, which were subject to a tax paid to the palace in hides. The Pylos tablets mention herds as large as 90 head of cattle and record the collection of over 200 oxhides in a single year.[76] This was carefully organized and administered, as is reflected in the Ma series of tablets from Pylos, which records the assessment and payment of tribute from eighteen communities from both the Hither and Further Provinces of the Pylian kingdom.[77] On each of the tablets oxhides are listed as the fourth of six standardized entries, and as with other palace accounts, scribes noted when any village failed to fulfill its obligation. The collection of hides as tax was therefore a regular feature of life in the kingdom of Pylos. Although the numbers involved are not large, averaging just over eleven hides per village, they imply considerably larger herds if, as the records suggest, the communities were expected to fill the same quotas every year. Hide taxes were collected from all corners of the kingdom, so it follows that herds were also found in every territory under Pylian control. Each of these herds was owned by a community, the *da-mo* mentioned in a number of tablets, and were no doubt grazed on communally owned land, *ke-ke-me-na*, which seems to have been distinguished from the land allocated to the *wanax* and *lawagetas*.[78] The presentation of skins given as tax payments to the palace may be depicted on the Chieftain's Vase, which shows a procession in which three men carrying hides are led before a nobleman, perhaps an official whose job is to receive and catalogue the skins as palace property.[79]

The herds that were taxed as communal property should be kept distinct from the herds owned by powerful individuals who were taxed separately. Private herds were no doubt restricted to the elite of Pylian society. The tax assessment for a certain Dunios of Pylos, recorded on PY Un 138, reveals his extensive land and stock holdings: 2220 liters of barley, 526 liters of olives, 468 liters of wine, fifteen rams, eight yearlings, one ewe, thirteen goats, twelve pigs, one porker, one cow, and two bulls. It has been suggested that Dunios was required to provide the animals and dry goods for a ceremonial feast.[80] He may have had trouble meeting his obligations, since tablet PY Ae 03 records the seizure of Dunios' cattle by another Pylian named Philaios. Many individuals owned or leased private land, called *ki-ti-me-na*, but the amounts of seed and grain recorded in relation to most of these plots are so small that they suggest that they were not big enough to allow cattle grazing. It is more likely that the community allocated the task of herding to herders such as the *qo-u-ko-ro* from *Ko-ri-to* (Korinthos?) on tablet Nm 831 who then moved the animals away from the land under cultivation—perhaps to places like Glykys Limin in Epiros, where Tartaron has found evidence for the processing of oxhides in the Mycenaean period.[81]

The Pylos tablets show that cattle were brought to the palace not only as tribute hides but on the hoof as well. One tablet Cn 23 records a consignment from an individual named *We-u-da-ne-u*, the details of which hint at the fate of the animals: the three oxen listed are described as "uniformly white," suggesting their suitability for sacrifice. As at Knossos while under Mycenaean control, the Pylos cattle were not only sent to the palace but were also dispatched from the palace. Cn 22 records the assignment of oxen to an inspector named Diwieus whose job is to send the oxen on to various groups:

A-ra-tu-a and Oikhalia: one ox
The *ku-re-we* men at *Pi-ru*: one ox
The men of Iwasos at *E-na-po-ro*: one ox
The men of Olympia at Erumanthos: one ox
The . . . of the Olympian: one ox

The locations on this tablet coincide closely with the toponyms on another tablet, An 43, which lists military garrisons of thirty, fifty, and one hundred and ten men. Ventris and Chadwick interpreted the Diwieus tablet as a simple record of food supplies being sent to the troops, but Leonard Palmer suggested that it was a list of animals being dispatched for sacrifice on behalf of the soldiers prior to battle.[82] We have already seen that at Knossos some cattle were probably distributed for sacrifice in outlying areas. The same practice evidently took place on the mainland. Whether this particular sacrifice was to ensure victory we cannot say with certainty, but sacrifices and banquets were held to celebrate the investiture of new officeholders.[83] Take, for example, Pylos Ta 711: "*Pu₂?-ke-qi-ri* made inspection, on the occasion when the king appointed Sigewas to be *damokoros*."

On one such occasion, recorded in PY Un 2, an official called the Overseer of Works assembles provisions for a feast following the initiation of a *wanax*. The supplies include 1,574 liters of flour, 115 liters of spelt, and 585 liters of wine, as well as prodigious amounts of olives, honey, and figs. Among the livestock killed for the celebration are twenty-six rams, six ewes, four goats, seven pigs, and an ox. These numbers alone demonstrate that thousands of people were directly involved in the sacrifice and feast. The same tablet reveals that the ceremony was to take place at a spot called *Pa-ki-ja* (Sphagianes), "the place of slaughter," also mentioned in another tablet, Tn 316, as the location of a holy ritual in honour of Potnia and other deities. A similar form, *sa-pa-ke-te-ri-ja*, is used in tablets from Knossos to designate animals for slaughter.[84] Sacrifices of large animals are also attested, as we would expect, in honor of the gods, one of whom is named *Bowia* (Cattle Lady, Tn 316 v.3). PY Un 718 records separate donations by officials, *damos*, and the workers (*worgioneion kama*) to a communal sacrifice in honor of Poseidon, an especially important god at Pylos. PY Un 6, unfortunately fragmentary, lists at least twenty sheep, pigs, and cattle being offered to Poseidon and another god.

Halstead has estimated that over 2,200 animals are mentioned in tablets from Pylos and Knossos in contexts related to consumption. Given that the total corpus of Linear B is fewer than 5,000 tablets, the tablets confirm that animal husbandry was of central importance to Mycenaean society.[85] However, the numbers tell only part of the story. The Linear B tablets show how inextricably economic procedures and religion were interwoven. For example, many tablets, including six from the most recent finds from Thebes, use the term *o-pa* in contexts referring to an animal and a human, named in the genitive. John Killen has demonstrated that the expression refers to the process of fattening an animal before sacrifice. Some animals were designated as payment for a religious obligation or debt (*qe-te-o*) and many came from herds referred to as *a-ko-ra* (cf. *agora*), "collections" under the control, and perhaps ownership, of important individuals known as Collectors.[86] These men and women constitute the group we can identify most closely with a Mycenaean nobility. The names of some show up in records from more than one palace, suggesting that their interests and influence extended across different regions. In fact, the predominance of bovine figurines in deposits of Mycenaean pottery from Italy to Syria would suggest that their influence was international.[87] Kings, their officials, the "Collectors," and communities throughout the Mycenaean world were bound together by the sacred bonds of stock breeding, sacrifice, and feasting. In short, Mycenaean society operated on the basis of a ritual economy. This does not mean that the palace controlled all resources or that Mycenae functioned along the lines of a temple–state—a category, in any case, of dubious value. Nor is this to deny that parts of the Mycenaean economy may have functioned as a true market economy. Farming activities, however, were structured around not only the seasons but also according to calendars of sacrifice. Food production was geared not toward markets but toward the palace, which served as a center for ritual activities such as sacrifice and feasting. These ritual performances redeployed the economic resources of the palace and its dependencies, transforming simple commodities like olives, wine, and meat into the stuff of socially significant transactions such as gifts, donations, and benefactions. While Pylos' control of the western Peloponnese was probably never complete, and required the willing participation of non-palace groups, the bureaucratic and organizational potential of the palace placed it at an advantage. Small villages could establish their place in broad networks of exchange with a few specialties such as salt production, but more complex manufacturing such as perfume required a degree of control over resources and labor that only a centralized hierarchical state could manage.[88] The palace could do this, and became the focal point for all the activities—social, economic, symbolic, religious—that regulated Mycenaean social order. Even if private ownership and entrepreneurial activities such as trading had a place in the Mycenaean economy, prestige goods were processed, manufactured, stored, and bartered in the palaces, which regulated the relations between these luxury goods and staples, thus permitting different types of economic activity to intersect.

This would prove to be crucial to Greek societies of later periods. From a strictly economic point of view, much of Greek society in the Classical age was made up of modest agrarian communities whose insistence on local autonomy was built on the ideal of the autarchic household. However, layered over this fundamentally agrarian economy would be other types of economic activity. Trade and manufacturing are the best known of these. But another economic model rarely studied is the ritual economy whose elements were piety, pilgrimage, dedications, and sacrifices. The killing of hundreds of animals, the burning of long bones wrapped in fat, and the supplication of the gods seem more fit as subjects of religious than economic history, yet these practices come down to the Greeks of the fifth century from their Mycenaean ancestors, whose legacy was to take cattle production out of a host of economic activities and to make it a central feature of social and political organization. At the end of the Bronze Age, the palace at Pylos would cease offering cattle to Poseidon at Sphagianes and would stop manufacturing perfume. But the end of the palace system would not mean an end to the ritual economy. In place of the palace would arise the sanctuary, still a locus for sacrifice. Few sites would continue to serve as the physical settings for rituals of slaughter, supplication, and feasting. Instead the model of ritual behavior was preserved for those who came after the Bronze Age by their poets. The *wanax* of *Pu-ro* would be remembered as Homer's Nestor of Pylos.

After the Fall

Both the Minoan and Mycenaean systems had collapsed by 1050 BC, but the subsequent periods of Greek history continued to reflect the influence of the complex systems of husbandry so carefully regulated by the palaces. The material changes in life in the Early Iron Age are still a matter of considerable debate.[89] The view once widely held that the end of the Bronze Age was a collapse ushering in a Dark Age has been modified in recent years as evidence has come to light on both the mainland and on Crete that the degree of change may have been less catastrophic than was once thought. Excavations at Lefkandi have brought to light a chieftain's residence used as the location of a heroic burial, while the goods from this burial and others at the nearby site of Toumba attest to a level of material culture and especially trade that would once have been thought impossible in the tenth and ninth centuries. Similarly, excavations at Elateia in Phokis also have demonstrated that the residents of central Greece in the so-called Dark Ages were connected to a trade network that extended across the eastern Mediterranean. A great many sites on Crete, especially the sites in eastern Crete such as Pseira, Mochlos, and Kavousi where there has been a high level of excavation and survey work done over the last twenty years, have produced evidence for virtually continuous settlement across the end of the Bronze Age and into the Iron Age.

If it is an exaggeration to resort to the apocalyptic language that once characterized studies of the Late Bronze Age, the new danger may be to overemphasize continuity. There were significant changes in the tenth and ninth centuries as power collapsed at the center of the Mycenaean world. The first of these changes concerns the location of settlement. At many Cretan sites there are signs that the population moved from lower-level coastal sites to more protected highland sites. At Kavousi, for example, the Middle Minoan settlement at Vronda appears to have moved uphill in the Late Minoan period to the site of Kastro, where there is continuous occupation from LM IIIC through all the early phases of the Iron Age. Krzysztof Nowicki has identified close to two hundred sites across Crete in the early Iron Age that replicate the pattern to be seen at Karphi, where the population enjoyed the natural protection of an isolated refuge site in the hills.[90] The reliance on refuge sites on Crete alerts us to one type of settlement change, as people moved away from the coast and sought protection inland. Another shift in settlement patterns points to a realignment of power in the post-Mycenaean world. It is noticeable that the places where signs of decline are greatest in the tenth century are at the old centers of power: Mycenae, Tiryns, Pylos, Thebes, and Gla. The regions where the scale of decline was least dramatic were away from the heartland of the Mycenaean world. Euboia and Phokis were not affected in the same way as Boiotia and the Argolid, in part because there were no palaces here in the first place. Instead, as the palatial system faltered, unable to maintain the delicate circulation of staple and luxury goods whose control was critical to the palace's existence, outlying regions emerged from the shadow of the palace. While squatters occupied Tiryns, a chieftain was building himself a longhall at Lefkandi. As Mycenae was abandoned, Thermon in Akharnania, peripheral to the old Mycenaean kingdoms, produced a succession of buildings bigger than anything built over any Mycenaean site. As central power diminished dramatically, chiefdoms and new regional centers emerged on the margins—which is not to say that they were as powerful as the Mycenaean kingdoms. The chieftain's hall at Lefkandi may be the largest tenth-century structure yet found, but it cannot stand comparison with the palace of Knossos or the citadel of Tiryns, in terms of architecture or social complexity. The Big Man of the Dark Ages still ruled in a more modest setting than the lords of Mycenae.[91]

The realignment of power and settlement in the tenth and ninth centuries was matched by a shift in husbandry. John Bintliff has observed that the cyclical fluctuations in population suggested by recent surveys is consistent with a shift from the intensive land use of the Bronze Age to a new period in which small communities controlled the fields close to their villages but left the outfields in the hands of the "big men" who could make "more effective use of the outfield (perhaps notably with stock)."[92] One of the best places where this shift can be seen is Nichoria, a site that was part of the Mycenaean kingdom of Pylos but far enough from the palace to have survived the end of the Bronze Age. The excavators of the site argue that the Late Mycenaean mixed economy of farming gave way to an Early Iron Age economy more focused on cattle herding.[93] The evidence for this consists of two data sets. The first includes the numbers and percentages of animal bones from Nichoria classified by period (Table 3.1). In the case of domestic cattle, the growing percentage of cattle bones relative to all others reflects an increasing reliance on herding. Compare the figures for cattle with numbers for sheep and goats. From a Late Helladic ratio of just under 11 percent cattle bones to just under 47 percent ovicaprids one finds a startling switch by the second half of the ninth

TABLE 3.1
Nichoria Faunal Remains

Species	*LHIIIB2 1250–1200 BC*	*DAI 1050–975 BC*	*DAII 975–850 BC*	*DAIII 850–775 BC*	*Total*
Bos taurus	7/10.9%	17/28.8%	32/35.1%	27/40.3%	277/19.7%
Ovis aries	1/1.6%	—	—	—	13/0.9%
Capra hircus	4/6.2%	—	2/2.2%	1/1.5%	31/2.2%
Ovicaprid	25/39.1%	16/25.4%	23/25.3%	16/23.8%	459/32.7%
Cervus	4/6.2%	5/8.5%	3/3.3%	3/4.5%	67/4.8%
Capreolus	7/10.9%	2/3.4%	8/8.8%	—	59/4.1%
Sus scrofa	14/21.9%	14/23.7%	16/17.6%	16/23.8%	412/29.3%
Equus	1/1.6%	1/1.7%	1/1.1%	2/3.0%	14/1%
Asinus	—	1/1.7%	2/2.2%	—	10/0.7%
Canis	1/1.6%	3/5.1%	4/4.4%	2/3.0%	54/3.8%
Total	64	59	91	67	1405
Hunting	11/17.1%	7/11.9%	11/12%	3/4.5%	136/9.6%

Source: Sloan and Duncan 1978, 63 (table 6.1)

century (DAIII): over 40 percent cattle to approximately 25 percent sheep. Nor do these numbers tell the whole story.

These cattle weighed between 180 and 200 kilograms, and provided much more animal protein per head than either sheep or goats. Robert Sloan and Mary Ann Duncan estimate that in the Early Iron Age cows supplied as much as 63 percent of the animal protein in the diet of Nichoria. This estimate is based on the numbers and percentages of bones, but also reflects an important second data set. The size of second upper molars among the preserved cattle bones is an indication of the average age of slaughter, which appears to have dropped from a Middle and Late Bronze Age high of ten years to an Iron Age low of five and a half years. The shift reflects a change in the way the animals were used. Cows stop lactating around twelve years of age and may live another ten years. A high age of slaughter reflects a cattle system in which animals are raised as dairy producers. Although bull calves may be slaughtered for their meat, the average age of slaughter will be high, reflecting the longer life of milk cows. A lower average age of slaughter points to a meat regime in which there is no particular bias toward maintaining animals as milk producers or as yoke animals. The distinction is reflected in references to the age of cattle in early Greek poetry. When Autolykos celebrates the arrival of his grandson, Odysseus, now come to claim his inheritance, he embraces the young hero and bids his sons butcher a bull, five years old, on which they feast.[94] When Hesiod offers advice about plowing, he recommends two oxen of nine years of age: their strength is unspent.[95]

The Nichoria evidence has been used by some historians to assert an Early Iron Age trend toward pastoralism throughout Greece. Anthony Snodgrass, for example, has suggested that "we may tentatively conclude that the Greeks had continued to subsist by stock breeding rather than by arable farming after the fall of Mycenaean civilization."[96] He has also assembled other evidence to support the notion of a move toward pastoralism. Among these are the small bronzes dedicated to Artemis of Pherai by seasonal shepherds; ephemeral structures that underlie more permanent later buildings at Eretria and Lefkandi, and have been identified as signs of occupation by short-term visitors; sites such as Vitsa with bones scattered all over a seasonal settlement situated on a major transhumant route across the Epiros mountains; the predominance of pulses (for fodder) in the Protogeometric palynology of Iolkos; and Olympia animal figurines, which reflect the importance of domesticated animals to the dedicating communities in the tenth and ninth centuries.[97]

Despite the arguments in favor of a significant shift toward herding in the Iron Age economy there are dissenting voices. Halstead has questioned the evidence for specialized pastoralism in Greek prehistory and has drawn attention to the close connections that exist between pastoralists and small-scale farmers. In Halstead's model, large-scale pastoralism is a rarity in the Greek world, and even at the height of the Bronze Age could only have taken place as a component of large, mixed farming estates.[98] In some respects the debate has become one of degree rather than type. Few people would now claim that any community in ancient Greece was entirely pastoral, in the sense of being primarily structured around stock breeding. Studies of modern transhumant communities such as the Sarakatsanoi in northern Greece have shown that even the most pastoral groups rely on the market economy of towns and villagers. Furthermore, John Cherry has noted that the confusion over pastoralism arises in part from an imprecise use of terminology. Some scholars use "pastoralism" to refer to specialized economic activity involving seasonal transhumance by entire communities; others use it to refer to husbandry of any sort, including the herding of small numbers of sheep or goats, a practice easily integrated into a regime of arable farming.[99]

So what does the Nichoria evidence mean? A clue may lie in the results of recent excavations at Kavousi in eastern Crete. Here the faunal evidence produced some unusual results.[100] While ovicaprid remains dominated the faunal remains both at Vronda (70 percent) and on the Kastro (77.9 percent), the cattle bones defied expectations. Five percent of the faunal remains from Vronda consisted of cattle bones. Since the site is located at a much lower altitude than the Kastro and has better forage available, it is an area that would seem to be a better environment for husbandry. Yet it was the Kastro, on a ridge high above Vronda and backing onto a dry upland plateau with only meager pasture, that has both a higher aggregate number of cattle bones (230 vs. 59) and a higher percentage of cattle bones within the overall record (8.8 percent). These figures must be put into chronological context: Vronda is a Bronze Age site and the Kastro is primarily an Iron Age site (though its origins lie at the end of the Bronze Age, when the Vronda community moved up from the lowlands). While neither site would ever qualify as a transhumant settlement, they represent two very different approaches to mixed agriculture. In the case of Vronda, cattle will have been kept as a complement to the agrarian farming conducted by households. Perhaps used for plowing or kept for milk, the cattle were in any case raised in small enough numbers to be part of the household. In the unsettled conditions of the tenth and ninth centuries, however, as central authority broke down, the community relocated to the hills and at the same time paid more attention to its movable wealth. A lower population would also help to explain a shift of resources to herding since, as James Whitley has recently noted, "herding may have been a more practical economic strategy when labour is short but land is plentiful."[101] Nichoria may represent a more extreme version of the same process. The enormous disparity in the number of cattle bones versus sheep and goats suggests a small population with a great deal of land at its disposal.

There are other factors that should be kept in mind when trying to interpret the finds from Nichoria. The first is that local conditions were better suited to stock raising here than in many other parts of Greece; we should not assume, therefore, that the same trend to pastoralism was widely copied everywhere else at this time. Regions respond to dramatic external changes in different ways, according to local conditions. So, for example, when world wheat prices collapsed in the 1870s and 1880s, European producers reacted in different ways. In Italy the response was emigration; in Denmark farmers turned from grain production to animal husbandry.[102] The second factor to consider is that agriculture was not abandoned at Nichoria, so that even at its most developed the husbandry here was still integrated into a mixed regime. The third factor is that Nichoria did not flourish beyond the early Iron Age, so that the intense emphasis on cattle breeding may have been an aberration that could not be sustained in the long term. If so, instead of simply postulating Nichoria as an example of an economically rational trend, according to which husbandry is a sensible adaption to the unsettled circumstances of the tenth and ninth centuries, we might instead be witnessing a response to crisis, the administrative breakdown of the palaces.

Why would the inhabitants of Iron Age Nichoria have experimented with intense stock raising? The answer to this can only be a guess, but even speculation may be productive. Claudia Chang and Harold Koster suggest that in the Bronze Age palatial herds will have competed with privately owned animals for access to pasture and water.[103] With the end of the palatial system came the end of that competition. One of the most dramatic differences between Greece in the thirteenth and Greece in the tenth centuries must surely have been the void created by the disappearance of palatially controlled herds. It is not difficult to imagine Dark Age communities filling that void where possible with their own cattle, moving onto pasture from which they had been barred. One might even imagine an unemployed *boukolos* appropriating herds abandoned by his absent masters, as Philoitios contemplates doing in the *Odyssey.* In fact, just as brigandage and banditry have been close cousins to modern pastoral systems, so too the presence of these elements in the heroic poetry of the tenth and ninth centuries suggests that pastoralism and raiding flourished at this time.[104] At the same time, the rituals most closely associated with cattle—sacrifice and feasting—were irrevocably established at the heart of Greek culture. As the Bronze Age of palaces was transformed into a golden age of heroes, feasting would remain a central social institution. In heroic poetry it is a forum for aggressively masculine performances of eating and boasting, just as similar performances characterize the ideology of recent Cretan herders.[105] Networks of communities would emerge between 1200 and 700 BC, and with them local elites, for whom feasting and the conspicuous consumption of meat were ways of asserting status and defining alliances.[106] In short, the world of the palaces would give way to the world of the *oikos*, yet in both societies cattle retained their status as the single most valuable commodity. The bull had served the lord of the Labyrinth. It would continue to serve the hero of Lefkandi.[107]

CHAPTER 4

Epic Consumption

> Where does Homer ever speak of any of the Achaians eating fish?
>
> —Euboulos, frag. 120KA

Transformations

The chieftain buried at Lefkandi was called a hero by the excavators, and rightfully so, since it was between 1000 and 800 BC that the notion of a hero became firmly rooted in the consciousness of the Greeks.[1] The bovine idiom—that cluster of values, institutions, and ideas that centered on cattle—helped shape the emerging notion of the hero. Indeed, a comparison of the Greek phrase at the heart of the heroic ideal, *kleos aphthiton* ("undying glory"), with its Vedic original reveals the degree to which the Greek hero is the heir to a pastoral tradition. "Indra," prayed the warriors of ancient India, "as everlasting one grant us fame *rich in cattle* and prizes, wide, lofty, imperishable."[2] The hero would steal cattle by raiding, would sacrifice them to show reverence to the gods, would swear mighty oaths over their dismembered carcasses, and would use feasts to entertain guests and strangers.[3] He offers them as prizes in competition and pays them over as compensation for his wrongdoings.[4] From these practices (and their opposites) would emerge very clear notions of piety, generosity, and self-restraint. The cattle experience would shape not just how the Greeks understood wealth, but how it should be used. Theirs, as we shall see, was the ideology of stock breeders.

The emerging notion of the hero was embedded in a relationship with the past. Heroes were thought to have existed before the present degenerate age, and all around there were physical reminders of that glorious past: the ruins at Mycenae, Tiryns, and Knossos, as well as hundreds of tombs. For the people of the Iron Age to fashion their own identity, they had to define a relationship with the past—often by asserting a genealogical link between present rulers and past heroes—while at the same time distinguishing themselves from it. The break was made by demonizing Crete, and especially Knossos, the most glorious palace of the Bronze Age. We have already seen that the Minotaur becomes a popular motif in Cretan glyptic at precisely the time when actual bull jumping was fading out. The Mycenaean presence on Crete appears to have precipitated a significant shift in the bull cult practiced in the palaces, with a new emphasis on sacrifice rather than athletic performance. The Minotaur looks very much like an attempt to fashion a new shamanistic connection with the bull-god. Yet versions of the Minotaur myth after the Bronze Age have little in common with the Minotaur as shaman. Instead, in later accounts the Minotaur was conceived of as a monstrosity, the offspring of an unholy union between Pasiphae and a bull sent by Poseidon as punishment for Minos' having withheld a promised sacrifice. In this, the version that goes back to Euripides' *The Cretans* and remained popular with Roman authors, any suggestion of legitimate cult activity has been replaced by perversion and monstrosity, or at best a parody of motherhood.[5]

It is difficult to believe that the earlier versions of the stories of Europa, Zeus, and their son Minos (and the doublet of this group, Minos, Pasiphae, and the Minotaur) were stories that dwelt on bestiality. They were, in fact, as P.B.S. Andrews demonstrated nearly forty years ago, stories that rendered into narrative form astronomical observations regarding the relationship of the new moon to the heliacal rising of various stars, especially the Hyades and Pleiades, two clusters that lie within the constellation Taurus (Fig. 4.1).[6] It is legitimate to infer that in the Bronze Age these astronomical events were marked by ritual celebrations, and that the songs and dances accompanying these rendered the cosmic events into narrative form. In later versions, however, sacred lore has been reduced to storytelling that was still capable of inventing richly imagined stories but was only tenuously connected to any specific ritual or sacred activity. Ovid, for example, refers to the *cornua Tauri* (Bull's Horns) that Phoebus must navigate past everyday as he crosses the sky, in an allusion to the constellation Taurus, but it would be hard to make the case that the *Metamorphoses* should be read as a cultic text in the same way as, for example, the *Enuma Elish*.[7] This is not to say that the Greeks no longer recognized the Pleiades, or that they were incapable of astronomical observations. Quite the contrary. Hesiod identified the rising of the Pleiades (in May) as the time to start harvesting, but there is a gradual shift away from mythology as a set of stories connected with a coherent religious system of ritual and belief, designed to render the cosmos into an orderly narrative, and toward a mythology that was a repository of stories only tangentially connected to the cosmos.[8] Stripped of a ritual setting, in which astronomical events were watched and celebrated as part of a religious calendar, the stories lose their cultic significance and serve instead as artistic tropes or templates, to be used and altered by poets and dramatists as they see fit. The stars still shine, but the song does not remain the same.

The complex and shifting relationship between myth, trope, and astronomy is nicely illustrated by the constellation Taurus and its represen-

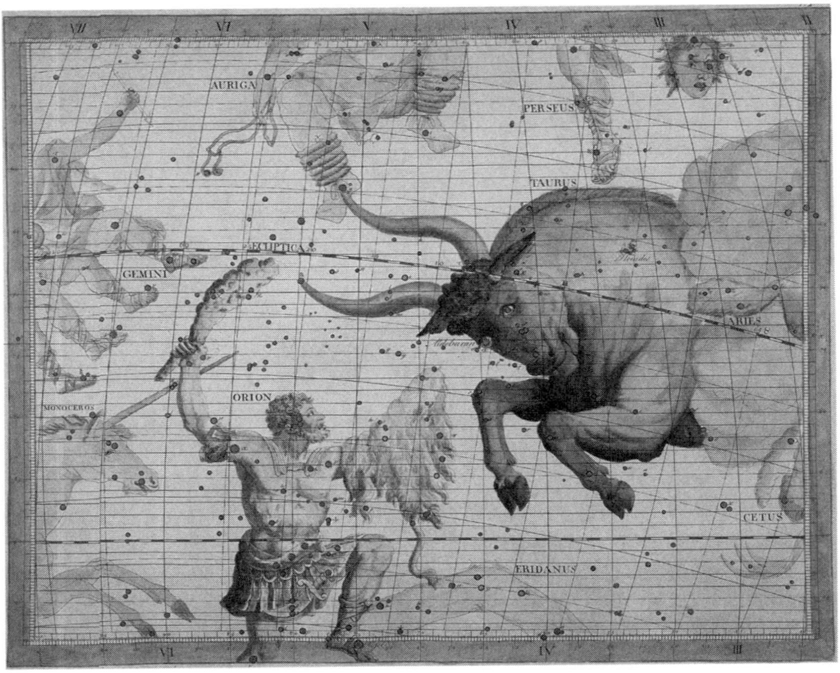

Figure 4.1 Sir James Thornhill, Constellation of Taurus, 1729. Photo credit: The Stapleton Collection/Art Resource, NY.

tation in art. Zeus created the constellation to mark the abduction of Europa, and depictions of the episode, from Roman mosaics to the paintings of Titian and Rembrandt, observe the same narrative detail: the bull's powerful front legs lift him clear of the water even as his hindquarters are hidden by Europa and shadows.[9] It is a curious detail, but corresponds to an observable feature of the constellation: its huge horns and charging front legs are all that can be seen. Its hindquarters are invisible. The many stars clustered around the bull, such as the Hyades and Pleiades, could be figured as nymphs, an imaginative step that made sense of the many lesser lights to be seen close to Aldebaran, the "eye of the bull."[10] But the story of Europa is not simply a narrative attached to the constellation.[11] It has a distinctive form arising directly from the cattle breeder's practice of putting a succession of cows to the same stud bull. Europa, the Pleiades and Hyades, and in other contexts Hera and Io, are all, before anything else, part of Zeus' herd. Hera's epithet, *boôpis* (cow-faced), Europa's name ("broad-faced"), Io's transformed state as heifer—all resonate with associations deriving from the domesticated herd, where one bull services many cows. Priests and astronomers of the great ancient civilizations, from Babylon to Egypt and Crete, used stories like these to render the observation of the stars into sacred lore, that is, into narratives in which the stories were themselves either allegories or metaphors for astronomical observations. But with the end of the Bronze Age the systems that made these stories significant were largely lost. Once disembedded, sacred narratives became merely tales. Thus, what was once a *hieros gamos* between the deities of sun and moon in Cretan cosmology, rendered in a narrative form involving Pasiphae and an incarnation of the sun god, was transformed by succeeding generations into a perverse tale in which Pasiphae conceived a disgusting desire to mate with a bull.[12]

Only vestiges of cosmology remained in the post–Bronze Age versions of these stories: Pasiphae was remembered as the daughter of Helios, and the Minotaur was sometimes known as Asterios, but the memory of the astronomy and cosmology that shaped the stories soon faded.[13] Like Pasiphae, Minos was demonized, transformed from a sun king into a disgusting creature, ejaculating scorpions and millipedes into his conquests, while Talos, the magical protector of the island, was refashioned from a bull into an automaton who killed trespassers by embracing them in his burning arms.[14] This is history rendered mythopoeically. Later myths, divorced from their Bronze Age Cretan setting and devoid of any ritual significance, sound like folktales. It is important to investigate the degradation of myth that has occurred here, because there persists a strong tendency to regard Greek myths as so perfect in their canonical form—Apollodoran or Ovidian—that any investigation of their origins is construed as an assault on their privileged status.[15]

The story of Glaukos illustrates how the transformation of Cretan myth into story could sap its richness. In the versions known to us, Glaukos, the child of Minos and Pasiphae, goes missing. Minos and Pasiphae are then told either by the Kouretes or by Apollo at Delphi (!) that the child would be found by the man who could solve the riddle of a curious creature in their possession. It so happened that they had a calf (or heifer) that changed color every day. Polyeidos of Argos correctly likens this to a mulberry, and then goes on to restore Glaukos to life after he is found drowned in a vat of honey. The story was popular in later times, figuring in plays by each of the major dramatists and preserved by compilers such as Apollodoros and Hyginus.[16] None of these versions, however, seems to have treated the story of the mulberry as anything more than the prosaic answer to a very odd puzzle whose only relevance to the story is to establish the credibility of Polyeidos. It has taken modern scholarship to understand the story of the mulberry is itself a metaphor of maturation and that it was connected to a specific ritual: adolescent initiation.[17] The changing color of the mulberry stands for the changing appearance of the child passing through puberty and approaching adulthood.[18] The connection between the tricolored calf and the mulberry—a clever but casual analogy in the hands of Polyeidos, since there is no explicit connection between the fruit and the animal—was significant in its earlier context because Cretan youths were thought of as calves and were organized into age classes called *agelai* (herds).[19] The story accompanied the initiation of young Cretan men of the elite, but the story as it comes down to us has lost its institutional context. Instead, the hero of the tale becomes the outsider Polyeidos (Know-All) from the mainland. As in the case of Theseus and the Minotaur, divorced from its Cretan setting in the Bronze Age the story can no longer convey its original meaning or serve its original purpose. Just so, in the story of Theseus in the labyrinth the bull remains a component of the story, but far from being a god or a shamanistic priest he is merely a monster lurking in the shadows, awaiting his death at the hands of Theseus and his resurrection at the hands of Picasso.

If the allegorical meaning of the stories of Europa and Pasiphae were lost, others stories emerged from the bovine milieu to give expression to the mores of Iron Age society, none more dramatic than that of the cow-maiden Io. Elements of her myth unmistakably identify its provenance in the world of the pastoralist: she is desired by Zeus, who assumes the shape of a bull to mate with her, she is transformed into a heifer by a jealous Hera, and her punishment, consisting of both madness and physical transformation, is made tangible through the continuous biting of the gadfly, an affliction that Hera also sends to disrupt Herakles' journey back to Eurystheus with the cattle of Geryon.[20] Despite the dominance of bovine motifs in Io's story, they alone do not explain what the myth is about. In the first place, Io's wanderings are part of a group of myths that arise out of the Greek attempt to fix genealogical and topographical relations between the Greeks and those whom they encountered around the Mediterranean: from Io are descended Aigyptos, Libye, Arabos, Phoenix, Europa, Cilix, and Minos, all heroes or nymphs whose descent from Io brings them and the territories associated with them into the Greek orbit. Not that the relationship guaranteed amity: Lycophron refers to the kidnapping of Io by Phoenician "sailor dogs" and "merchant wolves" who brought the "bull's maid" to Memphis, thereby kindling "the torch of hatred between two continents."[21] In fact, the Egyptian setting of the story's climax, since it is to Egypt she is driven, is linked to another important bovine element in the story: it is here that she gives birth to Epaphos, the black bull. Herodotos equates this with the Apis bull, implicitly attributing a Greek origin to one of the most famous Egyptian cults.[22] The line of transmission is more likely to have gone in the other direction: the Egyptian prototypes of Io and Epaphos were Isis and Apis, and Io's name is itself a Greek rendering of an Egyptian word for the moon.[23]

Yet even if Io's origins lie in Egyptian lunar observations, this is still not enough to explain the Greek versions of her story. To understand Io's story more fully, we must recognize that the Greek story has a very precise antecedent in the Near Eastern story of Geme-Sîn, and that the Neo-Assyrian version of this tale was told in the specific context of birth incantations. This is evident in the wording of the Neo-Assyrian incantation:

> Incantation. There was a cow of Sîn, Geme-Sîn by name. With ornaments decorated, tempting of shape she was. Sîn saw her and fell in love with her. The brilliance of Sîn he laid (. . . ?) upon her. He appointed her at the head of the herd, the herdsman followed her. In the lushest grasses she grazed, at the abundant well they watered her. Hidden from the herd boys, not seen by the herdsman, the wild bull mounted the cow; he lifted her tail (?). When her days came to an end, her months were finished, the cow trembled and terrified her herdsman. His head was bowed, all the herd boys lamented with him. At her crying, at her screaming in labour, Nannaru (Sîn) was downcast. Sîn heard her screaming in heaven and lifted high his hand. Two Lamassus descended from heaven. One of them carried "oil-from-the-jar," while the other brought down "water-of-labour." With "oil-from-the-jar" she touched her forehead; with "water-of-labour" she sprinkled her whole body. When she touched for the third time, the calf fell down on the ground like a gazelle's young. "Milk-calf" she called the calf. Just as Geme-Sîn gave birth normally, may also this girl in labour give birth. Let not the midwife tarry, let the pregnant one be all right. ("Cow of Sîn" 10–35)[24]

Although there are enough similarities between the Neo-Assyrian version and the Greek versions of the Io story in Hesiod and Aeschylus to show the story's Near Eastern ancestry, most commentators have treated the connections in terms of images and themes, or have concentrated on the discrepancies between the different versions.[25] Recently, however, Mary Bachvarova has argued that the Greeks, like the Sumerians, Hittites, and Persians, also employed the myth of the cow-girl and her miraculous birth in incantations uttered to comfort women as they underwent the pangs of childbirth.[26] The incantations prayed for a successful outcome and used the story of the sexual union between divine bull and cow to invoke good luck on the unfolding birth. This may well be so—the similarities noted by Bachvarova are very numerous—but once again similarities may reveal less than the differences between versions. In the Greek version of the tale Io is originally a girl, not a cow but the priestess of cow-eyed Hera. She is persecuted by a vengeful goddess, Hera, and her lowing is not just the screaming of an animal in labor but the cries of a girl driven mad. The Greek version of the story, despite its similarity to the Near Eastern antecedents, is much grimmer, placing a great deal of emphasis on Io's punishment. This is the Io of *Prometheus Bound*, who appears briefly to commiserate with another victim of divine anger, Prometheus, before the madness returns and drives her screaming from the stage. If, like the story of Geme-Sîn, the myth of Io was told as part of an incantation that accompanied labor, then it offered scant consolation to the suffering woman. Furthermore, Io's story had a postscript that underscored its bleakness. Io was the ancestor of the Danaids, whose story is the subject of Aeschylus' *Suppliant Women*. Forced to marry the sons of Aigyptos, the fifty daughters of Danaos slaughtered all but one of their fifty husbands and were punished in Hades by being required to carry water in broken vessels, a fitting metaphor for their own status. Their story was, in typical Greek fashion, a parable with a twist: their story reinforced the social imperative that girls must accept marriage, no matter how hideous the prospect. In supplicating Aigyptos before the fateful wedding night they retell Io's story, but in their version the Danaids edit out any explicit reference to the sex that has taken place between Zeus and Io. For Bacharova, there is an implicit moral in the Io episode that the Danaids, out of their deep fear of sex, have missed: "even if the woman is unwilling at first to be married, the resulting child is recompense for her suffering. If [the Danaids] had realized this, then they would have realized that they should lose their petition to avoid marrying their cousins."[27] Marriage, sex, and childbirth are women's common fate, according to this mindset. The Io and Danaid episodes are excellent examples of the distance any myth, including a bovine myth, can travel. What was a divine metaphor of motherhood in Egypt and Babylon became for the Greeks a means of reconciling young women to the grim reality of marriage.

Heroes and Feasts

If the transformation myths involving Europa, Pasiphae, and Io mark the loss of old meanings and the forging of new ones, it is in epic that the world of the Iron Age fully finds its voice. The traditional associations of cattle with wealth and status would allow Homer to investigate the values of the hero in settings that resonated with his audience. Cattle would serve as similes but also as the very objects whose treatment defines correct and incorrect human behavior.[28] In the *Odyssey*, for example, it is the consumption of the cattle of the Sun that will condemn Odysseus to his ten years of wandering, just as it is the consumption of *his* cattle that will mark the suitors for destruction. The suitors, in fact, eat every type of animal in Odysseus' herds and flocks, their taste for fatted hogs marking them as particularly swinish, but here again the symbolic importance of cattle outweighs their mere numbers. When called to supper by Medon they slaughter sheep, goats, hogs, and one heifer, reflecting an ascending order of value.[29] Beef is the top choice.

In the *Iliad*, cattle are less central to the action but are still important points of reference. Take, for example, the feasts in the *Iliad*: Agamemnon slaughters a five-year-old bull to feast the Argive princes and saves a special cut from the chine as a mark of distinction for Ajax (7.324). Phoinix speaks of his relations killing a succession of swine, sheep, and cattle to keep him engaged in a nine-day feast and to distract him from his murderous hatred of his father. Both instances suggest that Homer and his audience were alert to the significance of the feast as a vehicle for channeling rivalry and potential violence into a constructive ritual. There may be a correlation between these ideas and the archaeological record: at the very time that the epic poems were reaching their monumental form in the Iron Age, sanctuaries such as Kato Syme on Crete and Apollo Maleatas on Kynortion reveal strata rich in black, fatty earth, full of animal bones, which, as Birgitta Berquist has provocatively suggested, may not be simply the remains of burned animal offerings but "the remains of sacral meals of the worshippers and cult personnel."[30] Berquist goes on to draw attention to the evidence both for *hestiatoria* and temporary shelters that housed feasts at early Greek sanctuaries. Similarly, Uta Kron has drawn attention to the plentiful evidence for feasting at the Heraion on Samos, plausibly suggesting that the characteristic pottery found in the so-called *Pilgerschutt* was a type specifically manufactured for cult meals held within the sanctuary.[31] Yannis Hamilakis and Eleni Konsolaki have recently examined evidence for similar feasting, on a modest scale, at the Mycenaean site of Ayios Konstantinos and speak of the "privileged access to the cosmological powers that the active participation in the sacrificial and feasting rituals of the sanctuary would have conferred."[32] Much of the power associated with these rituals derived from not only their exclusivity, but from the fact that together sacrifice and feast formed a praxis, a coherent combination of theory and practice that transformed killing, a socially threatening act, into commensality, a reaffirmation of social bonds.[33] The power of the feast derives from the value of the animals involved and is recursively linked to the status of the participants as donors and consumers.

The relationship of meat eating to status is well established in the poem, as when Sarpedon rhetorically asks Glaukos, "Do you know why we are honored with the choicest cuts?" The answer is that they lead in battle. Similarly Agamemnon upbraids the Greek heroes at the ships by reminding them of their boasts:

What happened to all our boasts, all the big talk
That we were the best? Remember Lemnos?
All the beef and wine you ate there? Remember saying
That each of you could take on a hundred Trojans,
Two hundred?

(Homer, *Il.* 8.231–35, trans. Lombardo)[34]

When Agamemnon hosts a banquet for all these leaders he must make sure that they get equal shares, lest an inferior cut suggest a lower status (7.320). The connection between heroes and meat was sufficiently fixed for it to be mocked by Euboulos:

Where does Homer ever speak of any of the Achaians eating fish?
And their meat they roast, since he never has anyone boiling it.
Not one of them saw a courtesan either
But spent ten long years pulling each other.
A hard campaign they witnessed, those men who
Captured a city and came home with arseholes
Bigger than the breach they made in Troy's walls.

(Euboulos, frag. 120KA = Athenaios 1.25C)

The heroes participating in the feast come from a world that already computes value in terms of cattle. Athena's aegis has one hundred gold tassels, each worth one hundred oxen. Similarly, Lykaon, the hapless Trojan killed by Achilles in his rage, pleads for his life saying that when he was sold into slavery he was valued at one hundred oxen. The tripod offered by Achilles as first prize in the wrestling contest is worth twelve oxen (23.705), while the woman given to the runner-up is worth four (23.707). So fixed are cattle as the standard of wealth that when Diomedes and Glaukos famously exchange armor, the one for gold and the other for bronze, the poet computes the inequality of value in terms of cattle: the gold armor is worth one hundred cows, the bronze only nine.[35]

If cattle are the measure of value and status in the world of the heroes, they also manage to retain some very unheroic associations that the poet uses to bridge the gap between the heroic world and the world of Homer's audience. During the funeral games for Patroklos, for example, Ajax slips in the gore left from the sacrifices made by Achilles and ends up with a mouthful of cow dung (23.772–80). Similes of plowing and threshing are a little less comic, but do merge the heroic and the mundane. The two Ajaxes, for example, are described as standing side by side in battle like a yoke of oxen:

A team of oxen with faces dark as wine
Strain together at the jointed plow, pulling
Through a field that has not been worked in years.

Sweat oozes up from the roots of their horns,
And only the polished yoke holds them apart
As they cut the furrow to the field's far edge.
The two Ajaxes stood side by side in battle.
(Homer, *Il.* 13.743–50, trans. Lombardo)

In this episode Homer has already used similes of falcons, lions, and torrents, all of which are suitable for larger-than-life warriors, but the plowing simile is especially vivid, evoking not just their strength and their inseparable tenacity, but also their physical exertion. The sweat dripping down their horns gives the image an astonishing immediacy. In fact, just as cattle may be the food of heroes or beasts of burden in this world, so too plowing straddles both worlds. It is labor, the work of ordinary men, yet it requires strength and skill, and may be a test of manhood not unlike battle or athletics. When the braggart Eurymachos tries to humiliate the disguised Odysseus by offering him a job as a *thes*, a hired hand, Odysseus responds by saying he wishes they could have a contest cutting hay on a hot summer's day:

Or how about this?
We could drive oxen, the best there are,
Big and tawny, both well fed with grass,
The same age, yoked the same way, tireless animals—
And each with four acres of rich soil to plough.
Then you'd see if I can cut a straight furrow
Clear to the end.
(Homer, *Od.* 20.405–10, trans. Lombardo)

Although plowing on its own hardly seems heroic, it is here transformed into an *agon* with heroic overtones.[36] Other references are less charged, yet still allow the poet to break down the barrier between the heroic and the mundane. In the Doloneia, for example, Odysseus catches sight of Dolon trying to sneak into the Greek camp. He warns Diomedes and together they let him enter before springing their trap. The poet's description goes as follows:

. . . Dolon ran by unaware.
And had gone about the length of a furrow
A mule plows—they are better than oxen
At pulling the plow in deep fallow land—
When the two Greeks ran after him.
(Homer, *Il.* 10.364–68, trans. Lombardo)

Like the Thersites episode in Book 2, the Doloneia has elements that run counter to and thereby highlight the more conventional aspects of the

Iliad as hero saga. Diomedes will be likened to a lion and accomplish great deeds, but cutting the throat of the hapless Dolon is hardly one of them. Dolon is no hero and killing him brings Diomedes no glory. Rather, this episode and the oddly intrusive simile of the mule-plowed furrow are examples of places in epic where the world of little people and their ordinary activities—the world of the audience—keeps irrupting into the world of the heroes.[37] Even in the most intense battle scenes a simile from the agrarian world links the heroic to the familiar:

> . . . the black earth ran with blood.
> A team of broad-browed oxen has been yoked
> And is now treading white barley
> On a solid threshing floor. It does not take long
> For the bellowing bulls to tread out the grain.
> So the hooves of Achilles' horses trampled
> Dead bodies, shields.
> (Homer, *Il.* 20.515–521, trans. Lombardo)

In that more mundane world of plowing and threshing, sacrifice and feast do exist, but they are hardly the daily occurrences they are in the world of the heroes. Furthermore, in the world of Homer's audience cattle are also valuable for their traction power. Oxen do exist for Homer's listeners as sacrificial animals, but the preponderance of feasting in the imaginary world becomes all the more charged precisely because the animals have other, practical uses in the real world.

These moments when the contemporary world breaks through the fabric of the heroic world function in a way very much like similes, deepening and enriching the narrative texture of the epic. So, for example, Homer describes the battle in Book 16 as going on with death and slaughter evenly distributed on either side as long as the sun was still in midheaven. He continues,

> But when the sun moved down the sky and men
> All over the earth were unyoking their oxen,
> The Greeks' success exceeded their destiny.
> They pulled Cebriones from the Trojan lines
> And out of range, and stripped his armor.
> (Homer, *Il.* 16.817–21, trans. Lombardo)

Shortly afterward the death of Patroklos occurs. At this turning point in the poem, an extra layer of pathos is achieved by locating these events at the end of the day when work, whether farming or killing, ought to be over. Instead the battle is about to reach a new level of ferocity. It is not that cattle are unique in bridging the heroic and the mundane; in the *Odyssey*, for example, Odysseus associates the resurfacing of his shattered ship timbers with the time of day when a judge finishes work and goes to dinner. But cattle do have a particularly rich range of associations, since as beasts of burden they connote the hard work of farmers, while as objects of consumption they evoke the massive appetites of heroes.

The place where the poet most deliberately evokes the world recognizable to the audience in counterpoint to the world of the heroes is in his description of two cities on the shield of Achilles in Book 18. Cattle figure prominently here and the descriptions reflect experiences familiar to the audience. In the case of the city under siege, the besieged choose to go on the attack and ambush flocks and herds, presumably of the besieging army. At the sound of cattle mooing, the besiegers break off their battle council, rush into the fray and join battle. Beyond the cities the shield also shows fields under cultivation: a thrice-plowed field in which the plowmen are shown being given a drink at the end of each furrow. Lastly, the shield depicts a herd of cattle being released from their byre to graze on the reeds growing beside a river. These three categories surely reflect a world familiar to Homer's audience. While they listened to songs that repeatedly depicted a world in which cattle were sacrificed and consumed in epic quantities, they themselves will have been more occupied with tending herds, plowing with yokes of oxen, and protecting their cattle from raids. At times the distance between the heroic and the mundane contracts entirely: Andromache laments the loss of her seven brothers, all shepherds and cowherds killed by Achilles.

If the *Iliad* uses scattered and occasional references to farming activities as a way of connecting the heroic world to the world of the audience, the *Odyssey* goes much further in exploiting the cattle systems of both heroic and Iron Age Greece in order to explore both value and values. The *Iliad* requires its heroes to be warriors before all else, but the *Odyssey* reflects a world in which men are measured not only by their deeds but also by their means. It is the prosperity of the *oikos* that remains the central concern of this society. For this reason Homer takes pains to give as full a picture as possible of the wealth of Odysseus.[38] Before the hero exacts his revenge the poet has him meet his three herders: the faithful Eumaios and Philoitios, and the wicked Melanthios. Philoitios explains,

> He put me in charge of his cattle when I was a boy
> In the land of the Cephallenians, and now
> Those cattle are past counting. No breed
> Ever flourished like that for any mortal man.
> But now other men order me to drive them
> So they can eat them with no regard at all
> For the son of the house, or the gods' wrath.
>
> (Homer, *Od.* 20.224–30, trans. Lombardo)

Here is one of the anxieties at the heart of the poem: how to protect, increase, and pass on one's wealth to the next generation. Cattle are the most vivid embodiment of that wealth, because of their size, their value, and their vulnerability, so that the reckless consumption of Odysseus' cattle wealth is an attack on his *oikos*. Odysseus' entire household, in fact, falls into two simple categories: those who preserve his wealth (loyal servants like Eumaios and Eurykleia), and those who squander it by siding with the suitors (such as Melanthios and the twelve servant girls). Philoitios' dilemma arises from the fact that with Odysseus gone he must either stay and see the herds under his care consumed by the suitors, or preserve the herds by taking them elsewhere, and thereby betray his master's son. Penelope too makes it clear that the suitors' pursuit of her is fundamentally an attack on Odysseus' property. Once reunited, Odysseus and Penelope take turns telling each other stories:

She told him
All that she had to endure as the fair lady
In the palace, looking upon the loathsome throng
Of suitors, who used her as an excuse
To kill many cattle, whole flocks of sheep,
And to empty the cellar of much of its wine.
(Homer, *Od.* 23.308–13, trans. Lombardo)

While it is tempting to read the *Odyssey* with modern psychology in mind, and thereby to see Penelope as the true object of the suitors' desire (as is suggested, for example, by Menelaus' outburst at 4.357–58: "Those dogs! . . . Wanting to sleep in the bed of a hero!"), the poem consistently pairs the courting of Penelope with the consumption of Odysseus' cattle wealth. Penelope and the cattle of Odysseus are interchangeable, as in so many Indo-European myths where "the lady and the cow are one."[39] In fact, if we accept the fundamental importance of cattle wealth to heroic society then it follows that the suitors' consumption of meat is not a displacement of their desire for the woman. What they truly lust for is Odysseus' cattle wealth.

To get permanent control of this, however, will require a marriage, agreed to by Penelope and arranged by her father, Ikarios. It would be a mistake to read the *Odyssey* as a legal manual on the niceties of bride price and dowry, but it is possible to read the poem keeping in mind the practices with which Homer's audience was familiar.[40] Customarily a prospective groom would secure a betrothal by handing over a bride price. In the *Catalogue of Women*, for example, Hesiod describes the wooing of Eurynome, whom "Glaukos sought to win by Athena's advising, and so he drove oxen for her."[41] The bride would then bring a dowry to the marriage, a gift from her father, the use of which would fall to the groom. The two exchanges are not identical, since the bride price is an exchange between groom and father-in-law, while the dowry is a gift from father to bride, but in practice it was the circulation of cattle wealth between families that counted. Pausanias remarks that Neleus demanded the Thessalian cattle of Iphikles "as bride gifts for his daughter from her suitors."[42] The transaction was not necessarily smooth. According to Hesiod, Amphitryon killed his father-in-law Elektryon in a dispute over cattle.[43] In the case of the suitors' courting of Penelope, however, the intended bride has already been married, and so the prospective groom would not expect to pay a bride price. He would, on the other hand, expect a dowry. Furthermore, since Telemachos has made no claim to take up his inheritance, the successful suitor could also look forward to getting control over all of Odysseus' estate, whatever its legal status.[44] In its understanding of marriage as a set of gift exchanges, the *Odyssey* reveals a remarkable affinity to traditional pastoral societies and the emphasis they place on the exchange and distribution of cattle that constitute what Evans-Pritchard termed bride wealth. The acquisition of such wealth is of paramount importance. Michael Fleisher has demonstrated that among the Kuria of Tanzania the pressure to amass enough cattle wealth to afford a bride can be overpowering for young men who are "sister-poor," that is, from families in which the numbers of sons who must pay bride price greatly outnumbers the number of daughters who attract cattle wealth. As a result, many turn to cattle raiding, another point of connection, as we shall see, between the worlds of the African herder and the Greek hero.[45] It is, therefore, not only its roots in Indo-European mythology that account for the close identification of Penelope with the cattle of Odysseus, but also a set of values that the heroic world shares with the world of the pastoralist.

Impossible Worlds

The strict requirements of elite *xenia* are established early in the *Odyssey* in the description of Telemachos' stay with Nestor at Pylos.[46] (Is it nostalgia that leads Homer to render the Mycenaean palace nearest Ithaca a place of perfect courtesy?) As he arrives, nine companies of five hundred men are each sacrificing nine black bulls to Poseidon. In the ensuing feast the protocols of aristocratic *xenia* are followed exactly: the guests are greeted and feted without question. They are seated near Nestor, in places of honor, are given choice cuts, and receive the golden cup to pour libations. Peisistratos' speech is a model of politesse, inviting Athena (as Mentor) to take precedence over Telemachos. Only when the food has been eaten does Nestor even ask Telemachos who he is and what his business is.

An emphasis on correct treatment of the guest pervades the entire episode. Nestor will not hear of Telemachos sleeping on board his ship, but insists on giving him blankets and fleeces to sleep upon in his own halls. Furthermore, Nestor thinks of *xenia* not as simply a form of politeness in the moment, but as the expression of a relationship between families spanning generations:

> I have plenty of cloaks and blankets,
> And Odysseus' son shall never lie down
> On the deck of a ship while I am alive,
> Or any child of mine is left in the halls
> To entertain strangers who come to my house.
>
> (Homer, *Od.* 3.388–92, trans. Lombardo)

His generosity is inseparable from his piety, and his prosperity. Just as Nestor is constantly sacrificing bulls to Poseidon while returning from Troy and offers a heifer to Athena the instant he realizes it is she who has accompanied Telemachos, so too his piety is pleasing to Athena, who prays for him to win renown and grace for the men of Pylos, a prayer that she herself grants.

By putting such emphasis on the proper form of *xenia* at this early point on the poem, Homer is establishing a model against which the subsequent episodes can be measured. The differences become apparent as soon as Telemachos reaches the court of Menelaus at Sparta, where hospitality is far from automatic: Eteoneus specifically asks his master whether they should entertain the strangers or send them elsewhere. Similarly, even though Menelaus shows proper manners in postponing questions till after the feast, Helen's entry leads straight to this outburst:

> Do we know, Menelaus, who our guests
> Claim to be? Shall I speak my mind or not?
> My heart urges me to speak.
>
> (Homer, *Od.* 4.145–47, trans. Lombardo)

That she is right to guess that her guest is the son of Odysseus does not alter the fact that her forwardness pushes the boundaries of acceptable behavior.

Measured against Nestor's Pylos, Sparta's shortcomings become more pronounced. At the moment of Telemachos' arrival there is a twin wedding feast going on, as Menelaus celebrates the nuptials of both his daughter, Hermione, and his son Megapenthes, yet neither marriage promises happiness. Hermione's marriage is ill-fated as she will be fought over by Orestes and Neoptolemos in a replay of the feud between their fathers, Agamemnon and Achilles. We know nothing further of Megapenthes but his very name (Great Sorrow) is inauspicious. The wedding feast is unsatisfactory in other respects: while Homer emphasizes the generous portions given to Telemachos and Peisistratos, the food and drink is served without any libation, as it is in the matching Pylian scene. It is also notable that there is no mention of a sacrifice at Menelaus' banquet. This is a significant omission, because just as Nestor is distinguished by his sacrificial piety, Menelaus is conspicuously his opposite. The contrast is developed in their differing accounts of their return. Nestor's *nostos* keeps up a constant refrain of sacrificing bulls to Poseidon, while Menelaus stands out for his failure to appease the gods:

> I was in Egypt, held up by the gods
> Because I failed to offer them sacrifice.
> The gods never allow us to forget them.
> (Homer, *Od.* 4.376–78, trans. Lombardo)

Accordingly, once Menelaus has captured Proteus and forced him to reveal what he must do, he learns that he should have made a sacrifice to Zeus if he wanted to get home without trouble. Now he must expiate his oversight by returning to Egypt and offering hecatombs to the immortal gods.[47]

The two older men therefore represent different models of piety and hospitality. Nestor's willingness to share his cattle with the gods is mirrored in his punctilious behavior toward his friend's son. Menelaus' behavior consistently strikes the wrong note. He forgets the gods, reduces his guests to tears, talks for so long that Telemachos begs to be allowed to go to bed, and offers Telemachos gifts that are of no use to him.[48] The contrast between the two also extends to their wealth. Nestor mentions "the goods and softly belted women" that he took from Troy, but his wealth is modest. He has enough blankets and cloaks to be able to offer hospitality to his guests, and he offers Telemachos horse and chariots to get to Sparta, but aside from these the only precious item we hear of is his famous cup and the gold used to plate the horns of the heifer he sacrifices to Athena. There is no display of wealth at Pylos, and Telemachos leaves without any extravagant gifts. Nestor's only conspicuous wealth is in cattle: when we first see Pylos, the community is sacrificing more than eighty bulls to Poseidon by the seashore. Menelaus, on the other hand, is consistently associated with material wealth. Nestor refers to Menelaus wandering along the coast of Egypt as "a stranger in a strange land, amassing / a fortune in gold and goods" (*Od.* 3.334–35). At the wedding banquet the guests are served from golden pitchers and in basins of silver, part of the wealth that Menelaus says he amassed as he wandered around the lands of the Mediterranean. Menelaus speaks of this wealth guiltily, as if the getting of it cost him Agamemnon:

While I wandered through those lands amassing wealth
My brother was murdered, caught off guard
By treachery and the guile of his accursed wife.
So I do not enjoy being lord of this wealth.
You may have heard all this from your fathers,
Whoever they may be, for I suffered greatly
And saw my house ruined, with all its treasures.
(Homer, *Od.* 4.94–100, trans. Lombardo)

The Pylos and Sparta scenes reflect a coherent attitude toward wealth and its uses: cattle wealth is acceptable since it activates a set of institutions that establish good relations with the gods (sacrifice) and between humans, whether as a community (feast) or even with strangers (*xenia*). Cattle wealth increases without a great deal of effort on the part of humans, and it can be put to good use. Wealth in the form of material goods, on the other hand, requires a good deal of effort to acquire, but it is effort wasted since the property does not lead to any socially beneficial end. Only cattle wealth is truly honorable, while non-cattle wealth is suspect.

This may be termed the heroic understanding of wealth, and it may seem that it is no more than an aspect of the imaginative, and imaginary, world of Homer's heroes. It is, however, fundamentally no different from the view of many contemporary pastoralists, and the comparison suggests that Homer's imagined society reflects the mental landscape of Early Archaic Greece.[49] Among the Nuer, for example, the "cattle of money" (wealth created by market activity, labor, exchange, and sale—in short, Menelaus' kind of money) is constantly being transformed into the "cattle of girls," that is, real wealth measured in head of cattle, put into circulation through bride price and bride gift.[50] In similar fashion, the *Iliad* 18.593 refers to girls as "earners of oxen." These attitudes are deeply conservative, representing the atavistic hold of cattle on the imagination of the Greeks. As we have seen, this is in part because the habits of mind connected with stock breeding were deeply ingrained, and in part because this was a legacy of the Bronze Age. Yet Homer's world, the age of the eighth and seventh centuries, was also a world of rapidly expanding horizons, as Greeks moved across the Mediterranean as traders and colonizers. This would prove a challenge to the conservative vision of an ideal world in which all was well ordered, pious, and prosperous. The world of the heroes had rarely been thus, as the frequent references to Agamemnon's fate remind us, and the new world emerging around Homer was even less so. Increasingly, the ideal world characterized by generous hosts, pious sacrifices, and orderly social life seemed impossible both in the heroic past and in the shocking present.[51] The *Odyssey* draws attention to this impossibility by having Odysseus recount his adventures to the Phaiakians, who are themselves caught between the past and the present. At one level the Phaiakians seem perfect: the royal couple is of impeccable genealogy, descended from heroes and a god, Poseidon; the queen dispenses justice and is revered by her people; the forthright princess is nubile and, better still, available; the city has superb harbors, stout walls, and well-made assembly places; the bronze palace is tricked out inside with blue enamel, gold and silver finishings, and magical watchdogs made by Hephaistos, and outside, with an evergreen orchard hung with all manner of fruit irrigated by two gentle streams. Yet the behavior of the Phaiakians belies the image of perfection. Nausikaa warns Odysseus that they do not like strangers, and, as many commentators have remarked, Odysseus has to win the inward-looking Phaiakians over.[52] Although Alkinous is courteous and decent, considering the Phaiakians bound to help the suppliant as soon as he seeks help, there are jarring incidents during Odysseus' stay. Demodokos' song reduces Odysseus to tears, recalling Telemachos' tears at the song of Phemios at the court of Menelaus. Alkinous is generous, but he is a little too forward in offering Odysseus his daughter in marriage, an unwelcome gift that again reminds us of Menelaus' offer of horses to Telemachos. Particularly unfortunate is Euryalos' famous challenge to Odysseus during the athletic contest. When Odysseus demurs after being asked to compete, Euryalos replies,

> You know, stranger, I've seen a lot of sportsmen,
> And you don't look like one to me at all.
> You look more like the captain of a merchant ship
> Plying the seas with a crew of hired hands
> And keeping a sharp eye on his cargo,
> Greedy for profit. No, you're no athlete.
> (Homer, *Od.* 8.174–79, trans. Lombardo)

The insult is curious since the Phaiakians, despite their heroic posturing, are not in fact a part of the world of the heroes who fought at Troy. Odysseus himself draws attention to this when, having outdistanced all the Phaiakians with the discus, he angrily challenges any of them to a contest with the bow. Only Philoktetes had outdistanced him at Troy, he says. The Phaiakians are stunned into silence. They cannot respond to a real hero's challenge, as Alkinous immediately concedes. They do not excel at boxing or wrestling,

> But we are swift of foot, and the best sailors,
> And we love feasts and the lyre and dancing,
> Fresh clothes, warm baths and soft beds.
> (Homer, *Od.* 8.266–68, trans. Lombardo)

This is no longer the world of the *Iliad.* Soft and pleasure-loving, the Phaiakians may treat Odysseus well, showing him respect and showering him with gifts before ferrying him home, but they really have no place in the heroic world since they are not even athletes, much less warriors. They are seafarers, with their magic ships that pilot themselves automatically across "great gulfs with the greatest speed." But if they are scornful of trade, then what is the purpose of their seafaring? They have no place in the world of commerce either. In one sense the Phaiakians represent what happens when the aristocratic code of honor is grafted onto the emerging reality of seafaring commerce. The result is an impossibility, a world that looks too good to be true, and one that is doomed to disappear.[53]

The episode that presents the clearest inversion of the Phaiakian chapter is Odysseus' time in the Cyclops' cave. If the Phaiakians travel between the cities of men until they are condemned to isolation by an angry Poseidon, Polyphemos, a child of Poseidon, already lives apart, both from the rest of mankind but even from other Cyclopes. The Phaiakians seem to have everything at their fingertips and enjoy an idyllic existence. Polyphemos is a herder and must eke out a crude existence based on milk and raw meat. Phaiakians observe the niceties of *xenia* and help Odysseus, while Polyphemos displays nothing but contempt for the code.[54] Not only does he ask who they are before they can supplicate him, but when they do he dismisses them:

> You're dumb, stranger, or from far away,
> If you ask me to fear the gods. Cyclopes
> Don't care about Zeus or his aegis
> Or the blessed gods, since we are much stronger.
> (Homer, *Od.* 9.265–68, trans. Lombardo)

The pairing of the Phaiakian and Cyclops episodes has been interpreted as an exploration of the anxieties rooted in the experience of colonization. What kind of people were "out there"?[55] The two episodes suggest that the outside world wavers between savagery and luxury. Against this backdrop, Homer's lack of interest in trade is striking. Aside from a handful of exceptions, Homer seems unwilling to contemplate a world in which people produce goods and exchange them.[56] The second half of the poem, however, will bring Odysseus back to Ithaca, and so the failed or flawed communities Odysseus has experienced while coming home serve to highlight the crisis on Ithaca. Odysseus has seen Cyclopes living without assemblies, laws, shipbuilding, or skills of any sort. In short, they have no culture and fail to understand concepts such as hospitality or reciprocity. The Phaiakians have culture but lack purpose. They face no challenges, no conflict, no discord. Theirs is a stultifying perfection.[57] With these models before him, Odysseus will face a challenge on his return to Ithaca greater than simply getting rid of the suitors. The poem implicitly asks under what conditions communal life is possible. Can violence and aggression, requirements of the heroic code, be harnessed in a society at peace, or will they inevitably be turned inward and destroy the community?

The *Odyssey* offers no easy answer to this dilemma, but it does display a profound understanding of the contradiction of building a civil society whose dominant ideology is a warrior's code. In order to explore this contradiction the poem establishes a parallel between the killing of the cattle of the Sun and the slaughter of the suitors. From the point of view of the plot, Odysseus must be brought to Ithaca alone, and the cattle of the Sun episode explains the loss of his crew, who are killed for their impiety. At exactly the midpoint of the poem, at the end of Book 12, we find Odysseus on the high seas, adrift on the wreckage of his ship, symbolically reduced to nothing but his elemental self. But the needs of the plot are not enough to explain the prominence of the cattle of the Sun. The episode also marks an important change in Odysseus, for whom the time on Thrinakia is a test not of cunning but of respect for the gods. This is apparent from the moment he meets Teiresias in the Underworld. The seer tells him explicitly that the key to his survival involves not a quest but a temptation:

> You might just get home, though not without pain,
> You and your men if you curb your spirit,
> And theirs too, when you beach your ship
> On Thrinakia. You will be marooned on that island
> In the violet sea, and find there the cattle
> Of Helios the Sun, and his sheep too, grazing.
> Leave these unharmed, keep your mind on your homecoming
> And you may still reach Ithaca, not without pain.
> But if you harm them I foretell doom for you,
> Your ship and your crew.
>
> (Homer, *Od.* 11.102–11, trans. Lombardo)

At first this looks a little like telling Adam and Eve not to eat from the Tree, and both admonitions are equally unsuccessful, but Teiresias continues his prophecy in such a way that the consumption of the cattle of the Sun is linked to the consumption of Odysseus' wealth:

> And you shall find
> Trouble in your house, arrogant men,
> Devouring your wealth and courting your wife.
> Yet vengeance will be yours . . .
>
> (Homer, *Od.* 11.114–17, trans. Lombardo)

The encounter with the cattle is also referred to in Circe's speech to Odysseus (12.131–48), in which we are told that the cattle are watched over by Phaithusa and Lampetie, the daughters of Helios. Once again, Circe foretells disaster for Odysseus and his crew if the cattle are harmed. The poem is clearly setting up the visit to Thrinakia as a turning point, since it is immediately after this that Odysseus will lose his entire crew and arrive by himself on Ithaca.

What, then, is signified by the killing of the cattle of the Sun? It is hard to see the episode as a failure on the part of Odysseus, either in terms of piety or leadership. In the first place, Odysseus is well aware of the dangers posed by stopping at Thrinakia. He first tries to avoid the island altogether, only to face opposition from his exhausted men. He shares the prophecies of Teiresias and Circe with his men, but, unable to convince them to sail on, he has them swear not to kill a single animal. When the crew is caught on the island because of contrary winds, Odysseus keeps reminding them not to touch the cattle. Eventually their supplies run out, they are reduced to catching birds and fishing, and they face starvation. Only then, while Odysseus is asleep, does Eurylochos convince the others to slaughter the cattle. Even at this point, the crew members behave as respectfully as possible. Eurylochos argues that if they survive they can put up a temple to the Sun, and that if the gods are angry it is better to die quickly from drowning than slowly from hunger. When they do succumb to their hunger, they prepare the animals exactly as if for sacrifice, substituting leaves for barley and salt water for wine, before burning the thighbone wrapped, as required, in fat. Both Odysseus and his men have been driven to this by desperation and cling to reverence for the gods to the extent possible.

The twin prophecies give the slaughter of the cattle an aura of inevitability, yet neither Teiresias nor Circe actually says that the Greeks will slaughter the cattle, only that if they do, disaster will follow. It is a test of their ability to restrain themselves, to curb their spirits, as Teiresias says. It is a brutal test, to be sure, since they face starvation, yet the stakes only make the lesson more compelling. The need for self-restraint is absolute. The crew fails, and when news reaches Helios he calls on Zeus to punish them and make them atone for their wrongdoing. Zeus' justice is immediate: he promises to destroy their ship.[58] The gods advertise their displeasure in a remarkable omen, causing the hides of the slaughtered cattle to crawl and the meat on the spits to low, as if still alive. This episode foreshadows the punishment of the suitors, with Odysseus in the role of Zeus. In both episodes it is the abuse of cattle wealth that triggers violent retribution (*tisis*). Even if the crew begins by framing their deed as a sacrifice, order quickly breaks down. They do not content themselves with killing a single beast. Instead,

Each day for six days my men slaughtered oxen
From Helios' herd and gorged on the meat.
(Homer, *Od.* 12.409–10, trans. Lombardo)

They only stop their carnage on the seventh day because Zeus sends a favorable wind. Their gluttony signifies their lack of restraint. Similarly the suitors are described as both impious and wasteful. Eumaios complains to Odysseus,

The fatted hogs the suitors eat,
Men who have no fear of the gods, and feel no pity.
They just lounge about squandering our goods,
Sparing nothing in their insolence.
Every day and night they slaughter our animals—
Not just one or two either—and waste our wine.
(Homer, *Od.* 14.93–94, 104–7, trans. Lombardo)

The gluttony of crew members and suitors alike is the outward sign of their lack of restraint. There are, however, important differences between the episodes. Zeus' justice is swift and sure. Within fifty lines his promise to Helios has been carried out: the ship has been shattered by a thunderbolt and the men swept overboard. By contrast, Odysseus' restoration of his *oikos* will be teased out and requires help from both the gods and other mortals. Homer exploits the pattern established in the first half of the poem to reveal the full range of the suitors' shortcomings. It is Eumaios, for example, who offers the disguised Odysseus proper hospitality, welcoming him, as Odysseus says, "with an open heart." The suitors, instead of bringing the stranger to their table, arrange for a fight between Odysseus and the beggar Irus, the winner of which will dine with them "to the exclusion of all other beggars." Their hospitality is conditional, as imperfect as their gluttony. Homer describes the blow Odysseus delivers in the ensuing fight in such a way as to suggest a sacrificial killing:

Odysseus caught him just beneath the ear,
Crushing the jawbone. Blood ran from his mouth,
And he fell to the dust snorting like an ox
And gnashing his teeth, his heels kicking the ground.
(Homer, *Od.* 18.103–6, trans. Lombardo)

The evocation of a sacrifice is interesting for two reasons.[59] The first is that the suitors eat without making a sacrifice. They offer a stuffed paunch as the prize to the winner, but do not select a live animal as does Eumaios. The second is that Odysseus does not, in fact, kill Irus. In his deliberations before the fight Odysseus has decided only to knock him out rather than to kill him "stone cold dead." Even as the suitors are yielding to ever greater degrees of indulgence—their gluttony will soon be complemented by an outright lust for Penelope, and their feast threatens to end in a brawl—Odysseus is exercising a certain restraint.[60]

There is no question but that the suitors deserve their punishment for dishonoring Odysseus by wasting his property. Even the most casual asides reinforce their guilt, as when Odysseus lies down the night before the slaughter on a bed of fleeces "from the many sheep that were always being slaughtered," but the killing of the suitors does not result in a restoration of balance, as does Zeus' justice. When Helios demands atonement from the crew, Zeus' response is measured: the wrongdoing is swiftly punished and order is restored. Odysseus cannot accomplish this. In fact his imitation of Zeus only reveals the impossibility of absolute justice in the human realm.[61] Odysseus' vengeance opens up endless possibilities for retribution. This is understood by Laertes, who, when he realizes that Odysseus has returned, exclaims,

> Father Zeus, there are still gods on high Olympus,
> If the suitors have really paid the price!
> But now I have a terrible fear
> That all of Ithaca will be upon us soon
> And word will have gone out to Cephallenia too.
> (Homer, *Od.* 24.358–62, trans. Lombardo)

Divine justice may restore balance, but its human counterpart is no recipe for harmony. Odysseus' justice restores the integrity of his *oikos*, but the poem reflects uncertainty about whether the vigorous protection of the household's prosperity can be reconciled with the good of the greater community. Can human justice ever be complete? Certainly not if humans act as if they were Zeus. But isn't that exactly what a hero does?

CHAPTER 5

Heroes and Gods

> It is better to be killed stealing cows than committing adultery.
>
> —Kuria saying[1]

Cattle Raids

It is clear that epic drew on patterns of behavior and social practice deeply embedded in both the actual past of the Bronze Age and the imagined past of the heroic age. Reformulated by Homer, together these supplied the Greeks of the Archaic period with a complex model of social order and personal ethics that found expression in a range of coherent activities constituting the nexus of early Greek social practice: sacrifice, feasting, gift giving, and hospitality, in particular. We have also seen that the pastoral milieu favored these practices and values, inclining the Greeks toward a particular and characteristic way of imagining wealth and value and framing exchange. Yet the "bovine idiom," to use Evans-Pritchard's evocative phrase, does not guarantee that the social practices clustered around cattle breeding will always be useful or socially constructive. Consider, for example, a characteristic institution of one pastoral society, the Dafla, whose complete identification with their cattle, the mithan, we have already discussed. Among the Dafla there also existed, until recent times, another institution that, from a purely economic point of view, was catastrophic: the *lisudu*.[2] This practice, usually translated as "competitive destruction," consisted of escalating episodes of slaughtering mithan in response to a real or imagined slight. Honor was equated with the individual's willingness to kill more animals than his opponent, who was forced to match and outdo his opponent until either one side conceded or a settlement was negotiated. In one case, the offended party was about to kill eighty mithan until a group of arbitrators intervened and managed to convince him to kill only sixty, matching his opponent and calling it quits. When one takes into account that the competition had already escalated through three or four rounds of killing, the final count comes out around 200 head of cattle killed to maintain the honor of two men. This was a vastly greater number than could be eaten by the families involved. From a purely economic point of view, the institution is utterly irrational.

Pastoralism, by wedding the fortunes of a society to its cattle, can in fact foster values and practices that threaten the very society it defines. The contradiction between guaranteeing the prosperity of the household yet encouraging an activity that undercuts its security is especially marked in the institution of raiding. Here again, viewing early Greece as an example of a pastoral society, at least measured by its values and habits, if not its actual social organization, one can see the complex legacy of pastoralism. Consider, for example, the astonishing remark Odysseus makes in telling Penelope not to fret over the livestock consumed by the suitors: he says he will replenish their losses by raiding and by exacting tribute "until the pens are full again."[3] Protecting his goods has just led to the killing of dozens of rivals and the prospect of untold numbers of vendettas. Won't these raids simply incite more violence, triggering off further instances of retribution? Laertes, in fact, raises this very point, so the poet and audience are well aware of the dreadful ambiguity of Odysseus' triumph, but heroic society is characterized by a cycle of recurring violence in which raiding is catalytic: Katreus is mistaken for a raider and killed by his son, Althaimenes; Herakles kills Geryon and is attacked by Alebion, Derkynis, Cacus, and Eryx; Kastor is slain by Idas during a cattle raid; Amphitryon slays Elektryon for his "wide-browed cattle"; Odysseus' men are killed by the Kikones during a raid; Autolykos' theft of the cattle of Eurytos leads to Herakles' killing of Iphitos; Alkyoneus the herdsman is killed by Herakles.[4] The list is endless. Nestor's lengthy reminiscence in the *Iliad* makes the point explicitly:

> Oh to be young again, with my strength firm,
> As I was in the cattle wars with the Eleans
> When I killed Itymoneus, the valiant son
> Of Hypeirochus, a man of Elis,
> During the drive back . . .
> The spoils we corralled from out of the plain!
> Fifty herds of cattle and as many flocks of sheep, as many droves of pigs, as many herds of goats.
>
> (Homer, *Il.* 11.714–23, trans. Lombardo)

Itymoneus dies protecting his own property, as Nestor seizes cattle as reparation for cattle already stolen. Nestor's raid is therefore only the more recent episode in a history of violence between Pylos and the Eleans over their cattle. Furthermore, Nestor goes on to tell how his father tried to keep him out of the battle because all eleven of his older brothers had been killed by Herakles. Nestor does not elaborate on this, but in Isocrates' *Archidamos* we learn that Herakles attacked the sons of Neleus because they had stolen the cattle of Geryon from him as he was returning to the court of Eurytheus. Nestor's family, it appears, are a bunch of cattle rustlers and have been almost wiped out as a result. Nestor's entire existence, in other words, is defined by an unending cycle of cattle thefts, in which a perverse version of reciprocity, which Marshall Sahlins would define as negative reciprocity, produces an unending cycle of tit-for-tat violence.[5] For a hero longing for *kleos* this is fine, but how can a household benefit from raiding when it threatens to impoverish the *oikos* and expose it to ruinous retribution? In framing the cattle raid as inherently contradictory, Homer is moving beyond the raid as a trope of heroic activity and using it to explore the limitations of the heroic code.

In doing so Homer reflects the tumultuous changes of the eighth century. In more recent times, societies in which cattle raiding is an established institution have witnessed dramatic changes in the significance of the cattle raid at the very time when cattle have been transformed from producers to product. Among the Kuria, a pastoralist people whose territory straddles the border of Kenya and Tanzania, the commoditization of cattle in the last one hundred years has seen the earlier, looser forms of cattle raid give way to more regular, planned and violent forms of gang warfare.[6] What was once intimately connected to coming-of-age rituals, the acquisition of bride wealth, and irregular competition between various clans has been replaced by systematic theft as part of an underground economy responding to market forces that emerged in the postcolonial period. As a result, vigilantism, an alternative form of judicial procedure for hearing complaints and prompting restitution, has also sprung up. Like any other institution, the cattle raid is not fixed and unchanging, but can reflect the deeper transformation of the society in which it exists. It is particularly telling that the *Odyssey* relies on a divine intervention to stop the violence initiated by Odysseus from spiraling out of control. On the one hand, Homer anticipates the kind of solution that later Aeschylus and Euripides will rely on in the *Eumenides* and the *Iphigeneia in Tauris*, when only the appearance of a goddess can break the cycle. On the other, he reflects the heroic world, where raiding is ubiquitous. When Ajax, for example, needs to raise a bride price he casually promises to strip the flocks and herds from all the territories around Salamis.[7] Achilles' rebuke to Agamemnon reflects the same assumption:

> You shameless, profiteering excuse for a commander!
> How are you going to get any Greek warrior
> To follow you into battle again? You know,
> *I* don't have any quarrel with the Trojans,
> They didn't do anything to *me* to make me

To come over here and fight, didn't run off *my* cattle and horses
Or ruin my farmland back in Phthia, not with all
The shadowy mountains and moaning seas between.
(Homer, *Il.* 1.159–66, trans. Lombardo)

The prevalence of raiding in the value system of the heroes derives in part from the place of cattle raiding in Indo-European mythology, in which the cattle raid is matched to a combat between the hero and a monster. In Homer, however, raiding has lost much of its original mythic significance, and to understand raiding in the epic universe as an example of deeply embedded myth structures risks confusing origin with function.[8] Consider, for example, Odysseus' visit to Hades, where he sees Neleus' daughter Pero, the girl whose father would give her in marriage only to the man who could raid the cattle of Iphikles.[9] The story reflects the commonplace equivalence of cattle with women, but there is not even the vestige of an encounter with a monster. To reduce the cattle raids in Homer—and indeed in the rest of Greek myth—to simple reflexes of an earlier, original myth blinds us to the various ways the cattle raid is woven into the Homeric discourse on wealth, violence, and courage. So, for example, the *Odyssey* is careful to distinguish between raiding and the outright pillaging of Odysseus' *oikos* by the suitors. In Book 14 Eumaios expresses disgust at the wastefulness of the suitors who kill at least one animal every day (as do Odysseus' men when they slaughter the cattle of the Sun):

Even men who wage war in a foreign land
And sail for home with their ships filled with loot—
Even men like that fear the wrath of the gods.
(Homer, *Od.* 14.97–99, trans. Lombardo)

As Eumaios sees it, raiding is commonplace, but it is also morally ambiguous. A successful raid can only occur because Zeus wills it, so it is sanctioned, but there is no disguising the fact that it is theft. Accordingly the raiders should at least have the decency to feel guilty. The suitors do not. The same tension concerning the legitimacy of raiding comes up in Odysseus' tale of his attack on the Kikones:

From Ilion the wind took me to the Kikones.
In Ismaros. I pillaged the town and killed the men.
The women and treasure that we took out
I divided as fairly as I could among all hands
And then gave the order to pull out fast.
That was my order, but the fools wouldn't listen.
They drank a lot of wine and slaughtered
A lot of sheep and cattle on the shore.
(Homer, *Od.* 9.42–49, trans. Lombardo)

The Kikones rally, and in the ensuing battle Zeus brings them victory, forcing the Greeks to flee, only to be hit by a storm.

As in the episode of the cattle of the Sun, and as in the case of the suitors, the raid gives way to an orgy of eating and drinking. This emphasis on overindulgence arises not from any moral concerns over gluttony, but from an anxiety over the acquisition and use of wealth. The hero is supposed to raid, and at a deeply embedded level the myth of the cattle raid equates cattle wealth with the reward for the hero's *arete*. Raiding, therefore, is legitimate: it increases one's holdings and one's prosperity. At the same time it is theft, and in the real world it is finally counterproductive, undermining the stability of the *oikos*.

Furthermore, in the eighth and seventh centuries the contact between Greeks and non-Greeks put traditional notions of the hero under considerable strain. Trading forced a reevaluation of raiding. There were other ways of acquiring wealth that were less dangerous. This is reflected in the description of his activities that Odysseus, in the guise of the Cretan prince Aethon, relates to Penelope. Odysseus, he says, is alive in Thesprotia:

> And he is bringing home treasures,
> Seeking gifts, and getting them, throughout the land.
> . . . And he would have been here
> By now, but he thought it more profitable
> To gather wealth by roaming the land . . .
> But before I left, Pheidon showed me
> All the treasure Odysseus had amassed,
> Bronze, gold and wrought iron, enough to feed
> His children's children for ten generations.
> (Homer, *Od.* 19.298–99; 310–12; 321–24, trans. Lombardo)

We know that Odysseus is a sacker of cities and has amassed wealth by raiding. We also know that everything he had prior to his arrival among the Phaiakians was destroyed by Poseidon. So what is the precise nature of the wealth he has (ficticiously) amassed in Thesprotia?[10] Gifts from willing hosts? One-man raids, such as he envisages for restocking his farm? Or wealth gathered by barter and exchange? The reference to seeking and getting gifts locates Odysseus' activities in a heroic world of honor, where gifts of gold and bronze are standard, but no gift exchange described in the poem has ever resulted in a hero getting wrought iron. Once again, the present is intruding into the heroic world. Heroic ways of accumulating wealth, through gift exchange and cattle raid, are up against more prosaic methods: trade and piracy, the specialty, in the eyes of Odysseus, of the perfidious Phoenicians. It need hardly be said that these poles were more imagined than real, but the mindset of the Dark

Ages, if Hesiod is any guide, saw the present as an inevitable deterioration from an age of heroes to an age of iron.

How the West Was Won

Looking back a millennium later, Pausanias did not see the tension between past and present that Homer explored. Instead he simply equated the heroic world with endless competition over herds: Melampous stealing the herds of Iphikles, Eurystheus sending Herakles after the cattle of Geryon, Iphidamas paying one hundred head of cattle to his prospective father-in-law. For him the association between heroes and cattle was self-evident. He concludes, "This confirms my argument that the men of those days took the greatest pleasure in cattle."[11] Of the raids known to Pausanias, the most famous was without doubt Herakles' tenth labor: the stealing of the cattle of Geryon. The tale was the subject of a poem by Stesichoros, the *Geryoneis*, and was popular with sixth-century artists.[12] An examination of the myth reveals how the repertoire of cattle stories could be employed by the Greeks to process the anxieties of an age of far-ranging travel and contact with non-Greek people. Herakles' cattle raid is, in fact, an example of a myth that was based in the habits of a pastoral society and that did cultural work for the Greeks, allowing them to render into a comprehensible narrative the inchoate sensations of anxiety, misunderstanding, and fear that accompany exploration and colonization. If the Mediterranean contained the threat of Cyclopes and Lystraigonians, a Greek hero could always be relied upon to assert order, even if the pattern for doing this involved theft and trickery. This is quite different from the way that a similar story of cattle raiding, the *Táin Bó Cúailnge*, served the Irish. As we shall see, Herakles and Cúchulainn may both have engaged in cattle raids, but the similarity of plot disguises the fact that the stories served very different purposes for the societies that told them.

According to Apollodoros' summary, Eurystheus requires Herakles to travel far to the west, to the island of Erytheia, near the entrance to the Atlantic. Here lives Geryon, the triple-bodied, triple-headed son of Khrysaor and Okeanos' daughter, Kallirrhoe. He possesses a herd of crimson-colored cows, watched over by the herdsman Eurytion and his dog Orthos, a brother of Kerberos. On his way there Herakles threatens to shoot Helios and is given a golden goblet that allows him to cross the water. Upon reaching Erytheia, Herakles kills both watchman and dog and begins driving the cattle home. Geryon, hearing of the theft, overtakes him at the Athemos river and the two do battle. Geryon is slain by a poisoned arrow, whereupon Herakles makes off with the cattle.

How is the myth to be read? It is clear that it arises in a pastoral milieu, since the object of the raid is to get Geryon's cattle. Some scholars interpret the story as growing out of a shamanistic struggle with the Lord of the Animals, by which humans wrest control of the animals away from the god. Walter Burkert sees the story originating here, but argues that what was originally a hunter's myth was adopted by pastoralists who transformed the spirit of the wild animals revered by hunters into a tutelary deity for their herds and flocks.[13] Burkert takes this further and argues that the myth continued to evolve into an allegory concerned with mastering old age and death.[14] The raid has become a quest, leading the hero to the Underworld. The katabatic elements in the story are prominent: the color of the cattle and the name of the island, Erytheia (Red One), both point to the setting of the sun. The golden goblet in which Herakles crosses the western waters suggests the disk of the setting sun reflected in the waters. Herakles' journey takes him to the end of the world, where night and day meet, a magical place that leads to the Underworld. At one level, then, the myth reports a confrontation with the lord of the Underworld. His cattle are no longer just the animals given to us by a supreme being, but are a metaphorical representation of the souls of the dead.[15] Reading the myth as Herakles' voyage to the Underworld makes sense of many details in the story: the hound Orthos recalls his more famous brother, Kerberos, and the news of Herakles' theft is reported to Geryon by Menoites, the herdsman of Hades' cattle.[16] A doublet of the entire episode occurs in Latin in the story of Cacus, where Virgil makes the connection with the Underworld explicit:

> So Cacus' lair
> lay shelterless, and naked to the day
> the gloomy caverns of his vast abode
> stood open, deeply yawning, just as if
> the riven earth should crack, and open wide
> th' infernal world and fearful kingdoms pale,
> which gods abhor; and to the realms on high
> the measureless abyss should be laid bare,
> and pale ghosts shrink before the entering sun.
> (Virgil, *Aen.* 8.247–56, trans. Williams)[17]

The underworld elements in the story emerge in sharper relief if we compare the story as it comes to us from Stesichoros to Virgil with a variant reported by Arrian, and deriving from the early historian Hekataios. According to Hekataios, the entire episode had nothing to do with the Iberians and took place in Epiros.[18] This is an attempt to rationalize the story by diminishing the scale of the deed—Epiros is much closer to the Peloponnese than is Tartessos—but, like other rationalizations, stripped of its otherwordly themes the story loses its power. It is likely that Hekataios was trying to reconcile the western setting of the Geryon story with a local tradition that the fat cattle of northern Greece were descended from Geryon's herd, a tradition preserved in a number of sources.[19] Similarly, the myth could be reapplied to other, more local settings. A great many depictions of the episode on Attic vases, for example, give Geryon emblems strongly associated with Chalkis, a city hostile to Athens. The villain of myth thus becomes the local enemy.[20]

The Tartessos episode is interpreted by some as an example of the Indo-European cattle-raid myth, variants of which can be found as far apart as the *Rg Veda* and the *Táin Bó Cúailnge*.[21] In the basic form of the myth (as it is reconstructed by Bruce Lincoln) an Indo-European hero suffers at the hands of a monster, a three-headed serpent explicitly identified with the aborigines encountered by the Indo-Europeans: "[i]n the first encounter this serpent stole some cattle belonging to the hero or someone close to him, but in a second meeting he defeated the monster and recovered the cattle."[22] Lincoln sees Herakles' tenth labor broadly conforming to the pattern: Geryon is triple-bodied and triple-headed, while Cacus, the monster in the Latin version of the story, is both triple-headed and steals Herakles' cattle. Lincoln locates the original myth among the Indo-European peoples and reads the story as a justification for a type of imperialism. His conclusions deserve attention:

> Finally, an ethical concern seems to be present in our myth, for it must be noted that *Trito's raid was not unprovoked aggression, but followed upon the tricephal's earlier theft. It is thus justified, for the I-E hero is only taking back that which rightfully belongs to his people. Moreover, he uses open force to regain his stock, in contrast to what must have been regarded as the despicable stealth of the tricephal. The myth is an imperialistic myth, it is true, but even imperialists need their rationalizations.[23]

Having distilled later versions and reconstructed the original myth, Lincoln then adduces an even more specific function, connecting it to the initiation of young warriors. They battle a triple-headed creature and are told the story of the first cattle raid, "the prototype for all subsequent cattle raids."[24] Peter Walcot argues that the cattle raid was part of the ritual initiation of young men in early Greek society and that the episodes of raiding recounted in Homer are reflexes of this.[25] He particularly draws attention to Nestor's status as an untried warrior in the episode recounted by Nestor in *Iliad* 11.

These attempts to distill cattle-raiding myths down to an Indo-European prototype and to connect them with a loosely defined initiation ritual should be treated with caution. In Irish storytelling, for example, cattle raiding is a commonplace mytheme, reflecting the centrality of cattle husbandry, but is not associated with a specific ritual.[26] Furthermore, the correspondence between the story of the cattle of Geryon and Lincoln's reconstructed Indo-European *Ur*-myth is dubious. There is no sign that Herakles ever carried the name *Trito, which Lincoln sees as a central feature of the myth. More significantly, Geryon had not stolen Herakles' cattle, so that the ethical dimension identified by Lincoln is wholly missing. There is no justification for Herakles' theft, a point recognized as early as Pindar and Plato.[27] In order to make Herakles' theft justifiable, Lincoln speculates that in Hesiod's version Geryon had himself just stolen the cattle before Herakles' arrival, thus making him a villain. The lines in Hesiod are as follows:

> Now the force of Herakles killed him
> Beside the shuffling cattle in sea-girt Erytheia
> On that very day when he drove the broad-faced cattle
> To holy Tiryns, having passed over the ford of Okeanos . . .
> (Hesiod, *Th.* 289–92)

Nothing in Hesiod's Greek supports the view that Geryon had stolen the cattle, and, in any case, Geryon had no need to steal from Herakles since in all versions of the story he was known for breeding fine cattle.[28] Lincoln's case is also weakened by the fact that his reading of the Graeco-Roman texts is restricted to Hesiod and Propertius. He makes no reference to either Stesichoros or Apollodoros. In fact, Denys Page offers excellent reasons for supposing that Stesichoros' Geryon was an honorable and sympathetic character.[29] Quite aside from these textual difficulties, there is no evidence for any actual initiation ritual in Greece based on cattle raiding, at least not for boys.[30] The Athenian *ephebeia* and the Spartan *krypteia* do not involve cattle raids. The sole cattle connection to the *ephebeia* comes in the ephebes' practice of "raising the ox" at Eleusis in emulation of Theseus' feat of having hurled an ox over the roof of a temple.[31] Even if cattle raids were a part of life in Archaic Greece, one cannot demonstrate, as one can for the Maasai, that "a young male achieve[d] manhood by killing a warrior from another tribe during a cattle raid."[32] Plutarch claims that Peirithous stole the cattle of Theseus in order to test his reputation for strength and bravery, but the confrontation ends in their becoming bosom buddies, and in any case is known from literature alone, not as the charter myth of an actual cult.[33] In fact, the only evidence for the symbolic performance of cattle raiding in later Greek society is Xenophon's description of a folk dance performed by some of his Thessalian troops:

> After this, some Aenianians and Magnesians got up and started to dance the Karpeia, as it is called, in armor. The dance goes like this: one man puts his weapons aside and starts driving a pair of oxen, plowing and sowing. He keeps looking behind him as if he were afraid of something; up comes a cattle stealer, but as soon as the plowman sees him he grabs his weapons and confronts him. They fight in front of the oxen, accompanied by a tune played on the pipe. Finally the thief ties the farmer up and drives off the oxen, or else the plowman ties up the thief, in which he puts him into the stocks next to the oxen, with his hands tied behind his back, and off he goes. (Xenophon, *An.* 6.1.7)

There is quite a difference between dancing the Karpeia and "playing the bear" for Artemis, a ritual for Athenian girls conducted in the goddess' sanctuary at Brauron. The former can be performed anywhere, while the latter was a rite marking the transition from childhood to marriageability. The difference between the two shows that we should expect more evidence of an initiatory cult than just the traces of an Indo-European myth.[34] There ought to be some literary, epigraphic, or other material evidence connecting a specific cult to cattle raiding. Finally, while the cattle raid may be part of the repertoire of stories shared by Indo-European peoples, it is in its variations, not as a putative charter myth, that the cattle raid reveals its significance.

Attempts, then, to explain the myth of Herakles and Geryon by identifying the template on which it is based take us no further than confirming what we already knew: Greek myths shared much in common with other Indo-European systems of myth. What is harder to explain are the circumstances that made these stories meaningful in a specific time and place. Every telling of a myth reflects its *Sitz im Leben*, in this case the Greek encounter with the non-Greek world of the western Mediterranean. Thus the trope of the raid has a good deal in common with other cultural products designed to make indigenous people either monsters or threats to the Greeks.[35] The metopes from the temple of Hera at Foce del Sele, for example, consistently emphasize the barbarian and bestial characteristics of non-Greeks.[36] As so often, the way was paved by epic, in this instance Homer's depiction of the Cyclopes. In one tradition, indigenous to Sicily, they are music-loving shepherds, but in Homer they are presented as flesh-eating, lawless, and incapable of social intercourse.[37] As with giants, so too with pygmies, whose popularity in the sixth century suggests that they too served as models of what the Greeks were not. Maurizio Harari refers to pygmies personifying "a status of irreducible geographic, cultural, racial and anatomic alienism."[38] In the opening of the western Mediterranean to increasing numbers of Greeks, as Phokaians sailed to Massalia and established a Greek presence there, stories of Herakles' cattle raiding shaped the imaginative response of the Greeks to their New World experience. The Underworld aspect of his journey enriches the story with undertones of the heroic quest as a spiritual journey, but it too is made possible by the fundamental geographic reality underlying the whole process: the Greeks following Herakles were going ever westward, toward the setting sun. Without that basic fact the raid of Herakles would never have become a myth of the triumph over death.

One interpretation of Herakles' cattle raid that does place it in the context of contact with non-Greeks is Bruno d'Agostino's suggestion that the myth encodes the introduction of Greek modes of sacrifice and culture to non-Greek lands.[39] Herakles, however, does not stop to make sacrifice in the places he visits. Sacrifices, rather, are made to commemorate his passing, while the cattle are finally brought all the way back to the Peloponnese. As a charter myth for the "hellenizing" of the western Mediterranean, then, the story of Herakles is hardly compelling, although it was a suitable vehicle for exploring Rome's Greek roots. According to Strabo, Coelius Antipater interpreted the fact that the hereditary sacrifice to Herakles at Rome was conducted according to Greek practice as evidence that Rome had been founded by Greeks.[40] But the story of the cattle of Geryon takes Herakles far beyond Italy, and is against this broader geographic backdrop that the story must be read. Herakles' journey takes him around the Mediterranean coast from western Greece to Gibraltar and back, the details of which would have been meaningful to Greek sailors hugging the northern coast of the Mediterranean. The Greeks sailed for the most part by line of sight, so that being able to fix one's location by the shape of a headland was at least as important as being able to navigate by the stars.[41] For this reason maps were not a fundamental part of navigation. Instead the Greeks conceptualized space as a network constructed out of a series of steps moving from point to point. Such a conception of "odological" space, in which geography is neither theoretical nor static but constructed from a series of movements from place to place, lends itself to narrative treatments, descriptions of what is to be seen as one goes from point A to point B, fleshed out by accounts of what happened at each of these places. Pausanias' technique is a good example of this approach at its most developed. In this respect, a cycle of stories, such as Herakles' labors in the west or Theseus' labors around the Saronic Gulf has more in common with Aboriginal song lines, the songs that turned space into narrative, than with a map, a silent, two dimensional, non-narrative representation of space.

Traces of the odological dimension in the stories of Herakles are scattered through early Greek poetry and left traces in the epichoric traditions of the western Mediterranean. Hesiod had referred to the Garden of the Hesperides in the west, and poets before Hekataios had sung of the wonders encountered by heroes in their travels to the region of Okeanos.[42] Herakles' labors were the subject not just of Stesichoros' *Geryoneis*, but also figured the seventh-century *Titanomachy* and Peisander's *Labors of Herakles*. The stories had particular appeal to western Greek poets, such as Peisander (from Kamiros) and Stesichoros (from Himera), who helped crystallize the close association between Herakles and the western Greeks. As Thomas Braun puts it, "When the Phocaeans with their penteconters opened up the Adriatic and Tyrrhenian seas, Iberia and Tartessus . . . they cannot have doubted they were going where Herakles had gone before."[43]

Herakles was especially associated with the Tartessos region, where Geryon was born. The area was famous for its silver and gave the Greeks access to the tin mines of Cornwall, though they themselves rarely sailed beyond the Mediterranean. This boundary was set by the Pillars of Herakles, which marked the boundaries of the earth beyond which Herakles could not (and Greek sailors would not) go.[44] Pindar warned sailors to turn their ships back if they had come this far.[45] The pillars also neatly separated the northern side of the Mediterranean with its Greek colonies and emporia from the Carthaginian-dominated zone on the southern side. But right at the opening to the Atlantic the two cultures intermingled. The *archegetes* was worshipped as Herakles and Melqart, a fusion helped by the fact that the Greeks recognized a special relationship between Herakles and the Phoenicians.[46] Herodotos, for example, claimed that the cult of Herakles on Thasos had been founded by the Phoenicians.[47] The pillars were as much a feature of Phoenician cosmology as Greek.[48] Herakles led the cattle of Geryon back around the coast of Provence and into Italy, and stories attached to his journey reveal him taming the hostile locals. Not far from what would become Massalia Herakles is attacked by Alebion and Derkynis, who try to steal the cattle from him. Herakles kills them and is then attacked by the local people. When his arrows run out Herakles uses boulders to scatter the Ligurians, a story giving an *interpretatio Graeca* to an oddity of local geology: the rocks of the Plaine de la Crau.[49] Coming closer to Italy along the same coast, he is worshipped at Monaco as Herakles Monoikos in a temple above the small harbor.[50] Safe havens such as Monaco were invaluable to sailors hugging the Côte d'Azur because of the unpredictable and dangerous winds prevailing in the Gulf of Lyons and the Gulf of Genoa: the marin, an onshore wind that can quickly assume gale force anytime from late summer to spring; the mistral, a wind that blows offshore from the Rhone valley; and the Tramontana, a wind fed by cold mountain air in the Gulf of Genoa, an area of active cyclone formation in the winter.

After his arrival in Italy, Herakles has three adventures connected to the cattle of Geryon, each reflecting the position of Italy in the changing geographic imagination of the Greeks. The first, preserved in Dionysios of Halikarnassos, ultimately goes back to Hellanikos of Lesbos in the fifth century:

> When Herakles was driving Geryon's cattle to Argos and was come to Italy, a calf escaped from the herd and in its flight wandered the whole length of the coast and then, swimming across the intervening strait of the sea, came into Sicily. Herakles, following the calf, inquired of the inhabitants wherever he came if anyone had seen it anywhere, and when the people of the island, who understood but little Greek and used their own speech when indicating the animal, called it "vitulus" (the name by which it is still known), he, in memory of the calf, called all the country it had wandered over Vitulia. (Dionysios of Halikarnassos, *Ant. Rom.* 1, trans. Cary)[51]

While not formally making Herakles a founder-hero of Italy, the story does bring Italy into the orbit of the Greek world by making it a region that received its name from a Greek hero. Furthermore, variations of the story are widely attested in both Greek and Latin sources. Aulus Gellius, for example, citing both Timaios and Varro, picks up the cattle connection, claiming that Italy was named from the Greek word *italoi* ("bulls") because of the vast number of cows in Italy and prevalence of cattle grazing.[52] Varro's treatise on agriculture mentions both Timaios and the Herakles story preserved in Hellanikos. He notes that Italy was thought to have taken its name from cattle:

> In ancient Greece, according to Timaios, bulls were called "itali," so that Italia took its name from its many beautiful, and fertile, cattle. There are also accounts that Hercules pursued a handsome bull called an "italus" from Sicily to Italy. (Varro, *Rust.* 2.5.3)

Even Aristotle's account of the foundation of the Italian tribes, though based on a tradition that derived Italia from an eponymous founder, Italos, retained the connection between Italy and cattle herding:

> It was this Italos then who according to tradition converted the Oinotrians from a pastoral life to one of agriculture and gave them various ordinances, being the first to institute their system of common meals; hence the common meals and some of his laws are still observed by certain of his successors even today. (Aristotle, *Pol.* 7. 1329b, trans. Rackham)[53]

Hellanikos' curious story and its variants deserve comment. In the first place, while this looks like an early instance of colonialist nomenclature, it is noteworthy that Hellanikos believed that the name actually derived from the language of the indigenous people. In fact, *italos* ("bull") is etymologically related to Latin *vitulus* ("calf"), so Hellanikos was not far off the mark. Yet it is not a straightforward matter of Herakles' adopting a local toponym either. It is in Sicily that Herakles hears the word *vitulus*, but "Vitulia" is applied to "all the country [the calf] had wandered over," namely, Italy. In Hellanikos' account, therefore, Herakles is responsible for creating Italy as a coherent geographic space. In this story Sicily and Italy are understood to be geographically distinct, but it is unclear to what degree Hellanikos or his audience distinguished between the various indigenous people within the peninsula. "Vitulia" subsumes Ligurians, Latins, and all the other Italic peoples into one territory under one name, a phenomenon that does not reflect the political or cultural unity of the peninsula but does reflect the colonizer's practice of lumping disparate indigenous populations under one heading. The story, then, reflects how the Greeks arriving on the Italian peninsula saw the land. What they saw as they moved from the drier climate of Sicily up into the more fertile and well-watered coastal plains of Latium and Campania was cattle country. In fact, in the Pianura Padana of northern Italy, there was sufficient land available for pastoralism to be conducted alongside agriculture, rather than in competition with it, down into the imperial period.[54] To settlers coming from places in Greece called Boiotia ("The Land of the Cow") and Euboia ("Good Cattle Land"), this was a landscape ideally suited to cattle husbandry. Rivers flowed down from the Alps and Apennines, creating a broad, well-watered river plain. The eagerness of the Greeks to bring this rich pasture land into the metageography of the Greeks illustrates that trade and agriculture were not the only stimulants to the great wave of western colonization. Once colonies were established in Italy they prided themselves on the richness of their grazing land and advertised this on their coinage. As John Papadopoulos has recently argued, "By representing the bull, Sybaris stressed its extensive grazing lands."[55] Italy was thus made an extension of Greece, only bigger and better: Megale Hellas, Vitulia, the land of the Calf.

The second episode connecting Herakles to Italy is the story of Cacus, a doublet of the original theft of the cattle from Geryon. Having come as far south as Rome, Herakles goes to sleep on the Aventine Hill. Cacus takes this opportunity to steal some of the cattle, and, in a detail borrowed from the *Homeric Hymn to Hermes*, he leads the cattle backwards to his cave so as to leave no trail. The following day Herakles is leading the remaining cattle past the cave when the stolen cattle begin mooing. Breaking into the cave, Herakles kills Cacus, and reclaims the stolen animals. The story was enthusiastically taken up by Roman poets, and there are treatments of the myth by both Virgil and Propertius.[56] The Cacus episode brings Rome fully into the Greek cultural sphere. In Propertius' version, having killed Cacus, Hercules makes the future forum a blending of sacred Greek and Roman space:

> "Go, cattle, go, cattle of Hercules, last labour of my club, cattle twice my quest and twice my booty, and hallow with long-drawn lowing the Fields of Cattle: your pasture will be Rome's famous Forum."
>
> (Propertius 4.9.16–20, trans. Goold)[57]

Herakles thus becomes the tutelary deity of Rome's ancient cattle market, the Forum Boarium, which will in time house an Archaic altar and a temple to Hercules Invictus. The episode is an early example of the enthusiastic adoption by the Romans of Greek religious habits. From the point of view of cattle myth, Rome was Oxford-on-Tibur. Rome's cattle connection is not a matter of chance. Adalberto Giovannini and Alexandre Grandazzi have recently drawn attention to the enormous importance to Rome's prosperity of the saltfields of Ostia, "saltfields that attracted there and not elsewhere, herders and their herds."[58]

Before returning to Greece, Herakles undergoes one more adventure thanks to the cattle of Geryon. The bull that escapes from him in Italy and makes its way to Sicily is, in various versions of the story, first captured by Eryx, king of the Elymoi, who adds it to his herd. To regain the bull Herakles must wrestle the king, which he does, killing him in the process. In both Apollodoros' and Pausanias' accounts Eryx has staked his kingdom on the match.[59] Although Herakles does not stay to claim the throne, the tale was part of a cycle of stories that incorporated the contested area of western Sicily into the Greek sphere of influence. Eryx was himself the son of Aphrodite and Boutes, an Argonaut who was settled at Lilybaion by Aphrodite.[60] Boutes' name, which means "the Herdsman," again points to the attractive qualities of the pastureland encountered by the Greeks. Though the Elymoi were later regarded as a mixture of Sikanians, Trojans, and Phokaians, the genealogical connections with Aphrodite, Boutes, and Herakles amounted to a claim that the region they inhabited was neither Phoenician nor Sikel but legitimately Greek.[61] It is not difficult to guess when such a claim was made. In 510 BC the Spartan Dorieus led an expedition to colonize Sicily. According to Herodotos, Dorieus, the younger brother of the Spartan king Kleomenes, was encouraged to undertake the attempt by a seer named Antichares, who told him "that the whole land of Eryx belonged to the Heraklids, since Herakles himself had won it."[62] The attempt was repulsed by the combined forces of the Phoenicians and the men of Segesta, an Elymian city in western Sicily, and Dorieus was killed. But even though the episode ended in failure, it nonetheless demonstrates the flexibility of myth, and, in this instance, its ideological possibilities. Herakles' tenth labor took him to Gibraltar and back, but the tellings of these stories covered even more ground: the myth served as an allegory of the confrontation with death, rendered into narrative form the sailor's perilous journey along a hostile and dangerous coast, and expressed the complex relationship between Greeks and non-Greeks up and down the Italian peninsula. The only thread linking these diverse narratives was the structural pattern of repeated cattle thefts, by Herakles, by Cacus, and by Eryx. The cattle raid recurs repeatedly not because it was linked to adolescent initiation, but because it is a narrative trope embedded as deeply as possible among cultures that raised cattle.[63]

As a myth and as a way of understanding social relations, the cattle raid authorizes a permanent state of conflict. This is readily seen in societies that remain more fully pastoral, such as the Dinka and Nuer, who attribute the continuous warfare between them to a primeval raid when the first Dinka stole cattle intended as a gift from the sky god Kwoth to the Nuer.[64] Cattle raiding myth retained its power for the Greeks, not because they too lived entirely by pastoralism (since they did not), but because they did live in a state of permanent conflict, among themselves and with non-Greek people. The cattle raid encapsulates agonism. Its complements are those myths that substitute women for cattle, the other commodity that causes endless disputes. As Herodotos recognized, one could claim that at the heart of Greece's grand confrontation with Persia lay a series of reciprocal raids and the kidnapping of Io, Europa, Medea, and Helen.[65] With women as the focus of dispute, institutions at the intersection of gender, power, and conflict—that is, rape and marriage—were suitable vehicles for understanding the same conflicts between Greek and non-Greek. Hence, as Carol Dougherty has shown, the origins of Greek colonies are often given a narrative form based on wedding narratives and epithalamian poetry: Greek men rape or marry local women, before or after killing the local men.[66] In one final inflection of the Geryon myth, reported by Herodotos, the mythemes of cattle and marriage are finally interwoven. According to the Greeks of the Black Sea, Herakles drove the cattle of Geryon back from Gadiz to the Peloponnese by way of Skythia (!) where he lost the mares of his chariot team. To get them back he must sleep with a monstrous half-woman, half-snake, who tells him that she will now bear his child. In time the child grows to manhood, proves his worth by drawing the bow of Herakles, and becomes the founder of the royal dynasty. His name is Skythes, and thus are the kings of the Skythians descended from Herakles.[67]

Taking the Bull by the Horns

If cattle help shape the Greek hero, they are even more important to the Greek conception of their gods. Zeus, Poseidon, and Dionysos are all worshipped as bulls or described as assuming bull shape in myths recounting their deeds. Hera is frequently described as *boôpis* ("cow-eyed" or "cow-faced"), and there are close cultic associations between Athena and Artemis and cattle. This was a stratum of belief that was overlaid by the emergence of anthropomorphic deities more familiar to us: Apollo majestically raising his arm over fighting lapiths and centaurs, or Zeus carrying Ganymede off to Olympos to be his cupbearer. Bovine avatars come to be replaced by gods who are cowherds. Hades, Helios, and Apollo all own herds, and Apollo serves a period of indentured servitude as a herder to Admetos. The best example of the god as cattleman is, of course, Hermes the cattle rustler. This cattle nexus has not been more fully explored, probably because the notion that the Greek gods were, or had been, theriomorphic is not currently in favor. Since the Homeric pantheon is so recognizably anthropomorphic, locating a different stratum in the Greek conception of the gods smacks to many recent historians of excavating Greek religion stratigraphically in search of its earliest layers, rather than asking how it operated. The search for origins, especially by comparison with other cultures, is often viewed with great suspicion, associated as it is with the excesses of the Cambridge ritualist school of the early twentieth century. Scholars such as J. G. Frazer and, most notably, Jane Harrison drew on a vast range of comparative studies to uncover what they saw as the roots of Greek religion and to reconstruct the most basic layers of Greek religious practice.[68] The results were mixed. Brilliant insights arose side by side with idiosyncratic observations that relied more on imagination than evidence: Harrison's concept of the *eniautos daimon*, or Year-spirit, was (and remains) particularly contentious.[69] It is not uncommon for more recent scholars to acknowledge the adventurousness of this work, but the tone of such judgements is likely to damn with faint praise. A good example is Geoffrey Kirk's estimation of Harrison: "in spite of the brilliant texture of her *Themis* little that is original in it can also be said to be correct."[70] As a result, there remains a curious reluctance to acknowledge an animistic dimension to the gods of the Greeks. For example, in discussing the vexed question of epiphanies of gods as birds in epic Walter Burkert asserts, "The owl of Athena, the eagle of Zeus, and the peacock of Hera-Juno are little more than heraldic animals for the Greeks," and dismisses animal metamorphosis as nothing more than a playful characteristic of myth rather than a feature of actual worship.[71] But when Achilles threatens to eat Hector's flesh raw, is the poet merely employing a figure of speech to suggest Achilles' rage, or do he and his audience understand that for a moment Achilles is in essence a wild animal?[72] The latter is vastly more disturbing, yet recent studies of metamorphoses in Homer suggest that that is exactly how the Greeks saw their gods, their heroes, and presumably themselves.[73] When, therefore, Homer speaks of Athena *Glaukopis*, he is evoking a goddess who assumed the form of an owl. Artemis, too, was both the mistress of animals but also the she-bear.[74] And in speaking of Hera *Boôpis*, Homer is referring to a divinity who, as Rhys Carpenter puts it, "was in some sense a cow-goddess, not merely in that cows or white oxen were sacred to her, but that she manifested herself in that animal's form."[75] Similarly, there is nothing symbolic about Boreas' transformation into a stallion: it is in this form that he copulates with the mares of King Erichthonios, siring twelve foals.[76] We learn this from Aeneas, himself the son of a mortal and a goddess, as he speaks to Achilles, another son of a mortal and a goddess, in the middle of a recitation of the genealogy of the kings of Troy. Here heroic, divine, and animal ancestry all interweave with each other. Zeus' transformation into a bull to seduce Europa is thus not a metaphor occasioned by the similarity of god and bull in terms of power or virility, so much as a metamorphosis made possible by metonymy, an association based on contiguity rather than similarity.[77]

The distinction between metaphor and metonymy is an important one for our understanding of the Olympian gods. The bull is not merely a symbol. It remains part of the god's presence, and in the encounter with the divine the human still sees the bull in the god and vice versa. Greek religion was not, as the great scholar of hero cult Lewis Farnell asserted, "sober and sane."[78] Rather, it was bloody, passionate, and imaginatively rich. In it, the boundaries between the human and the divine were constantly breaking down.[79] Heroes showed that the line between men and gods was fluid, while gods showed that the line between human and animal was also opaque. As John Heath asserts in a recent study of animal imagery in the *Oresteia*, "The boundaries of humanity are too porous, allowing the beast to slip in and out with discomforting ease."[80] The categories of divine, human, and animal bleed into each other. When, therefore, Agamemnon strides forth among the Achaians going into battle, the poetic description deploys a welter of animal, human, and divine associations. The captains marshal their men like goatherds separating their animals into different flocks, while in supreme command is Agamemnon:

So the commanders drew up their troops
To enter battle, and Lord Agamemnon
Moved among them like Zeus himself,
The look in his eyes, the carriage of his head,
With the torso like Ares', or like Poseidon's.
Picture a bull that stands out from the herd
Head and horns above the milling cattle—
Zeus on that day made the son of Atreus
A man who stood out from the crowd of heroes.

(Homer, *Il.* 2.513–21, trans. Lombardo)

He is the chief herder, but at the same time his physical form evokes both the gods and a prize bull. It is this richness of association that makes Homeric similes so much more powerful than mere metaphors. When, for example, Menelaus recounts Proteus' description of the death of Agamemnon, the metonymic connection between Agamemnon and the bull permits a subtle shift. Aigisthos sets an ambush, after which

> He led Agamemnon up, not suspecting his death, and killed him
> After feasting him, as one would kill an ox at the trough.
> (Homer, *Od.* 4.534–35, trans. Lombardo)

No longer the bull, Agamemnon has become an ox. The bull has been castrated, evoking another aspect of Aigisthos' treachery: the usurper has cuckolded the king and now he will slaughter him.[81] At first the verb *anago* ("lead up") seems to point toward the killing as a sacrifice, since this is the word one uses to describe leading a bull to the altar, but then the simile strips the action of sacral overtones. Instead it is no more than the casual killing of a draft animal caught unawares, too old to be of any more use.

Like Agamemnon, the chief gods also have a bovine side, reflected in scattered references throughout poetry and art. As early as the middle of the sixth century Zeus was depicted on a metope of Temple C at Selinous as a bull carrying off Europa, while the Orphic Hymn to Zeus explores the Zeus bull mystically:

> His head and beauteous face the radiant heaven
> Reveals and round him float in shining waves
> The golden tresses of the twinkling stars.
> On either side bulls' horns of gold are seen,
> Sunrise and sunset, footpaths of the gods.
> His radiant body, boundless, undisturbed
> In strength of mighty limbs was formed thus:
> The god's broad-spreading shoulders, breast and back
> Air's wide expanse displays; on either side
> Grow wings, wherewith throughout all space he flies.
> Earth the all-mother, with her lofty hills,
> His sacred belly forms; the swelling flood
> Of hoarse resounding Ocean girds his waist.
> (Porphyry, *On Images* frag 3.11–15, 23–30, trans. Gifford)[82]

Porphyry concludes, "Zeus, therefore, is the whole world, animal of animals, and god of gods." His essence, then, is as much purely animal as it is purely divine. This is reflected in some of the cults associated with Zeus. The cult of Zeus Atabyrios, in particular, had a strong cattle associa-

Figure 5.1 Figurine of a bull. Greco-Roman. Bronze.15 cm. Museum of Fine Arts, Boston. Catharine Page Perkins Fund, 96.707. Photograph © 2010 Museum of Fine Arts, Boston.

tion. The cult is attested as far west as Akragas and as far north as Olbia.[83] The cult originated on Rhodes and was heavily influenced by the Phoenician worship of Baal Tabor.[84] Zeus' worshippers dedicated oxen to the god (Fig. 5.1), and bronze cows were supposed to low when danger threatened.[85]

There is also the suggestion of a connection between god and bull in the Archaic Athenian festival of the Bouphonia, in which an ox was sacrificed and skinned, after which the stuffed hide was set to a plow. The celebration was held in honor of Zeus Polieus, as protector of the city, but nothing in Porphyry's lengthy description of the festival points to an identification of the animal with the god.[86] The focus of the festival is on the ox's significance as the plow animal, making it indispensable to agriculture.

A second god who appears as a bull in myth and cult is Dionysos. He is referred to as "Bull" (*tauros*) or as "bull-headed" (*taurokephalos*).[87] Dionysos was called *taurophagon* because winners of the dithyrhambic competition at the City Dionysia sacrificed an ox to him, and at a festival in his honor in Arkadia men coated themselves with grease before carrying a bull to the altar for sacrifice.[88] At the *boukoleion* ("bull pen") in Athens the Basilinna engaged in a *hieros gamos* with Dionysos. A large marble relief depicting a bull once found in the Library of Hadrian may have originally adorned this building, especially if, as Stephen Miller has suggested, the *boukoleion* was a cult place of Dionysos.[89] In some places the bull god's priests were called *boukoloi* (cowboys). Like maenads they danced ecstatically in his honor and provided the chorus for Kratinos' play of the same name.[90]

The god most strongly identified as a bull in both literature and cult, however, is Poseidon. Hesiod refers to him as the Earthshaking Bull and he was worshipped in that form particularly in Thessaly, where it was believed that he had created the gap to the sea through which the Peneios flowed.[91] In Roman times the Thessalian cult of Poseidon Taureos included bull games in which young riders wrestled wild bulls to the ground; the depiction of the *taurokathapsia* on earlier coins from Thessaly suggests that the practice was ancient.[92] At Ephesos young unmarried men serving as wine carriers at sacrifices to Poseidon were called "Tauroi."[93] The more familiar face of Poseidon is the *Theogony*'s earthshaker, given dominion over the sea. He was associated with earthquakes and horses, but as with the other gods, identities and avatars elide, even to the point of complete inversion.[94] He is Poseidon Hippios, but also Taraxippos, "he who terrifies the horse." There has been considerable debate regarding this face of the god, but the well-known story of the death of Hippolytos, who is dragged to his death by his horses after they have been frightened by a bull sent by Poseidon, suggests that Taraxippos and Poseidon Bull were one and the same.[95] A similar motif of the bull as punishment occurs in the story of the animal sent by Poseidon to punish Minos for failing to sacrifice the bull he had vowed to the sea god. It is this Cretan bull who sired the Minotaur by Pasiphae, making the bull-man, in some sense, another avatar of the god. Metamorphosis is thus characteristic of Poseidon, whether of the god's own appearance, as when he becomes a stallion to mate with Demeter, or in his bestowing the power of transformation on others, such as Periklymenos and Mestra.[96]

In the *Odyssey* Poseidon appears as the father of another savage creature, Polyphemos, and is the recipient of all eight of the bull sacrifices mentioned in the poem.[97] This may reflect the fact that Poseidon is as hostile to Odysseus as Athena is well-disposed to the hero, but it may also arise from a particular reverence for Poseidon going back to Mycenaean times. Tablets from Pylos record communal sacrifices to Poseidon and underscore his importance.[98] A theriomorphic Poseidon has also been suggested in the cults of southern Italy. Coins from both Sybaris and the

Figure 5.2 Poseidonian stater, fifth/fourth century, showing Poseidon and bull. American Numismatic Society acc. no. 1944.100.5148.

colony named after him, Poseidonia, frequently depict Poseidon on the obverse as the sea god with his trident and show a bull on the reverse (Fig. 5.2).[99]

Domestic(ated) Goddess

Another deity whose divinity is closely connected to a cattle incarnation is Hera, who is referred to fourteen times in the *Iliad* as *boôpis*. This epithet, when taken to mean "cow-eyed," has been interpreted as a reference to the goddess' modesty. Unlike Athena *Glaukopis* with her bold and manly stare, like the wide-eyed gaze of an owl, Hera displays the downward glance appropriate to a sober matron.[100] Others attenuate the connection between goddess and cow even further: Burkert notes the epithet and adds parenthetically, "wide, fertile plains with grazing herds of cattle and cattle sacrifices are her special preserve."[101] Yet the bovine connection is more than casual. At the Heraion on Samos the cult statue of Hera wore a mitrê with horns, placing her in a tradition of Near Eastern nature deities who also displayed a strongly warlike character.[102] The most common sacrificial animal at the Heraion was the cow, which is not true of other goddesses such as Demeter or Aphrodite, to whom pigs, goats, and sheep were more commonly sacrificed.[103] In epic Hera's ferocity is sometimes reduced to semicomic nagging, but behind it lies a power to inflict punishment: Io turned into the "cow-horned Maiden"; the women of Kos given horns to punish them for abusing Herakles; the daughters of King Proitos of Tiryns, who run around madly mooing; and her eponymous hero, Herakles, the "Glory of Hera," whose cattle raid nearly ends in disaster when Hera sends a gadfly to prick the cattle of Geryon.[104] Furthermore, her very name is the female equivalent of "hero" and points to her role as a strong, protective deity.[105] As the female version of the hero protector, she is not Zeus' wife so much as his female counterpart. Indeed, one hundred years ago Albert Cook argued that Zeus and Hera were seen as the very model of a married couple at war with each other precisely because they derived from two separate traditions in which each had been the dominant figure within a divine couple.[106] Traces of this warrior queen persist, as for example in the *Iliad*, where her bloodlust inspires the Achaians to fight. Homer refers to her roaring with the voice of fifty men, an aspect of her power that Hesychios recognized when he glossed *boôpis* to mean *megalophonos* ("loud").[107] The epithet *boôpis* points, then, not toward her modest gaze but to her animal power. This is consistent with the pattern of the epithet's use in Homer. When Homer employs other animal imagery he refers to Hera as the "white-armed goddess" so as not to clutter the poem with competing animal associations. When, however, she is described in close proximity to Zeus the poet refers to her as *boôpis*, in deference to her queenly power and status.[108] The same equivalence between king and queen, cow and bull, was recognized at the feast of the Great Daidala celebrated on Mount Kithairon once every sixty years. Each participating city sacrificed a cow to Hera and a bull to Zeus.[109] The theme of the ritual was reconciliation, ostensibly that of Zeus and Hera through the celebration of a *hieros gamos*, but the political overtones for the Boiotians, whose federal league existed by voluntary ascription, are clear.[110] By the time of Hesychios in the fifth century AD it was possible to gloss *boôpis* metaphorically as *megalophthalmos* ("large-eyed") and *euophthalmos* ("gracefully eyed"), but behind Homer's cow-faced goddess stands not a coy matron but *Bowia*, the divine Cattle Lady attested in Linear B.[111]

The cattle connection carried over into Hera's cult.[112] On Samos, excavations at the Heraion have brought to light terracotta oxen, perhaps going back as early as the Mycenaean period, dedicated to the goddess who, according to the *Greek Anthology*, delighted in the sacrifice of calves.[113] Her special preference for cattle sacrifices is also demonstrated by the high concentration of cattle bones found in excavation, and the existence, still to be seen in Pausanias' day, of an ashen altar, which, like the Altis at Olympia, was comprised of the remains of the animals sacrificed to her.[114] Around 500 BC the butcher (*mageiros*) Kyniskos in one of the colonies of Magna Graecia dedicated a bronze ax head to Hera *en pedioi* (on the plain).[115] At Argos she was worshipped as the Goddess of the Yoke (*zeuxidia*).[116] The worship of Hera was as central to Argive identity as that of Athena at Athens, and her priestess was a woman of high status. Chrysis served as priestess for at least forty-eight years and Argive chronology was based on this priesthood.[117] Io, too, was remembered as a priestess of Hera, and the story of her punishment may have arisen in the merging of the familiar mytheme of the rape of the maiden by the god with an injunction against sex laid upon the servant of the goddess.[118] In Apollodoros' version of the myth Hera responds to Zeus' trick of turning Io into a cow by requesting that the animal be given to her, and as the story unfolds Hera and Zeus struggle over not just his adultery but who gets to keep the cow. Structured around the poles of here (Argos) and there (Egypt, where Io ends up giving birth to Epaphos), the story affirms Hera's primacy at home, defined as *oikos*, *gamos*, and *patris*. Hera's festival at Argos was noted for the sacrifice of one hundred cattle in her honor and was sometimes called the *Hekatombaia*.[119] The festival provides the setting for one of the best-known stories in Herodotos, the tale told by Solon to Kroisos of the happy death of Kleobis and Biton:

> They were of Argive stock, had enough to live on, and on top of this had great bodily strength. Both had won prizes in the athletic contests, and this story is told about them: there was a festival of Hera in Argos, and their mother absolutely had to be conveyed to the temple by a team of oxen. But their oxen had not come back from the fields in time, so the youths took the yoke upon their own shoulders under constraint of time. They drew the wagon, with their mother riding atop it, travelling five miles until they arrived at the temple. When they had done this and had been seen by the entire gathering, their lives came to an excellent end, and in their case the god made clear that for human beings it is a better thing to die than to live. The Argive men stood around the youths and congratulated them on their strength; the Argive women congratulated their mother for having borne such children. She was overjoyed at the feat and at the praise, so she stood before the image and prayed that the goddess might grant the best thing for man to her children Kleobis and Biton, who had given great honour to the goddess. After this prayer they sacrificed and feasted. The youths then lay down in the temple and went to sleep and never rose again; death held them there. The Argives made and dedicated at Delphi statues of them as being the best of men. (Herodotos 1.31.2–5, trans. Godley)[120]

This passage has elicited a number of rich analyses.[121] Otto Regenbogen noted that the reward of a painless death offered to a pair of pious men mirrors Apollo's gift to Trophonius and Agamedes, builders of his temple at Delphi, and argues that the story reflects a strong Delphic influence.[122] However, Richard Seaford has placed the episode back in the setting of the Heraia festival at Argos and sees the Kleobis and Biton tale as marking an important shift in the festival's function. Originally a festival to celebrate the taming of young women eligible to marry, a transition evoked in the yoking of cows, the festival instead became a full-fledged celebration of the ephebe's assumption of adult status.[123] Seaford notes that in taking up the *zugon* not only do the youths become substitutes for the oxen, but they submit to the yoke of manhood and assume their place as men in the line of battle. In similar fashion, Leslie Kurke sees the youths' reward as a painless attainment of a man's most glorious end: a death on behalf of the community, rewarded by civic commemoration.[124] Read in this light, Herodotos' story points toward the successful completion of male ritual initiation. The cattle symbolize the strength and service of the man to the community, while the sacrifice of the animals, reversing the original substitution of the young men for the oxen, solemnizes the ritual.[125] The final stage in this process was to dedicate a figurine of a bull, like Homoloichos' dedication to the Kabeiroi (Fig. 5.3), which commemorates his sacrifice and piety.

Figure 5.3 Votive bronze bull figurine dedicated by Homoloichos to the Kabeiroi. Greek, Late Archaic Period, about 490 BC. 5.3 x 8 cm. Museum of Fine Arts, Boston. Henry Lillie Pierce Fund, 98.663. Photograph © 2010 Museum of Fine Arts, Boston.

Just as the yoking of cattle could stand for the young men' assumption of civic responsibility, so too terracotta figurines (and in some cases cakes called "cows") could substitute for cattle or serve as dedications to commemorate the completion of the ephebes' rites.[126] Similar connections between coming-of-age rites and dedications of animal models have been proposed recently to explain the large number of bovine figurines found at the sanctuary located at Kato Syme on Crete. Dedications here include twenty-three bulls and seven rams from the end of the end of the tenth century, and forty-nine bulls and eighty-seven rams from the ninth.[127] The excavator of Kato Syme, Angeliki Lebessi, has offered a similar interpretation of the bronze bulls dedicated at the Kabeirion outside Thebes.[128] Whether the population participating in these rites was made up of pastoralists or not, their cultic practices reveal the continuing hold of the bovine mystique on the Greeks. Through their cattle the Greeks formulated their understanding of their gods, their heroes, and themselves.

CHAPTER 6

Gods, Cattle, and Space

> Here I intend to build a beautiful temple
> to be an oracle for men who shall always
> bring to this place unblemished hecatombs.
> —*Homeric Hymn to Apollo* 3.287–89,
> trans. Athanassakis[1]

Trouble in Paradise

In a much-quoted passage, Herodotos remarks, "It was Homer and Hesiod who composed a divine genealogy for the Greeks, and who gave the gods their titles, allocated to them their powers and fields of expertise, and made clear their forms."[2] Yet the cults of Zeus, Hera, and Poseidon revealed by archaeology and literary sources outside of epic reflect gods quite unlike their Homeric and Hesiodic incarnations. All three remain strongly associated with cattle, demonstrating the hold that cattle continued to have on the imagination of the Greeks, both as avatars of divine power and as objects of symbolic and economic value. An exploration of the gulf between the cults of these gods, especially local cults, and their presentation in epic is warranted, since it sheds light on the tensions that accompanied the emergence of a Olympian, and by extension Panhellenic, order.

An obvious starting point is the petulance shown by Zeus' companions in Homer's poems. In the *Iliad* Hera complains of Zeus' private conversation with Thetis;

> Who was that you were scheming with just now?
> You just love devising secret plots behind my back,
> Don't you? You can't bear to tell me what you're thinking,
> Or you don't dare. Never have and never will.
> (Homer, *Il.* 1.573–76, trans. Lombardo)

She is a shrew and a snob. In Book 24 she complains that Hector does not deserve the same honor as Achilles since Hector was only fed at the breast of a mortal woman.[3] She seems, in fact, jealous of other women and of other women's children. Yet this is the goddess who had been closely associated with the initiation of young women and whose worship was a focus for the territorial aspirations of the Argives.[4] There is a similar disjunction in the case of Poseidon, worshipped throughout Thessaly, at Onchestos in Boiotia, and at Helike on the north coast of the Peloponnese. He was the tutelary deity of at least four federal leagues: the Thessalian and Boiotian *koina*, the Kalaurian League, and the Achaean League.[5] Why then should he appear in epic literature primarily as an earthshaking sea god who is, like Hera, notable for his querulousness? "It's not fair!" he cries, complaining of his inferior status to Zeus.[6]

This discrepancy between the evidence of cult and the evidence of epic arises from a tension at the heart of Greek religious thought and practice in the Archaic period: the struggle between an emerging Panhellenic consciousness, identified with the will of Zeus, and the hold of epichoric customs, expressed in an attachment to local divinities often markedly different from their Homeric or Hesiodic incarnations, if they appear in epic at all. These local cults often demonstrated a surprising tenacity in the face of the emerging Olympian pantheon. On the border of Attic and Boiotia, for example, the cult of Amphiaraos continued to attract visitors throughout the Classical period, even though the nearby town of Oropos changed hands dozens of times. At Lebadia in Boiotia the oracle of Trophonios was still famous in Cicero's day, more so, in fact, than the impressive sanctuary of Zeus Lebadios.[7] In some instances the local god fused with an Olympian. The local Tegean goddess Alea was assimilated to Athena, and another Peloponnesian god, Maleatas, eventually merged with Apollo, receiving cult both above the Thyreatic Plain and at Epidauros.[8] These epichoric cults were inseparable from the landscape. As *lieux de mémoire*, local sanctuaries were both places fixed in memory and places that fixed memory, attaching stories of the past to specific locales, and making the landscape meaningful in the creation of identity. At Oropos one could see the very chasm into which Amphiaraos had fallen. Even if his cult was picked up in other locations and helped to define the borders of Attica, Oropos remained his central locale.[9] At Lebadia, to consult the oracle of Trophonios one had to be lowered into a hole, after which one experienced a sometimes unsettling epiphany. In some cases the topography already had an epic identification, as, for example, at the Menelaion outside Sparta, where as late as the sixth century a certain Euthikrenes could commemorate Menelaus with a bronze statuette.[10] For hundreds of years this site served as a place where memories attached to the heroic age shaped dedicatory behavior in the landscape. At Menidhi, where we do not know the hero's identity, a tholos tomb made a tangible link to the imagined, heroic past and so was a suitable place to leave gifts to a hero.[11] In all these cases, physical remains from an earlier time combined with oral and epic traditions to provide the setting and the stories that accompany religious performance. In discussing a contemporary example from southern Angola, the *okuyambera* ceremony, Christofer Wärnlöf notes, "the landscape is filled with mythological and cosmological contents relating the physical landscape to ancestors and spirits. . . . History is created in the ritual process which manifests itself in the landscape."[12] The same combination of specific location and ritual performance imbued the local cults of ancient Greece with significance. And in cases where the specific identity of the person for whom the ceremony was enacted was forgotten, the ritual could take on a life of its own. Pausanias notes that the people of Daulis did not agree on the identity of their founder–hero, but continued to offer him chthonic sacrifice every day. Similarly, while crossing Afghanistan in 2002, Rory Stewart observed the treatment of the graves of local people: "Tombs like this [the grave of a village schoolteacher's father] are frequently on the edges of settlements and most are revered as shrines, even when the occupants' names and deeds are forgotten."[13]

There is hardly a hint of this layer of epichoric religion in Homer and Hesiod, who describe a world in which the Olympians have triumphed, and where, as a result of Zeus' victory over chaos, the various gods have each been given their own discrete portfolios. Jane Harrison contrasted this with the profusion of cults in Athens and astutely noted, "It was only the great orthodox Olympians who in their ultimate supremacy could indulge in a perfect specialization of attributes."[14] Hesiod's neat teleology is a pleasant fiction, because in fact struggles continued. Zeus and his most loyal ally Apollo faced the threats not only of Titans without, but also of constant subversion from within, from a recalcitrant Poseidon, a jealous Hera, and warlike Athena.[15] In the *Iliad* there are hints of this, as, for example, when Achilles refers to the time that Hera, Poseidon, and Athena tied Zeus up until Thetis helped free him.[16] The episode is oddly left unexplained but is evidently located close to the critical time of Zeus' victory over the Titans. The subversive Olympians accept Zeus' rule against a common enemy, but the battle between Zeus and his family continues throughout the *Iliad*, particularly in Books 14 and 15, when with the help of Aphrodite, Hera seduces Zeus and removes him from the battlefield, allowing Poseidon to rally the Argives. Sleep stirs him to action, saying,

> Help the Greeks all you want now, Poseidon,
> And give them their glory, however brief,
> While Zeus still sleeps, for Hera has bedded him,
> And I have wrapped him in downy slumber.
>
> (Homer, *Il.* 14.363–66, trans. Lombardo)

Poseidon does so, only to be recalled by Zeus. Poseidon complies, but like a truculent younger sibling only does so grudgingly, with a mixture of complaints, threats, and scowls. Deceptive, conniving, biased, the gods in epic act in ways that were shocking to those who expected perfection. Xenophanes famously complained, "Homer and Hesiod have ascribed to the gods everything that is shameful and blameworthy for human beings: stealing, committing adultery, and deceiving one another."[17] The rancorous struggles on Olympos serve to remind us that Zeus' order is not only hard won but also only tenuously maintained. It is an attitude that mirrors the view prevailing in the Archaic period that the natural condition of human life was conflict. *Eris* ("Strife"), as Hesiod observed, existed all over the earth in two types: a positive form that led men to compete with each other, and an evil version that was the spirit of warfare.[18] The same seems to have applied to heaven. This is perhaps not entirely surprising since Xenophanes also understood that the Greeks created their gods in their own image.[19] The shadow of bull and cow would not disappear entirely, retaining more than a vestigial existence in myth and cult, but the Olympian pantheon was modeled on an entirely human family, its dramas reflecting in the divine realm a very human anxiety over conflict. Paradoxically, however, the cults of the Olympian gods played a somewhat different role, mediating conflict and channeling it into the socially constructive forms of agonistic competition. At Olympia and Delphi, Zeus and Apollo could exercise the kind of Olympian calm that eludes them before the walls of Troy. In their *temene* (sacred enclosures), cut off from ordinary space, during periods of truce distinct from ordinary time, their festivals were occasions for contestants and visitors to affirm their common Greek identity through competition with each other.[20] Thus, while the gods of epic reflect a society riven by conflict, their sanctuaries and festivals offered the means to make conflict productive.[21] As the opposite of local gods and heroes, the Olympian deities offer the opportunity for an entirely Panhellenic identity to emerge against the backdrop of local cults.[22]

The radical nature of this Archaic religious evolution has barely been explored, in part because archaeologists and historians persist in giving priority to evidence of cult continuity from the Bronze Age. Consider, for example, the approach of Anthony Snodgrass, who, having surveyed a number of sanctuaries asks, "The . . . sanctuaries . . . have produced some material which is probably of a date before 700, and much else that belong to Archaic and Classical times: *the great question is*, how far back into the dark age do their dedications extend?" (emphasis added).[23] Where changes have been noted, the explanations tend to oversimplify. For example, Christian le Roy notes, "La hiérarchie olympienne à prédominance masculine remplace la domination féminine symbolisée par la

Potnia."[24] In actuality the emergence of a coherent Olympian and Panhellenic order was much less neat and tidy than the imposition of a male hierarchy on the worship of "the Goddess" or the annexation of local cults. The Olympian religion, to use an old-fashioned term, was not based on the repetition of cult activity at a single place but on a cycle of performances—*theoria*, oracular consultation, sacrifice, festival, and contest, all regulated by a fixed calendar—repeated with variations at a number of sites. It was this fugue that allowed the sanctuaries of the Olympian deities to emerge from the background noise of local worship and to become focal places for those performances that created regional ties. Just as at the level of the polis there existed a calendar of sacrifices, rituals, and initiations that constituted the performance of communality, so too at the Olympic level the cycle of athletic games, musical contests, sacrifices, dedications, and consultations performed Greekness.

Panhellenism and Homeric Hymns

The Olympic pantheon differed from local cults in one important respect. There were many versions of the Olympian gods: Zeus might occupy Olympia, but he was also the god at Nemea and Lebadia, prophesied at Dodona, and was associated with Mount Ida on Crete. Accordingly, the god was connected with many places. Furthermore, few of the sites that emerged as the settings for the great Panhellenic sanctuaries had a Homeric pedigree. Homer's description of the geography of Pylos includes the region of Olympia, yet he says nothing of the sanctuary. The cult of Zeus at Nemea used the story of Opheltes being bitten by the *drakon* Nemeios in its ktistic myth, but Homer describes only the *drakon* at Aulis and makes no mention of the Opheltes story. Helikonian Poseidon is mentioned by Homer, but not the god's cult at the Isthmus. Only the oracle of Pytho is mentioned in terms suggesting the importance of the spot.[25] Nor did the Panhellenic sanctuaries have much in the way of physical remains to bolster their legitimacy with a heroic pedigree.[26] Instead, they had to create their own histories, and did so largely through the manufacturing of divine narratives telling of the birth, journeys, and deeds of their godly inhabitants. Hesiod and Homer may have allocated *timai* to the Olympian family, but it is the *Homeric Hymns* that found them a place in the landscape by supplying the cults of the Olympian gods with stories that could plug into Homer and Hesiod's master narrative. With these stories of the gods' births, wanderings, and ktistic performances, the *Hymns* contribute to the evolution of the Greek gods away from their pastoral, bovine associations. As the gods become more anthropomorphic, their space also becomes more distinctly Greek.

Unlike epic poems, which could be enjoyed in sympotic settings, the *Homeric Hymns* were composed for performance as part of cult festivities.[27] Some of the hymns, especially the shorter ones, make reference to a longer song to follow and can therefore be classified as *prooimia* (introductions).[28] In the case of the longer hymns the issues of performance and function have led to a debate characterized by more heat than light. Some scholars propose reading the hymns as cult histories, and note their strong aitiological tendency.[29] Others have emphasized a liturgical function that would seem incompatible with competition. By calling on the god and telling the god's story, the rhapsode summons the god into the worshippers' presence.[30] But it is doubtful whether this is a meaningful distinction. Nearly fifty years ago W. E. McLeod pointed out the overlap between oracles, epic, and hymns, noting that bardic art and mantic art were linked.[31] It is entirely likely that some hymns were composed to accompany a specific cultic event, just as Philodamos' paian welcomed the arrival of Dionysos to Delphi.[32] Herodotos speaks of Olen of Lykia composing hymns for the women of Delos, and these too appear to have been liturgical.[33] Others, however, were composed for contests, as was, for example, *Homeric Hymn VI to Aphrodite*, which ends with the poet's prayer, "Grant me victory in this contest and arrange my song" (20). The longer hymns, such as those to Demeter, Apollo, and Hermes, recount episodes in the lives of the gods that may have accompanied cult events or have been performed in competition, either on their own or introducing other recitations.

The three major topoi recounted in the hymns are the birth of the god, the wandering of the god, and the establishment of the god's cult in a specific location. Birthplace was the least important in terms of the god's veneration at a sanctuary to which he later came, and competing claims over origins were possible. *Hymn I to Dionysos*, for example, begins:

> Some, O divine Eiraphiotes, say that Drakanon was your birth place
> But others claim it was the wind-swept island of Ikaros, others at Naxos,
> And others by the deep-eddying river Alpheios
> That Semele conceived and bore you to Zeus who delights in thunder.
> (*Hymn. Hom. Bacch.* 1–4, trans. Athanassakis)

This last claim may have arisen in an attempt by Olympia to strengthen the connection between Dionysos and Zeus, which would also explain the repetition of the motif of Zeus' participation in a divine birth: just as Zeus produced Athena from his head by swallowing Metis, so too he bore Dionysos after sewing him up in his thigh. Depending on where a particular hymn was composed or performed, one tradition might come to dominate—but in the case of a god like Dionysos, who had no Bronze

Age roots in Greece, it was easier to emphasize his foreign origins and deny all local claims equally:

> And, O lord, some liars say you were born
> At Thebes when in truth the father of the gods and men
> Gave birth to you and kept you well out of sight of men
> and white-armed Hera.
> There is a certain Nysa, a lofty mountain overgrown with trees,
> Far from Phoinike and near the flowing stream of the Aigyptos.
> (*Hymn. Hom. Bacch.* 5–9, trans. Athanassakis)

The geography here is imprecise but the message is clear. The story of Semele may give Dionysos a special connection to Thebes, but he does not belong to the city.

When, on the other hand, there was little or no dispute over the god's birthplace, it was usually because it was believed to have occurred in isolated or neutral territory. Hermes, according *Hymn IV to Hermes* 1–12, was born on Kyllene after Zeus and Maia slept together in a cave. The description of the spot emphasizes its isolation: it is in Arkadia, abounding in sheep, where Maia had her refuge, since "she shunned the company of the blessed gods" (l.5). By Pausanias' day the temple of Hermes on Kyllene was in ruins and it seems never to have developed into a major sanctuary, even at the regional level. Hermes was much more commonly associated with herms or altars, which he often shared with other gods, and so his birth deep in pastoral Arkadia provoked no serious counterclaims.[34] Apollo's birth, too, occurred in a neutral setting. As an island, and an unimpressive one at that, all of "rocky Delos" could be consecrated to Apollo as his birthplace, a sacred spot that belonged equally to all Ionians and any others who sought to worship Apollo. The stories of Apollo's birth site deliberately juxtaposed the humble position of the island with the promise of future prosperity. In *Hymn III to Apollo* Delos expresses her anxiety to Leto, saying,

> Thus I dreadfully fear in my heart and soul
> lest, when he first sees the light of the sun,
> scorning an island whose ground is rocky,
> he overturn me with his feet and push me into the deep sea.
> (*Hymn. Hom. Ap.* 70–73, trans. Athanassakis)

Leto's response is to promise on oath,

> that here there shall always be a fragrant altar and temple
> for Phoibos and that he shall honour you above all others.
> (*Hymn. Hom. Ap.* 87–88, trans. Athanassakis)

A similar motif occurs in the Delphic half of the hymn, when the Cretans, newly landed at Delphi, ask,

> How are we now to live? This we bid you tell us.
> This charming place does not abound in vineyards or meadows
> From which we may live well and be in the service of men.
> (*Hymn. Hom. Ap.* 528–30, trans. Athanassakis)

The god's reply is to promise them an eternal supply of sheep brought to the sanctuary for slaughter by visitors from all over Greece. The inauspicious quality of the countryside, isolated and unpromising, is thus integral to the ktistic tradition of the sanctuary.[35]

Why should this be so? In the case of city-states, shrines were often set on the edges of the polis' territory in order to mark boundaries, a structural relationship Polignac has labeled the bi-polar city.[36] The great shrines, however, are generally not located on the periphery of a single city. They exist in an entirely different political and topographic milieu. Delphi, for example, was controlled by an amphiktyony of states for most of its history.[37] The Argive Heraion, Jonathan Hall has recently argued, similarly served as a type of confederate sanctuary for all the communities on the eastern side of the Argive plain.[38] In the case of Olympia, the victory lists from the Olympic games show that Olympia was an important regional sanctuary for the entire Peloponnese as early as the seventh century, but control of the shrine was left to the people of Elis, who, during the Archaic period, did not live in a single polis.[39] Taking Corinth, Argos, and Sparta as the most powerful states in the Peloponnese, Olympia could hardly be located in a more isolated spot. The third Panhellenic sanctuary, Nemea, is close to the centers of power in the northeastern Peloponnese, yet its topographical location preserves an almost mathematical equidistance both from its nearest neighbours, Kleonai and Phlious, and from two larger centers of power: Argos to the south and Corinth to the north. Only Isthmia seems to have been closely associated with a single city, Corinth, until the destruction of the city by Mummius prompted the Sikyonians to assume control of the shrine.

The anomalous position of Isthmia may be explained by the nature of the god worshipped here: Poseidon is himself something of an anomaly. He rarely receives cult within a city and the Earthshaker was not brought within the walls of Corinth. In fact Poseidon appears to have been systematically removed from polis centers as the Olympian order became established. At Athens he lost to the goddess whose gift of the olive was deemed more beneficial than the salt sea. The evidence for the competition was still to be seen hundreds of years later in the form of the sacred olive tree and the salt sea inside the Erechtheion. Poseidon seems to have had a cult on the site of the Erechtheion at one time, but his place was taken by

Zeus and his cult merged with that of the Zeus substitute, Erechtheus.[40] Certainly in the fourth century it was the tribe of the Erechtheidai who were most assiduous in guaranteeing that Poseidon continued to receive honors.[41] In Thessaly Poseidon had been the god worshipped at the Peloria festival, where he was the consort of Gaia and was celebrated for smashing an exit to the sea for the rivers of Thessaly through the vale of Tempe, but later, as the coins of Thessaly show, he lost his position to Zeus Peloris.[42] The pattern is consistent: Poseidon was associated with more elemental forces than with *metis* ("cunning intelligence"). As Rob Schumacher has recently pointed out, "Poseidon creates the horse but Athena invents the bit; Poseidon is lord of the seas, but Athena invents the first ship."[43] At the same time this devaluation focused on areas where the polis emerged as the dominant sociopolitical form. In Attica Poseidon is relegated to the distant promontory of Sounion. In the Corinthia he has no place in the city center but is pushed out to the edge of the territory by Isthmia, close to the water. Among the Boiotians his cult does not figure in Thebes but at Onchestos, a sanctuary not a town.[44] Poseidon is often relegated to deserted areas such as promontories or mountains. In *Hymn XXII* he is addressed as "the sea-god, who is lord of Helikon and broad Aigai," both mountains.[45] Shunning cities made him a suitable god for federal leagues, and, as noted earlier, he was frequently worshipped as the tutelary god of regional *koina*: at Helike by the Achaians, at Onchestos by the Boiotians, and at Kalauria by states of the Saronic Gulf. In the context of these associations with non-polis regions his sanctuary at the Isthmus therefore may represent a declaration of neutrality by the Corinthian aristocracy. This would make sense of Pausanias' story that Poseidon and Helios contested over the land of Corinth, and that in the negotiated settlement Poseidon received the Isthmus but Helios received Acrocorinth.[46] As in the contest over Athens, Poseidon missed out on the city and was relegated to the outskirts.

Creating Greek Space

Although the *Homeric Hymns* often tell of a god's birth, it is the god's travels that are a more prominent feature. One recent study interprets these divine itineraries in terms of a difference between mortal and godly movement: mortals wander helplessly, while the travels of the gods is purposeful.[47] This is partly, but not entirely, true. Anthony Bulloch, for example, appropriately terms the wandering of Leto in search of a place to give birth "geographic anarchy."[48] On the other hand, the apparently aimless wandering of cattle may hint at a divine purpose: both Kadmos at Thebes and Ilos at Troy found their cities after being instructed by the gods to follow a spotted cow until she lies down.[49] In contrast with these, Demeter's wanderings in *Hymn II* are truly aimless, reflecting her grief at the loss of her daughter. She ends up at Eleusis, but there is no suggestion in the poem that she went there intentionally to fulfill some divine purpose. Rather, she is as grief-stricken as a human mother, a point that makes her wanderings topographically aimless:

> She withdrew from the assembly of the gods and from lofty Olympos
> And went through the cities of men and the wealth of their labours
> Tearing at her fair form for a long time . . .
> (*Hymn. Hom. Dem.* 92–94, trans. Athanassakis)

The daughters of Keleos find her and persuade her to come home with them, but it is not until Demeter has been confronted by Metaneira that she reveals herself and orders the construction of her temple at Eleusis. The aetiological aspect of the poem, to establish the rites of Demeter at Eleusis, resides then, in the logic of the poem, in a response to Metaneira's foolishness, and is not the purpose of her wandering.[50] Other gods also move across the landscape without a clear aim. Artemis in *Hymn XXVII* comes to the house of her brother at Delphi, yet she is also a hunter who "roams all over destroying the brood of wild beasts" (line 10). Pan, too, appropriately for the god of shepherds, traverses the wilds erratically:

> He wanders all over through the thick brushwood,
> Now drawn to gently flowing streams,
> Now again making his way through to steep crags
> And climbing to the topmost peak overlooking the flocks.
> (*Homeric Hymn XIX to Pan* 8–11, trans. Athanassakis)

There is, however, a real difference between, on the one hand, the shepherd or hunter, whether human or divine, crossing the landscape with his flocks or stalking her prey, and on the other, the great journeys of the gods. Poseidon steps from Samothrace to Aigai on Euboia in four strides, but the difference is not solely a matter of scale and speed.[51] Mythical itineraries such as Io's in the *Prometheus Bound* supplied the Greeks with imaginative maps that placed them at the center of a world that grew increasingly marvelous and monstrous the farther one went from the *omphalos* of Delphi. The world contained one-eyed Aramaspians and rivers rushing with gold, and was full of dangerous tribes such as the well-armed Skythians or savage Chalybes, but putting these marvels into itineraries rendered the world orderly. Aeschylus' geography may not fit well on a map, but it offered a coherent shape to a world that included Egyptians and Phoenicians. Accounts of the travels of the gods should therefore be interpreted in the context of a time when the changing shape of the world was fueling a growth in periegetic literature, from the Hesiodic *Catalogue of Women* to Hekataios.[52]

It is in that setting that Jean Davison explicates the rhetoric of these myths of wandering, arguing that they are strongly polarized: "the course moves from 'near' to 'far,' from 'now' to 'then' and from 'identity' to 'alterity.' "[53] However, in the *Homeric Hymns* the movement is usually in the other direction. Demeter ends up at Eleusis, Leto at Delos, Apollo at Delphi. Similarly, the temporal movement is not into the past but from it into the present, as the hymn summons the god into the presence of the poet and his audience. Still, if the entire purpose of the journey narrative were simply to bring the god to his sanctuary, then the itinerary would be in itself insignificant. What distinguishes the itineraries of the gods from, say, the wanderings of Io is that the places visited by the latter are alien and dangerous, while places visited by the gods are not. Space is a set of culturally constructed relations, and the gods' travels signal a divine presence whose complement is the human presence of their worshippers.[54] The gods' journeys create Greek space by marking out specific places where common religious practices could be performed on a regular basis in ways that were meaningful to all participants.[55] But the Greek system was not composed simply of a ritual repeated with only minor repetitions in any setting, much as a Catholic mass might be conducted in Ireland or Venezuela. Each place had its own version of the god and a story specific to that place, making it both wholly Greek and wholly unique at the same time. Apollo might be worshipped in many places, but only at Delphi did he protect Orestes from the Furies.

Entire regions could thus be incorporated into a Greek sphere, in which each contiguous unit, whether mountain, spring, or *ethnos*, was wholly Greek, such as the Peloponnese or the islands of the Aegean. Yet because each had a version of Apollo or Artemis, or an episode in the god's life particular to that place, Greek topography was infused with stories linking place to place, such as the Arethusa fountain in Syracuse, which overflowed with dung from sacrificial animals after a great flood at Olympia.[56] There are also enclaves of hellenicity within alien spaces, such as Naukratis or the colonies of Magna Graecia, whither the Greeks are led by Apollo Archegetes. Here sanctuaries defiantly advertised a Greek presence, as at Selinus or Akragas, where they dominate the skyline, warning those passing by that the territory is occupied by Greeks. Finally, there are places at the edges of the known world, virtually unknown to the Greeks, yet assigned a place in the Greek topography as settings for divine action. These remain elemental places, the realms of Okeanos and Tethys, whom Hera proposes to visit, or are occupied by Aithiopians or Hyperboreans who, despite their foreignness, are brought into the Greek sphere by their recognition of Greek gods. The *Iliad* begins with Zeus and the Olympians feasting among the "worthy Aithiopians," a visit that, like Poseidon's visit to them in the *Odyssey* and Apollo's winter sojourn among the Hyperboreans, defines the edges of the world. They also reinforce the Greekness at its center. At Delos, for example, two Hyperborean maidens, Opis and Arge, were buried behind the temple of Artemis and honored by the women of Delos in hymns. Distance, maidenhood, Artemisian wildness all play in counterpoint to the male, Greek melody of Apollo.[57]

That tune runs through the itinerary in *Homeric Hymn III to Apollo*, in which Leto's anguished search for a place to give birth authorizes Apollo's dominion and the creation of an Ionian space centered on Delos and the sanctuaries of Apollo:

> Thence you arose to rule over all mortal men:
> Over the inhabitants of Crete and of the towns of Athens,
> Of Aigina and Euboia, famous for ships,
> Of Aigai and Eiresiai and Peparethos by the sea,
> Of Thracian Athos and Pelion's lofty peaks,
> Of Thracian Samos and Ida's shady mountains,
> Of Skyros and Phokaia and Autokane's steep heights,
> Of well built Imbros and Lemnos enveloped in haze,
> Of holy Lesbos, realm of Makar, son of Aiolos,
> Of Chios, brightest of all the islands lying in the sea,
> Of craggy Mimas and the lofy peaks of Korykos,
> Of shimmering Klaros and Aisagee's steep heights,
> Of well-watered Samos and Mykale's towering peaks,
> Of Miletos and Kos, city of Meropian men,
> Of rugged Knidos and wind-swept Karpathos,
> Of Naxos and Paros and rocky Rheneia.
> So many places did Leto visit, in travail with the far-shooter,
> Searching for a land which would give him a home.
> (*Homeric Hymn III to Apollo* 29–46, trans. Athanassakis)

The entire catalogue is replete with references to Apollo's cult places throughout the Aegean.[58] Leto's itinerary, in fact, illustrates the way that the *Homeric Hymns* define Greek space in a fashion analogous to the assertion of Greekness conveyed through genealogy. What the one maps in time, the other maps in space. The similarities go further. Genealogy is an open system: endless brothers, cousins, raped nymphs, and eponyms can be created back to Hellen and out to the margins. As Jonathan Hall has demonstrated, the problem is how and when to stop the infinite regression. This is done by establishing an autochthonous *Urvater* or an exogenous *Stammvater*.[59] There is evidence to suggest that Greek geographic thinking worked in a parallel fashion. Adalberto Giovaninni has argued that the Catalogue of Ships represents an attempt by the priests of Delphi to accommodate those states that participated in the sacred truce and Pythian games by fashioning a heroic pedigree that actually reflects the political geography of the seventh century.[60] Two of the longest extant hymns, those to Apollo (3) and Hermes (4), also foreground geography and suggest that Greek space was defined by negotiation between gods and their sanctuaries. The Delian half of the *Hymn to Apollo*, as we have seen, catalogued the territories affiliated by Leto's passing. The journey ends at Delos, whose centrality arises from the interweaving of territory, divine presence, worship, and competition:

> But it is in Delos, O Phoibos, that your heart delights the most,
> For Ionians with trailing garments gather there
> In your honor together with their children and modest wives.
> And with boxing matches, dancing and song,
> They delight you and remember you whenever they hold the contests.
> (*Hymn. Hom. Ap.* 146–50, trans. Athanassakis)

But the Olympian gods are Panhellenic, and a bridge is needed to connect Ionian Apollo to another sanctuary capable of advancing Apollo's claims beyond the regional appeal of Delos. In short, the god must get to Delphi. Before supplying a direct route from Olympos to Delphi, however, the hymn alludes to various accomplishments of Apollo in wooing girls and defeating competitors. The lines refer to a cycle of myths once well known but now usually dismissed as opaque:

> Or am I to sing of you as wooer and lover of maidens,
> sing how, wooing the daughter of Azan, you raced
> against godlike Ischys Elationides, possessed of good horses,
> or against Phorbas sprung from Triops or against Ereutheus?
> Or in the company of Leukippos and his wife,
> You on foot and he with his horses?
> (*Hymn. Hom. Ap.* 146–50, trans. Athanassakis, modified)[61]

Recovery of the mythic substrates is difficult but not impossible. In the first place, the names of Azan, Ischys, and Elatios point to a coherent genealogy with connections from Arkadia to Phokis. According to Pausanias, Azan and Elatos [*sic*] were brothers and sons of the nymph Erato and Arkas, the eponymous king of Arkadia. As a younger brother Elatos received no share of the kingdom of Arkadia. Instead, when Delphi was threatened by the Phlegyans, Elatos went to the god's aid and stayed in Phokis, founding Elateia, the largest town in the region. The genealogy links Delphic myth with an Arkadian cycle through the story of Elatos. This was elaborated in a second episode with the same characters. Ischys, the son of Elatos, became the husband of Koronis. She had already slept with Apollo and conceived Asklepios, but preferred her human consort. The story, at one level, asserts an equivalence between the house of Elatos and Apollo, even as it singles out Koronis for punishment for her madness. In the various versions of the story Koronis is always the daughter of Phlegyas, eponym for the lawless tribe that raided Delphi and was renowned for its recklessness.[62] The story, then, has elements that emerge from Delphic accounts of the many attacks on the sanctuary, attacks that not surprisingly either fail or are punished. But there are also elements with an Arkadian provenance. In recounting some of the genealogy of Arkas and Erato, Pausanias refers to the sanctuary of Despoina in Arkadia, which included a shrine of Pan where Erato was said to have served as Pan's oracular prophetess. It is therefore very probable that it was at this sanctuary that Pausanias learned of Apollo's competition with the Arkadian heroes.[63] Different threads of the story were teased out in different places and then rewoven into versions that bound different cult locations into a great tapestry of myth.

The second cycle alluded to in the coda from Delos to Delphi concerns Phorbas and Triops. Once again, connections between sanctuaries emerge from the haze of mythical references, hinting at a kind of diplomatic negotiation carried on through stories about the gods and heroes. Triops gave his name to the federal sanctuary of the Dorian Hexapolis of the Knidos peninsula. Phorbas, his son, was believed to have come to Rhodes from Thessaly and was worshipped there for having rid the island of snakes, a deed hinting once again at an equivalence between the hero and Apollo. This and a number of other stories involving Thessalian Phorbas and Triops were integrated into cycles of myth linking both Argos and the Dodecanese to Thessaly.[64] There is no concrete evidence to support a theory of any actual migrations, and whether the cult of Triops has an actual Thessalian origin is unknown, but the putative Thessalian origins of Triops and Phorbas supply all the later elaborations with a common ancestry of place as well as genealogy. On Rhodes, where Phorbas ends up, the story of his exile and arrival supplied the *aition* for a form of ritual begging in which masters of households, as opposed to women or servants, were expected to supply hospitality.[65] The *Homeric Hymn to Apollo* reveals no familiarity with this, but in any case, that is the local story about Phorbas, and not relevant to the composer of the hymn. His interest lay in integrating Phorbas and Triops into the network of places, specifically sanctuaries, affiliated with Apollo. The references to competitions between the god and Phorbas recognize Apollo's Dorian appeal and mark the move away from Apollo the exclusively Ionian god. The mechanism for this lies in the hymn's use of catalogue-style references to places and allusions to various episodes in the god's and hero's life. Narrative patterns behind these episodes were easily replicated: gods contending with men and competitions to woo a *parthenos*. When, then, the Athenians picked this up and elaborated an Athenian version of it, Phorbas was refashioned as Erysichthon, son of Triops, and his daughter was named Mestra ("She who is wooed"). Their story was told in the Hesiodic *Catalogue of Women* but surely was also told by the *genos* of hereditary priests of Apollo at Athens, the Erysichthonidai. Accordingly, in the *Homeric Hymn* we have but a sample of a story whose fabric changed from place to place and time to time. Some tellings were primarily for local consumption, attached to local cults, but in the *Homeric Hymn* we have stories woven together to suggest a larger pattern. In sponsoring these elaborate versions, sanctuaries and their dominant families made overtures to each other by acknowledging the legitimacy of each other's versions of the god. In this way the cultic, functional layer of Greek myth underlies the more literary manifestations of it. The *Homeric Hymns* draw on both. They hymn the gods, but the theology serves the needs of an age grasping for ways to build cross-regional networks.

At the same time, however, as the locus for this system building, sanctuaries were much more vulnerable than *poleis*. Their inviolability rested on little more than goodwill and the willingness of member states, when the sanctuary was integrated into a federal structure, to defend them. For this reason the stories employed to speak between sanctuaries also hinted at the threats posed by the sanctuaries' vulnerability. In this instance, any mention of a "Phorbas" was necessarily ambiguous. Phorbas, son of Triops, in the Dodecanese might be a worthy opponent, but the same name was carried by Phorbas the Phlegyan, who preyed upon pilgrims on the way to Delphi until defeated by the god.[66] So there are two contests with a human named Phorbas: a tale involving a competition with a Dorian hero from Rhodes, and the hint of a much darker story involving depredations committed on those coming to Apollo's sanctuary. A similar ambiguity may lie behind the reference to "Leukippos and his wife" if, as has been suggested, we are meant to take this as an allusion to Daphne, who was punished for fleeing the god's advances.[67] The best-known myth dealing with the vulnerability of sanctuaries was certainly the story of Herakles' and Apollo's struggle for possession of the Pythia's tripod, an episode whose popularity on vases peaked in the generation following the First Sacred War around 590 BC, when Delphi came under the control of an amphiktyony of central Greek states.[68]

In the Pythian section of the poem many of the themes identified before are elaborated, but in surprising ways.[69] The itinerary takes Apollo not directly from Delos to Delphi but by way of Olympos and Pieria, allowing the poet to mention locations that resonate with the myths of Apollo but without strong local cultic associations. Pieria, for example, includes the cattle lands north of Olympos where the god kept his herds. From here the god moves south to sandy Lektos among the Ainianes. This Thessalian *ethnos* sent a *theoria* to Delphi every four years to make sacrifices at the tomb of Pyrrhos, claiming descent from the Thessalian hero whose tomb was as important at Delphi as the hero Pelops' was at Olympia.[70] This is a different type of movement from Leto's itinerary. In that case, all places visited by the goddess are guilty for having turned her away and become Apollo sanctuaries in order to expiate that guilt. But the Pythian itinerary is more exclusively teleological: only Delphi will do, because now it is Apollo himself looking for a permanent abode.

Apollo has no cult associations in Pieria or Iolkos, the next stage of his itinerary. In fact, Apollo's connections to the region derive from a time before he comes into his power, when he serves Admetos, king of the Thessalian kingdom of Pherai. It is while herding Admetos' cattle in Pieria that Apollo sees Hymenaios, falls in love with him and refuses to leave the house of Magnes, whereupon Hermes steals his cattle.[71] Having served Admetos, Apollo twice intervenes to help him: in the yoking of a chariot with a boar and a lion to win the hand of Alcestis, and in gaining for Admetos a postponement of the day of his death.[72] These northern locations, far from celebrating the god's power, draw attention to a young Apollo, a cowherd and a hired man, who has not yet proved his power nor found his place. Apollo even serves as a cowherder when indentured to Laomedon and Admetos.[73] Appropriately, these territories were known as rich cattle land. Jason refers to Iolkos as *eubotos* ("well-pastured"), and a number of heroes and heroines from the area have names that recall the region's reputation for its livestock: Eumelos, son of Admetos; Polymele, mother of Jason; and Perimele, daughter of Admetos.[74] From Iolkos Apollo makes his way to Kenaion, the northernmost part of Euboia, so called by Zeus for its wealth in cattle.[75] During the Archaic period the straits between Euboia and the mainland, from the Petalai Islands to Cape Kenaion, were controlled by Eretria, whose most important sanctuary was that of Apollo Daphnephoros.[76] It comes as a shock, therefore, to find the hymn addressing Apollo thus:

> You stood on the Lelantine plain, but it did not please
> Your heart to build a temple with wooded groves.
> (*Hymn. Hom. Ap.* 220–21, trans. Athanassakis)

One wonders how these lines would have been received by an audience at Eretria. The dismissal of the Eretrian cult of Apollo is especially notable since Delphic tradition maintained that Apollo's first temple was made of laurel, a legend for which there is no archaeological evidence at Delphi but which exactly describes the late Geometric temple of Apollo Daphnephoros at Eretria. The *Homeric Hymn* represents a quite deliberate attack on Eretria's credentials as a sanctuary of Apollo, even to the extent of appropriating Eretria's specific emphasis on laurel, which subsequently became a cornerstone of Apollo's prophetic role at Delphi.[77]

Quitting Eretria, Apollo continues into eastern Boiotia, past Mykalessos and Teumessos, places whose mythic resonances are with cattle stories but are not associated with Apollo: Mykalessos took its name from the mooing of the cow that led Kadmos to Thebes, and Teumessos was where Zeus hid Europa.[78] Since Apollo has not yet slain the dragon or acquired the lyre, he is still little more than a cowboy, his wanderings a cosmic expansion of the pastoralist's passage across a landscape appropriately named Boiotia: cattle country. The god even acknowledges as much when he tells Hermes of learning divination from the three sisters dwelling in the folds of Parnassos, "when as a mere child I tended the cows."[79] Arcing southwest he comes to a Thebes as yet unpopulated by humans, still a primeval forest. These are necessary details because, as with Eretria, any audience from the Archaic period onwards might have objected that Thebes was home to a well-known Apollo sanctuary, a fact confirmed by Herodotos' report that in the sanctuary of Ismenian Apollo he saw a tripod inscribed with Kadmean letters. The god's epithet shows that the cult was connected to the river Ismenos, which ran past Thebes. Pausanias explains that Apollo fathered two sons by the nymph Melia: Teneros, to whom he gave skill in divination, and Ismenos, who gave his name to river flowing through Thebes.[80] The cult of Ismenian Apollo, then, was as important to the sacred topography of Thebes as was Apollo Daphnephoros to Eretria, but it is once again deliberately ignored, no doubt because of the threat of competition to Delphi. Apollo continues on to Onchestos, where a shrine and cultic activity are described by the poet, but since the place is identified as the "fair grove of Poseidon" it is safe to dwell on the place: Poseidon's ownership has been acknowledged.[81] From here he approaches Lake Copais by way of two sites, Okalea and Haliartos, which both figure in the Catalogue of Ships.[82] The geography of the itinerary at this point is oddly inaccurate: the god crosses the Kephissos prior to reaching Haliartos. This is impossible, since the river flows into the Copaic basin on the northwestern side, while Apollo approaches Haliartos from the southeast. Upon reaching Haliartos Apollo surprisingly announces his intention of building his temple here, and even goes so far as to lay the foundations. At this point the local nymph Telphousa speaks to Apollo, explaining that the sound of pounding hooves at her spring will annoy the god. At nearby Onchestos such traffic is already controlled, as the poem has suggested at lines 235–38. Andrew Miller has shown that Telphousa is able to persuade Apollo because the god's preference is for a place that will guarantee his freedom from disturbance, a goal he will achieve with the killing of the dragon at Delphi, but it is also worth noting that Apollo's search is for a place in a landscape that already has sanctuaries in it. This space is taken, explains Telphousa, but Krisa in the folds of Parnassos is not.[83] When Apollo subsequently realizes that he has been deceived, since Telphousa forgot to inform him of the dragon's existence, he returns in a rage, buries the Telphousa spring under "a shower of rocks," and builds himself an altar. In other words, Apollo will recognize the claims of Poseidon, to whom all of Boiotia was sacred, but will take sanctuary land for himself when such a seizure is justified as punishment for attempts to deceive the god. The entire narration of Apollo's journey to Delphi reflects a delicate balance between rival sanctuaries and powers: Poseidon is acknowledged, rivals to Phoibos are not.[84]

After the Telphousa episode the *Hymn to Apollo* describes the birth of Typhaon and the killing of the she-snake by Apollo. The mention of Typhaon casts Apollo's victory as a repeat of Zeus' victory, but whereas Zeus' victory is cosmic, Apollo's, like the victory of Kadmos over the snake at Thebes, is tied to a particular place. The triumph over a monster is a regular part of the foundation narrative of a town, and by modeling Apollo's arrival at Delphi on ktistic traditions the poem in essence creates an equation: sanctuary is to god as city is to men.[85] It is important to recognize this, since the traditions concerning the previous owners of the sanctuary have been interpreted as evidence for the suppression of older cults, dominated by female deities such as Gaia, by the new patriarchal order of the Olympians. Yet the only evidence for these earlier, female earth cults is the *Homeric Hymn* itself. What looks to be the suppression of older, epichoric religious practice may be something quite different: an invented tradition designed to give the victory of the Olympians—in this case Apollo—a teleological impetus.[86] The elevation of the locale to the special status of Panhellenic sanctuary requires that there be an earlier indigenous cult, the suppression of which is not only necessary for Apollo's triumph, but which must be inscribed in memory and topography:

> Rot now right here on the man-nourishing earth;
> You shall not ever again be an evil bane for living men
> Who eat the fruit of the earth that nurtures many
> And will bring to this place unblemished hecatombs,
> And not Typhoeus or ill-famed Chimaira
> Ward off woeful death for you, but right here
> The black earth and the flaming sun will make you rot.
> (*Hymn. Hom. Ap*. 363–69, trans. Athanassakis)

The double *figurae etymologicae*—both *puthô*, "to rot," and Python, the serpent's name, evoke the god's epiklesis, Pythian—serve to reinforce the connection between land, narrative, and memory.[87] The act of hymning the god's passage through the landscape transforms it into a memoryscape, connecting all places he has touched upon, fashioning them into a network of cultural relations that come together at Delphi. The hymn allows Delphi to emerge from the landscape as a *lieu de mémoire,* a central place whose significance is marked by the god's presence, his coming of age, and the commemoration of this in the hymn's performance.[88]

In narrating Apollo's time as a herder and his deeds before assuming his place as the mature, oracular god of Delphi, the Pythian hymn treats Apollo as a youth who must come of age before coming into his full power. A modern parallel from a contemporary pastoral society sheds light on Apollo's role, since the same combination of coming of age, herding, adventure, and storytelling can be seen in the Yaaral and Degal festivals, celebrated twice a year by the Fulani and Peul people of west Africa. Young Fulani men spend months herding other people's cattle across the Sahel before returning to Diafarabé in December. After driving their cattle across the river they celebrate the Cattle Crossing Festival. The first day of this popular community festival is called the Promenade des Jeunes, when the young men and women, beautifully dressed and with their hair carefully braided, mingle freely, while the young men recount their accomplishments.[89] How they and their stories are received influences their position as men (and especially as husbands) in Fulani society. Those with the fattest, best-kept herds are highly esteemed; the worst herder is given a peanut, and stands little chance of attracting a bride.[90]

Here is a living tradition from a pastoral culture that shows a much greater degree of similarity to the *Homeric Hymns* than any hymn from a monotheistic tradition, reflecting once again the continuing hold of pastoral practices on the mindset of the Greeks. At Delphi it is the rhapsodes who will perform in honor of the god, but rather than simply praise the god's omnipotence the hymns explore the tricks and battles of strength in which the god has been tested. In fact, as Marcel Detienne has shown, the Apollo worshipped at Delphi is a god "of murderous drives," knife in hand, whose thirst for blood and meat, far from evoking a distant Olympian calm like the god's portrait on the pediment at Olympia, instead "makes public the extreme fragility of a cultural frontier between the blood-crime and the sacrificial meal."[91] The performances will be witnessed by sacred ambassadors, whose own journey to Delphi imitates the god's. Embassy, journey, and bearing witness to the god's victory and presence are all aspects of the *theoria,* a recursive reinforcement of both the god's stature and the status of those who come as pilgrims and who are sanctioned to offer sacrifice and share meat.[92]

From Apollo's triumph over the serpent the hymn moves immediately to the story of Apollo bringing the Cretan priests to serve him at Delphi. A recent study of the hymn draws attention to the fact that the southern and western movement of the Cretan priests describes a broad arc that parallels the arrival of the god from the north and east, both arcs converging on Delphi.[93] It is also notable that the Cretans' exogenous origins parallel Apollo's, each reinforcing Delphi's claim to lie at the center of the Greek world, just as Delos lies at the center of the Ionian sphere. The claim of centrality is clearly critical to a sanctuary that boasted the *omphalos*, and was perhaps made pressing by the emergence of the Amphiktyony, the "Dwellers-Around" as the controlling power at Delphi. In fact, Beate Wagner-Hasel has recently argued for a broad correspondence between the places visited by the god coming from Pieria to Delphi and the states of the Amphiktyony. Wagner-Hasel also draws attention to the late date (sixth century) for the first epigraphic attestation of Apollo at Delphi, pointing out that there is no evidence for an Apollo cult here in the Mycenaean period. Moreover, the seasonal movement of Apollo, who quits the sanctuary in the winter to spend time among the Hyperboreans, parallels the seasonal rhythm of the Amphiktyony, which rotated its spring and fall meetings between Delphi and Anthela. All of this points to an Amphiktyonic setting for the Pythian hymn.[94] The connections should not be pressed too far, however. A number of Amphiktyonic states are not mentioned: the smaller ones include the Dorians, Phthiotians, Oitaians, and Malians, while the powerful Thessalians and Athenians are also missing.[95] On the other hand, the itinerary is heavy with Boiotian sites while dismissing Thebes. It would be a mistake, therefore, to read the Pythian hymn as a charter for Amphiktyonic control of Delphi any more than as an instance of Olympian propaganda at the expense of an earlier cult of the Earth-Mother. If there is a political message in the poem, it is surely in the downgrading of powerful political centers, such as Eretria and Thebes, and the dismissal of their Apollo cults. The hymn does not tolerate potential competitors to the primacy of Delphi.

Competition with Delphi, nevertheless, lies at the heart of the second hymn with a strong geographic component, the *Homeric Hymn IV to Hermes*, a poem that in many ways picks up the themes of the *Hymn to Apollo*. The hymn relates the story of the birth of Hermes to Maia on Mount Kyllenes before passing quickly on to the theft of Apollo's cattle. The god is suitably precocious and the hymn in his honor can be read as a complement to the *Hymn to Apollo*. Just as Apollo is a cow herder, Hermes is labeled a "cattle-rustling robber."[96] Apollo travels from Pieria to Onchestos to Delphi, while Hermes travels from Kyllene to Pieria via Onchestos. The hymn to Apollo emphasizes music as the god's special domain and speaks of him going to rocky Pytho to play his lyre and entertaining the gods on Olympos with his songs; Hermes' first deed is to address the tortoise he finds on the hillside, promising to do it honor, whereupon he immediately kills it and transforms its shell into a lyre. Even the asides in the *Hymn to Apollo* asides concerning the chariot ritual at Onchestos and the exchange with Telphousa, in which she alludes to Onchestos, are inverted: Hermes comes across an old man attending his vines and warns him,

> You are to be blind to what you saw and deaf to what you heard
> And silent too when no harm is done to what is your own.
> (*Hymn. Hom. Merc.* 92–93, trans. Athanassakis)

Returning from Pieria, where Apollo keeps his cattle, Hermes comes back not to Kyllene but to the Alpheios River in Elis, where he slaughters two of the stolen cows. It has been tentatively suggested that this itinerary recalls some early transhumant route across Greece, but the journey should probably not be read so literally.[97] Surely it is no coincidence that a hymn that so clearly and cleverly answers the hymn that places Apollo at Delphi should find Hermes slaughtering the stolen cattle on the banks of the Alpheios, the very river that ran past the sanctuary of Zeus at Olympia. The geography of the poem includes sanctuaries referred to either explicitly or obliquely: Zeus at Olympia, Hermes at Kyllene, Poseidon at Onchestos, and Apollo at Delphi. In short, employing the mytheme of the cattle raid with all of its heroic associations, the hymn represents a claim by Hermes for inclusion in the Panhellenic pantheon.[98] In narrative form Hermes is permitted to demonstrate that he merits this inclusion: the theft proves cunning, his desire for meat is heroic, his cooking of the meat shows he is civilized, and the fact that he does not eat the meat is evidence of his piety.[99]

Commentators have often noted the subversive quality of Hermes: thief and trickster, he aspires to be "chief of robbers." But attempts to historicize the poem and read it as an attack on the emerging Olympian order miss the mark. Norman Brown felt that Hermes represents "the aspirations and achievements of the Greek lower classes," challenging the aristocratic Apollo.[100] Morris Silver also finds a political consciousness in the poem and reads the myth as a code "concerned with the cultic symbols and social practice of corvée labor (*leitourgia*)."[101] Not only is this interpretation based on the most tenuous arguments—lyre playing equals building public works, turtle shells symbolize the bending of the back in labor, the whip is a scourge for beating recalcitrant workers—but it fails to adduce any evidence for the existence of such an institution in Archaic Greece. A god displaying cunning intelligence can hardly subvert order when such intelligence is a prized virtue in Greek culture. Furthermore, analyses focusing on the dispute between Hermes and Apollo must also account for the reconciliation between them. Hermes' theft and deceit must be seen in terms of the resolution of conflict that they lead to; it is the sharing of honors with Apollo that makes the earlier details both amusing and manageable. Nor is this simply a general interest in abstract notions of justice. Early on Hermes threatens to sack Apollo's great house at Pytho with its cauldrons and tripods, gold and iron, so the threat to order is framed specifically as a threat to the sanctuary of Apollo, an issue of immediate concern to Delphi. Pausanias records seven attempts to sack the sanctuary, and stories of the god's intervention on behalf of his temple and of the punishment of those who succeeded were disseminated by Delphi as a warning to those tempted to ignore the sanctuary's inviolability.[102] Yet, having raised the possibility of the sack of the god's sanctuary, the hymn deflects it through an elaborate exchange of gifts: Hermes' lyre and Apollo's shining bullwhip. By this exchange the two gods solemnize an arrangement that ends their antagonism. Henceforth, in fact, their opposition will be reformulated as a complementarity: Apollo may be famous for his lyre playing, but he will owe this to the gift of Hermes. Similarly, Hermes owes his position as god of herds to his older brother. Thematically the hymn moves from theft to restitution, from conflict to amity, from deceit to *timai*, shown by the most potent sign of compromise, the exchange of gifts. In fact their agreement is reached in two stages. In the first, Apollo listens to Hermes singing. His theme, appropriately, is how each god received his lot. Apollo offers to settle their dispute, explicitly linking Hermes' song to the first steps toward resolving their dispute:

> Scheming cattle slayer, industrious comrade of the feast,
> Your performance is worth fifty cows;
> I think we will settle our accounts at peace.
> (*Hymn. Hom. Merc.* 436–38, trans. Athanassakis)

Taking the hint, Hermes then offers to leave lyre playing to Apollo, and instead asserts that he will content himself with grazing cattle in the mountains, whereupon he offers Apollo the lyre.[103] In effect, Hermes has placated Apollo, and without conceding his guilt has made an offering to the god and compromised. In the second stage of the settlement, the two gods adjourn to Olympos and work out the final details. Apollo receives the lyre and a guarantee that his sanctuary will remain inviolate:

> And then the son of Maia with a nod of his head promised
> Never to steal away whatever the Far-Shooter possessed,
> And never to approach his sturdy house.
> (*Homeric Hymn IV to Hermes* 521–23, trans. Athanassakis)

In return, Hermes is showered with gifts, receiving the Thriai's power of divination, dominion over lions, boars, dogs, herds, and flocks, and is appointed the escort of souls to Hades. In short, his claim for inclusion has been acknowledged, and in return Delphi's status has been recognized.[104]

Hermes and Apollo are by no means the only Olympians to engage in a contest, nor is theirs the only contest for control of a sanctuary. Poseidon and Athena competed for the sovereignty over Athens and Troizen, while Apollo and Poseidon also assert competing claims to Delphi, Kalauria and other spots.[105] Most of these disagreements play out in two ways. Contests for a territory end up with a single winner and a clear loser: Athena wins Athens, Poseidon loses Athens, and Troizen, and Corinth.[106] At sanctuaries, however, the potential conflict results in compromise and either the sharing of a sanctuary or an exchange: Apollo gets Delphi, but Poseidon does not exactly lose. He is compensated by receiving Tainaron and Kalauria, and Apollo's victory also involves sharing Delphi with Dionysos.[107] In short, stories about the dealings between members of the Olympian family serve as models for different approaches to conflict resolution. Territory, that is, human space, is subject to competing claims that tend toward a zero-sum solution: only one god can possess Athens. Sanctuary-based resolution, on the other hand, is characterized by compromise. As the political landscape of Greece evolved throughout the Archaic period, then, sanctuaries were positioned to assume a position of critical importance to the Greeks not simply as the settings for a religious activities, but because they slowly came to offer models of communal interaction quite distinct from the institutions of the city-state. Sanctuaries formed a bridge between local, regional, and national (taking this term in a cultural rather than political sense). They provided locations that permitted a Greek identity, as opposed to a Achaean, Elian, or Thessalian identity, to express itself.

It is clear, then, that the religious changes occurring in Greece during the Archaic period were extraordinarily complex. Local sanctuaries adapted in a variety of ways, sometimes emphasizing their special place in the landscape, but also affiliating with the Panhellenic order. Panhellenic sanctuaries emerged as places of unique importance, sponsoring religious activities such as games, festivals, and oracles that established a pattern embedded in Greek culture at once complementary to but different from the culture of the polis. It remains now to examine the position of sanctuaries in the physical landscape of Greece, not in terms of the polarity of city and sanctuary, but in terms of the landscape as a terrain in which conflicting economic activities, such as farming and large-scale herding, were carried out. As we shall see, by functioning as the focal point for an emerging sacred economy, sanctuaries made possible the intersection of agriculture with an increasingly sacralized herding culture.

CHAPTER 7

Sacred Economics

> For we do not have need of the *thytes* for himself, but that through him we think we can reach an understanding of the future and the signs from heaven.
>
> —Arrian, *Epictetus* 1.18

Sacred Land

In the *Politics*, Aristotle specifies the different ways humans may get their food supply without resorting to exchange or commerce. He lists nomadism, hunting (which for Aristotle includes brigandage and fishing, as well as more traditional styles), and agriculture.[1] The last category, claims Aristotle, comprises by far the greatest number of people, namely, those who live off the land and its produce—in other words, farmers. In his scheme he does not reserve a particular place for raising cattle, and his description of nomads seems to refer to shepherding; he describes nomads as extremely lazy and being compelled to follow their flocks from field to field, "as if they were running a living farm."[2] His decision not to include ranching is significant and points to the fact that in the Classical world raising cattle was no longer the business of ordinary people. Out in Epiros King Pyrrhos might have his herd of four hundred prime head, but land was too precious in most parts of Greece to turn over to ranching. The Greeks had made a tentative move in that direction in the Geometric period, but the process was unsustainable despite the prestige of cattle and their importance in Greek religion. In fact, precisely because cattle were central to the entire system of sacrifice-feast-dedication-commemoration that constituted the key practices of Greek religion, the Classical period would witness the development of a sacred economy centered on herds and land that operated separately from the secular economy of agriculture, manufacturing, and trade. At times the two systems would intersect, but the sacred economy, as we shall see, operated according to its own laws of supply and demand and was not the secular economy in disguise.

Sanctuaries and the religious organizations that operated them were critical to the emergence of the sacred economy, providing the locales where sacrifice took place, and managing the business of stock raising that supplied sacrificial animals. This raises the question of the role played by sanctuaries in the broader economic life of the Greeks. Tullia Linders interprets the sanctuary in terms similar to Max Weber's conception of the ancient consumer city living off the produce of the countryside, but offers this modification: "[the sanctuary] lived on contributions and did not produce any material goods. I suggest, however, that what we may call its spiritual wares, the cult, created a demand for certain material commodities, which stimulated production and provided the producers with a market they might not otherwise have had."[3] In fact we can go further: the sacred economy operated as a sphere of economic activity according to a different set of procedures and with different outcomes from the agricultural-mercantile mixture that characterized secular activity. It was influenced by different patterns of consumption, supported by different practices of land use, operated according to a different calendar, and depended on a different set of values. Ideologically and conceptually, it was a different beast.

By definition, sanctuaries were classified differently from the surrounding territory since they were *temene* ("cut out"). The land on which the god's altar or temple was built therefore constituted the most basic type of sacred land. The notion of the *temenos* was itself based on two related ideas: first, that the earth itself was divine, and further, that the realms of gods and men were meant to be kept distinct.[4] This distinction still operated even when the sacred land was extended beyond the immediate confines of the sanctuary to include much larger swaths of land, and is reflected in the different terms used to describe the different types of terrain. Terms such as *hiera chora*, *hiera ge*, and *orgas* were used to designate specific pieces of sacred land such as the Krisaian plain below Delphi and the broad pastureland on the border of Athens and Megara.[5] On Delos *ge* referred to the god's land while *kepos* and *chorion* designated plots of land for human use.[6] Sacred land was meant to be kept free of the taint of impurity, so that activities such tanning hides might be banned from the vicinity, although this ban did not include manure, which usually could not be removed from a *temenos*.[7] Rarely did sacred land overlap with land permanently occupied by humans, although this practice is attested during the Roman period in Asia Minor.[8] Keeping people from permanently occupying sacred land was practical as well as pious, since it minimized the opportunity for disputes over ownership. In the Skorta plain, for example, on the border of Attic and Boiotia, the two states managed to avoid disputes over marginal land by agreeing to graze the land in common, an arrangement also tried between Ilion and the Skamandrioi.[9] Flocks do not need enclosed fields and boundaries, and allowing shepherds from both sides of Mount Kithairon to graze here was potentially a smart way of putting the land to use without having to resolve the question of ownership. It is just such shared pasture that Sophocles alludes to when the Corinthian shepherd describes being given the infant Oidipous by Laios' shepherd on Mount Kithairon.[10]

The extension of sacred land beyond the immediate vicinity of the temple reflects an increasing pressure on sanctuaries to service the needs of pilgrims, whose visits to a sanctuary typically involved a sacrifice. The case of the sacred meadow on the western side of Eleusis illustrates how concerns regarding the supply of sacrificial animals could overlap with other worries over the vulnerability of sanctuaries. The territory in question had been a subject of contention between Athens and Megara at the outbreak of the Peloponnesian War, when the Athenians charged that the Megarians were cultivating consecrated ground.[11] As at Delphi, the matter long remained unresolved. Then in 352/1 BC the Athenians put the matter in the hands of Apollo. The inscription that preserves details of this business, *IG* II^2 204, deals primarily with the procedure for determining the boundary in accordance with the will of the gods. Along with various officials "and any other Athenians who wish," the *epistates* was to take two tickets at random from a bronze vessel and place them in either a gold or silver vessel, which was then sealed. The tickets contained two options for the future use of the land. The first read:

> If it is preferable and better for the Athenian people that the *basileus* should rent out the parts of the sacred *orgas* currently being cultivated outside the boundaries, for the building of a colonnade and the equipping of the sanctuary of the two goddesses . . .

The second option was quite different:

> If it is preferable and better for the Athenian people that the parts of the sacred *orgas* currently being cultivated outside the boundaries be left to the two goddesses untilled.[12]

The options under consideration both concern the exploitation of land described as part of the sacred meadow yet outside the boundaries.[13] The first option under consideration in the Eleusinian inscription envisages the leasing of this land, which was already under cultivation, in order to raise revenues for the construction of buildings associated with the Mysteries. The second option involved removing the land from cultivation altogether. Delphi's decision is recorded by Androtion:

> They apportioned the lands in accordance with the determination of the boundary officials (*horistai*) [i.e., the *hierophant* and *diadouchos* from Eleusis]. The farther areas (*eschatiai*) such as were next to the Orgas they sanctified (*kathierosan*), after consulting the oracle and being told by the god that it was better and more profitable for the Athenians not to work the land. And by the decree of Philocrates it was bordered by a circuit of stone stelai. (Androtion, *FGrH* III B 324 F 30)

In other words, despite the fact that the land bordering the *temenos* at Eleusis was already under cultivation and could have been exploited for revenue by leasing, Delphi supported a decision that saw this land taken out of agricultural production. Giovanna Daverio-Rocchi has observed an increasing pressure to open up the border lands of Attica during the fifth and fourth centuries for cultivation, yet here the solution went in the other direction.[14] This may strike us as odd, but it made sense in the context of the competing demands on land in Classical Greece, where political borders, religious requirements, agriculture, and pastoralism were often at odds with each other. At Eleusis, consecrating the land as uncultivated removed it from use by either Megarian or Athenian farmers but did not render it unproductive: it transferred the principal exploitation of the land from private cropgrowers to sanctuary-based husbandry, a more suitable use of the land given the sanctuary's needs. It is important to recognize that a sanctuary might exert pressure for change on prevailing patterns of land use, since recent theories of economic development in the ancient world have tended to focus on the growth of urban centers as the decisive force for change.[15]

Yet the intensification of market activity produced by increasing urbanism is very different from the economic processes occurring at major sanctuaries. Sanctuary populations may swell, but the markets associated with this will fluctuate enormously according to a calendar of oracular consultation, initiation, and sacrifice. Accordingly, there can be no simple correlation between economic activity and sanctuary size, as may be the case with towns. In a settled community, periodic trading based on intermittent supplies and low levels of demand is subject to intensification, giving way to daily markets as supply, demand, coordination, and administration all grow at comparable rates. Sanctuaries, however, remain subject to enormous fluctuations, at one moment servicing a local community while at another time having to deal with a flood of visitors from all over the entire Greek world. Moreover, not only was the economic rhythm of sanctuary activity unlike that of urban communities, but the goods needed were not typical of the commerce associated with urban growth. As sacrificial sites, sanctuaries required a guaranteed supply of meat on a scale that made no sense in any rational model of economic development. The towns of Eleusis and Delphi needed far less meat than the sanctuaries located in their midst. The disproportionate pressure to supply sacrificial animals to sanctuaries also leaves a powerful mark in the material record. At Didyma Klaus Tuchelt notes that of 6,734 pieces of animal bone found in all contexts, 75 percent come from cattle.[16] Whatever the interplay between agrarian practices and city life may have been in Miletos, the sanctuary of Didyma in one sense functioned as the local abattoir and meat market.

Town and sanctuary thus had distinct and contrary needs. The town of Eleusis, for example, continued to require grain but the Athenians took care of this by building three granaries at Eleusis. These were filled with the first fruits of barley and wheat collected from each deme, from all the allies and from any other Greek states who wished to participate. Accordingly, the basic alimentary needs of the sanctuary and its visitors were satisfied by what amounted to a tax in kind on the agricultural produce of all Attica and allied territories. The land around Eleusis was thus freed up to serve as pasture for the considerable numbers of animals required by the rites of initiation. These can be reconstructed in broad outline from a number of sacred laws concerning the Eleusinian Mysteries. Regulations from ca. 435 BC specify sacrifice in three stages. The first was the offering of an ox with gilded horns to Demeter and another to Kore; the second stage was the sacrifice of unblemished animals "to the god, to the goddess and to Euboulos each," amounting therefore to another three animals; the last stage was the offering of another cow with gilded horns to Athena.[17] Estimating the amount of meat at 150 kilograms per cow, the sacrifice of six animals would produce 900 kilograms of meat, enough to supply a magnificent feast celebrating the Mysteries for initiates numbering in the thousands. According to another regulation of ca. 460 BC, initiates were also supposed to make an individual sacrifice at Eleusis inside the sanctuary.[18] Doubtless these were much smaller animals than were used for the state-sponsored cow sacrifices, but the total amount of meat produced as part of the celebration of the Mysteries was staggering.[19] In the law of 435 BC the animals designated for sacrifice are referred to as coming "from the *pelanos*" and "from the *krithos* and the *puros*," all varieties of grain, which suggests that the animals were purchased from funds generated by the sale of grain under the control of Eleusis. This was before the land beyond the *orgas* had been taken out of cultivation. By the middle of the fourth century, however, the pressure produced by the sacrificial calendar resulted in the decision to turn the sacred land over to pasture. Even at Eleusis, where the Greeks celebrated the invention of agriculture, the needs of the sacrificial community exerted an increasing pressure to turn a portion of local land over to stock raising.

Similar pressures were felt at Delphi. Here, after the First Sacred War (ca. 600–590 BC), the fertile alluvial fan of over 13 square kilometers between Delphi and the Gulf of Corinth was dedicated to the god.[20] From both a speech of Demosthenes and a dossier of inscriptions dealing with the administration of the land by Amphiktyonic commissioners we know that the land was supposed to be left uncultivated.[21] Fines were levied against anyone who worked the land, built a mill or kneading trough on it, or removed dung from it. The official explanation for Apollo's acquisition of the land was preserved in an oracle said to have been delivered to the Amphiktyons during the First Sacred War, when Krisa was encroaching upon the god's land:

> You will not take and dismantle this city
> Until blue-eyed Amphitrite breaks in surf
> Through my enclosure to the glittering sea.
> (Pausanias 10.37.4, trans. Levi)[22]

By the time Pausanias was told the story, Solon had been credited with the correct interpretation of the oracle, that all the land from the sanctuary to the shoreline was to be dedicated to Apollo. Earlier versions offer slightly different accounts, but agree in framing the seizure of the land as a defense of Delphi.[23] Strabo's version is that the Krisaians were exacting illegal taxes from the pilgrims coming to Delphi.[24] It is more than likely that all of these accounts are variations on a justification for the annexation of the Krisaian plain originating from Delphi, and it is highly likely that all go back no further than the Third and Fourth Sacred Wars, fought between 356 and 338 BC, when the sanctuary was seized by the Phokians and a segment of the Sacred Plain was cultivated illegally by the people of Amphissa.[25]

Traditions regarding the First Sacred War offer an unreliable explanation for Apollo's acquisition of the Sacred Plain, although they shed a good deal of light on Delphi's anxiety regarding its vulnerability. What is more certain is that the Amphiktyons took great pains to enforce the ban on cultivation. Pausanias found the plain bare and mused that it may have been under a curse, but his suspicion was unwarranted for a variety of reasons. Not only was the plain watered by two rivers, the Pleistos on the eastern side and the Hylaithos on the west, but recent study of the palynological record has revealed that the plain was well wooded in antiquity, with hazelnut, oak, and willow trees.[26] Today it still supports one of the richest olive groves in all of Greece. Clearly, then, this was attractive, productive land. The reason Pausanias found it bare was because the ban on cultivation was still in force. But to take alluvial land out of production in a country where such land is rare and highly prized was economically irrational, unless a more compelling condition was at work. If we dismiss the notion that the land was sanctified to protect Apollo's sanctuary as Delphic propaganda, then a more prosaic explanation presents itself: the growing popularity of Delphi beginning in the eighth century was putting the sanctuary under increasing pressure to guarantee a supply of sacrificial animals. In consecrating the plain of Krisa to Delphi, the Amphiktyons with one stroke seized a swath of land to provide pasture for the sacred herds of Apollo. The inscriptions from Delphi often contain honorific phrases praising the officials who served as ambassadors between the sanctuaries and states of Greece, but it is worth remembering the logistics that lie behind these honors, such as the inscription honouring Phormio and Herakleidas for the "one hundred–animal sacrifice, led by a cow" they conducted in honor of Apollo. Unless they had shipped the animals from the Pontic Chersonesos, they will have purchased the animals at Delphi from the flocks and herds that grazed below the sanctuary.[27] Dedications and treasuries were only one measure of Delphi's growing wealth. The other was the fantastic quantity of meat sacrificed on the gods' altars.

The presence of a hippodrome near Delphi also meant that land was needed for grazing horses, and a decree of the *hieromnemones* in 178 BC refers to a part of the *hiera chora* that had been set aside for both sacred cattle and horses. It is possible that Apollo bred not only cattle for pilgrims to sacrifice in his honor but horses for the elite to race in his honor as well. Horse breeding was also carried on in the sanctuary of Artemis at Ephesus, and at Delos the hippodrome was located on a plot of land purchased by Nicias for 10,000 drachmas and then given to Apollo in return for perennial sacrifices to be conducted in the donor's honor.[28] J. H. Kent reasonably suggested that in the years when the race course was not in use the land could have been exploited as pasture.[29]

Unfortunately the creation of a Sacred Plain did not put an end to disputes over Apollo's land. It was attractive enough to tempt the men of Amphissa to make repeated attempts to cultivate portions of the plain, precipitating the Fourth Sacred War in 338 BC.[30] Even when the ban on cultivation was enforced scupulously through periodic inspections by the *pylagorai* representing the Amphiktyony, private flocks and herds continued to encroach on the Sacred Plain. This led to a great deal of confusion over which animals were privately owned and which were Apollo's. The Amphiktyony addressed the matter repeatedly in its decrees, ultimately deciding to reserve a portion within the Sacred Plain for Apollo's cattle and horses, and adding the stipulation that if anyone did graze his animals within this area the animals were subject to seizure.[31] Animals grazing on land set aside for Apollo's herds became sacred property, a forfeiture that was also spelled out explicitly in regulations at the sanctuary of Zeus Temenites at Arkesine on Amorgos.[32] Here the regulations stipulated that sheep were not allowed to be brought into the *temenos* to graze, although a sacred herd must have been present, since the leasing arrangements of the sacred land required the lessee to provide 150 measured loads of manure when his lease was up. There was a close connection, then, between sacred land and sacred herds. This is what Plutarch has in mind when he relates a story of the first Pythia, who claimed that after her death from her body would spring various plants to nourish the sacred animals, so that when the victims were sacrificed men would be able to read portents of the future in the animals' entrails.[33] For Plutarch, a priest at Delphi toward the end of his career, animals were sacred not merely because they had been so designated, but because they were a link in the chain joining the human, the natural, and the divine.

Supplying the Sanctuary

It is not surprising that stipulations against cultivating sacred land are complemented by regulations designed to guarantee a supply of sacrificial animals, whether from the god's land or from other sources. This is particularly understandable because sanctuaries not only required supplies for their own use, but also because many sanctuaries were the focal point for state-sponsored sacrifices. We have seen that cattle were grazed on the Sacred Plain below Delphi, and many of them will have been kept as breeding stock. Other animals were purchased and donated, such as the fifty cattle from Oianthaia that the Lakedaimonian village of Tyritai gave to Apollo.[34] Such donations might be voluntary benefactions, as when a certain Diomedon bequeathed funds in his will for the purchase of sacrificial animals, or they might be public demonstrations of power relations, such as the Athenian requirement that their allies and colonies each send a cow and panoply to the Panathenaia.[35] Such livestock processions were spectacular and were widely copied. In the second century AD, C. Julius Demosthenes of Oinoanda established a festival in honor of Apollo Patroos. He used the opportunity to coordinate the sacrificial calendars of the surrounding villages and create a single procession in honor of the god: referring to the twenty-seven sacrificial animals sent from outlying *komai* (hamlets), his decree states, "when they are sent [the sacrificial animals] are also to participate in the *pompe* through the theatre and be announced. . . ."[36] These animals are designated *thusiai* (sacrificial animals) and will have been sacrificed and consumed during the festival, but in many cases animals were donated but then used as breeding stock. Two Hellenistic inscriptions from Morrylos are good examples. The first is a decree honoring a certain Paramonos for giving to the city, and specifically to the cult of Asklepios, a cow for the breeding of a sacred herd.[37] The inscription states that he "gave to the city and Asklepios a *bous agelaia*," a cow that produced many offspring over the next fifteen years. The editors translate *bous agelaia* as "une vache pâturante" to distinguish it from a cow designated for sacrifice, but the adjective can also refer to breeding stock, and that is its meaning here, as the rest of the sentence makes clear.[38] The second decree honors a certain Alketas for a similar act of generosity: endowing the citizens and the god with an unspecified number of cows.

As with other features of sanctuary organization, there was no single strategy for keeping sanctuaries stocked; different practices were employed in different places. What we find are variations on a theme: sacrifice was thoroughly ubiquitous in Greek life, so that the need to supply sacrificial animals was a constant, whether by breeding, donation, exaction, or negotiation. A law, for example, establishing the cult of Artemis Hyakinthotropos ca. 200 BC invited the people of Knidos to participate in the cult on Kos, provided they brought their sacrificial cattle with them.[39] Regulations governing the sanctuary of Alea also show that different arrangements operated at different times of year. The priest of the cult is permitted to graze twenty-five sheep, a yoke of oxen, and a goat, and is subject to *inphorbismos* (either seizure of the animals or a tax) if he exceeds that number. The priest in charge of sacrifice (*hieothytes*) is then instructed to graze only animals that are unblemished (*askethes*), and punishments are prescribed for breaching this ordinance as well. Both regulations pertain to the customary practices of the sanctuary, but the decree also stipulates that on the last days of the *tripanegyris* anyone can graze any animals they like, provided they do not enter the *perichoros*, the inner portion of the sanctuary. Similarly, no citizen or foreigner is permitted to graze animals in Alea unless coming to the sacrificial feast (*thoina*). The Alea regulations distinguish between the customary use of the land, in which even the presence of a sacred herd is carefully monitored, and the temporary exploitation of the land at festival time, when greater numbers of visitors were coming both to sacrifice and to feast.[40]

One of the most common arrangements for supplying the needs of the sanctuary was through the leasing of sacred land, but here too no single pattern can be discerned, although the aim remains the same: to guarantee revenue for the upkeep of the sanctuary and a supply of animals suitable for sacrifice.[41] The most simple arrangement involved leasing land that had been donated to the sanctuary and applying the income directly to pay for a sacrifice. This is illustrated by a third-century inscription from Kalauria that records the donation of land and money by Agasikles and Nikagora to support the cult of Poseidon. The citizens agreed to the following terms of the bequest:

> . . . Taking the interest from the money and the lease payments (*dotine*) from the land, let the *epimeletai* sacrifice a suitable victim to Poseidon and a suitable victim to Zeus Soter. (*IG* IV 841)

In another case, at Olymos in Karia, the community decided to purchase land on behalf of the state's tutelary gods, Apollo and Artemis, using sacred money. Those chosen by the demos to purchase the land were then entrusted with the task of leasing it on terms favorable to the gods.[42]

In the case of large sanctuaries with multiple holdings, this required close monitoring of sacred lands by sanctuary officials. The best-known example is that of the sacred lands belonging to the sanctuary of Apollo on Delos and published in a superb study in 1948.[43] In his analysis of the documents relating to the estates, J. H. Kent was able to clarify many issues concerning the location of the various properties and the conditions according to which they were leased. Kent detected a period of agitation in the rental market in the early third century and showed that the sanctuary officials devised a *hiera syngraphe* ("sacred contract") in order to standardize the conditions under which leases were given out.[44] Contracts were offered at auction, defaults were rare, guarantors were required, and heirs inherited obligations. Rentals rose by approximately 10 percent with each new contract period (every ten years). From these observations Kent concluded that the land was regarded as valuable; the sanctuary was successful in making the leasing of sacred land an attractive proposition.

It is interesting, then, to find that the *hiera syngraphe* assumes that much of the sacred land is being leased as grazing land. A number of clauses support this view and shed light on the sanctuary's role in modulating the cattle economy. The first clause is at lines 19–21:

> If lessees keep livestock, they are to pay in the month of Artemision an amount for each animal [which is to be counted as part] of their rental, for all the animals they keep. (*IDelos* 503.19–21)

Kent interpreted this as an attempt to forestall lessees squandering money raised from the sale of wool and spring lambs and calves, leaving them unable to pay when rent was due. Signe Isager also connects the payment schedule to livestock, noting that the rent payments came due at different times for breeders and farmers (lines 27–30). The latter paid rent in Metageitnion (August/September), after the harvest. Breeders paid twice, in Artemision (April/May) and Lenaion (January/February). The first of these dates corresponds to wool-shearing season and sale of lambs. By Lenaion, when the second payment was due, the previous year's lambs and goats were fully grown. This was a good time to reduce numbers, especially of males.[45] Thus the arrangements devised at Delos meshed with sensible flock management, setting the due date for rental payments at precisely the times when the lessees would be best able to pay.

Other clauses also suggest that there was more husbandry taking place on the sacred lands than on regular farms:

> The *hieropoioi* in the month of Galaxion are to take a census of the cattle according to custom . . . (*IDelos* 503.19–22)

The following lines reveal that there were two types of cattle grazing on these lands: (1) animals belonging to the lessees, and (2) branded cattle (*enkekaumena boskemata*), the sacred animals subject to the *hieropoioi*'s inspection. The cattle grazed side by side and clauses in the contract refer to lessees raising sacred animals and not being assessed for this. A major incentive for taking care of the god's cattle was that lessees enjoyed the usufruct of the sacred animals: wool, in the case of sheep, and milk and perhaps even their use as draft animals in the case of cows.

The distinction between private and sacred herds was further blurred by the fact that lessees were permitted to sell those branded cattle for whose replacement they were responsible, provided they supplied a guarantor. At first glance this is a puzzling clause. Why should lessees be permitted to sell the god's cattle? Kent argued that it could not be simply a matter of the lessee turning over the proceeds of the sale to the sanctuary at some future date, since an unscrupulous lessee could sell the animals for much less than its real worth, in order to raise a quick loan, and the sanctuary would be left to carry the loss. On the other hand, as Kent also observed, if it was simply a matter of replacing one animal with another, "there was nothing to prevent the lessee from selling a good animal and replacing it later with a poor one."[46] It may be, however, that the clause makes better sense if we consider the unique way that a sanctuary's needs and status affected its operation as a cattle market. Delos raised cattle for sacrifice on the altars of Apollo and the other gods of the sanctuary, but the management of the herds was left to the men who leased the sanctuary's land. The *hieropoioi* conducted periodic inspections, but did not take an active hand in breeding or raising Apollo's cattle. Furthermore, they did not sell cattle, since no revenue from the sale of cattle is listed in their very detailed accounts.[47] In leaving the actual husbandry to the lessees they had to offer incentives to offset the burden of paying both lease payments and tending the god's cattle alongside their own. By permitting the lessees to handle the sale of sacred cattle, the *hieropoioi* were essentially handing over the entire cattle trade from production to market to the lessees, subject to the supervision of the sanctuary. The lessee was required to replenish the sacred herd from his own stock whenever he sold a beast. This was mutually agreeable. The sanctuary could rely on a stable stock, since the sacred herd grew by natural reproduction and was supplemented by addition from the lessees' herds whenever animals were culled for sacrifice. The lessee profited by having a permanent and reliable market. Demand for cattle was guaranteed and could only be supplied by sacred cattle, since it is unimaginable that the *hieropoioi* would have permitted the lessees to sell both sacred and profane cattle for sacrifice. The system at Delos represents a balance between economic sense and religious scruples. Lessees raised herds which, to be suitable for sacrifice had to be branded as the god's cattle. Yet they were still a commercial commodity sold at market to those making sacrifice, thereby repaying the lessees for their investment of time and money. As religious items the cattle passed into the possession of the god once they were branded; as a source of revenue, however, they directly profited the lessees, not the sanctuary.

This may seem an odd arrangement, as if the sanctuary were forfeiting the opportunity to make as much money as possible from its potential monopoly of the cattle market, but in fact the *hiera syngraphe* reflects a different mindset altogether: the sanctuary's officials were less motivated by greed and more by concern for the prestige of their sanctuary, which depended on the sanctuary's ability to guarantee a constant supply of sacrificial animals.[48] This is confirmed by the sacred law of Ilion, from 77 BC, where the Ilian Confederation also leased sacred land attached to the sanctuary of Athena in order to guarantee the availability of sacrifical animals. After mandating a suspension of interest payments to the goddess from the member cities of the Confederation, whose ability to support the sanctuary had perhaps been upset by the turmoil of the Mithridatic War, the stipulations lay out what is to happen when payments resume:

> From the income from all the fields that are leased by the goddess let cattle be bought. Let them be sacrificed to the goddess according to the traditional law and let one other cow be sacrificed to Zeus Polias and all the gods equally, as has been decreed. Let the cost be paid by the current *agonothetai* and the representatives in charge of the future revenues. (*LSAM* 10 B)

As at Delos, the land of the goddess was leased, but the animals grazing on it had to be purchased before they could be sacrificed. Here too, then, the sanctuary preferred the reliability of lease payments despite the fact that it resulted in the apparently illogical process of the sanctuary having to purchase the animals that grazed on its own land.

If we recognize that the administration of sacred lands was not designed to maximize income, and that sanctuaries preferred regular lease payments to the uncertainties of a bull market, so to speak, we can better understand why Delos required a guarantor for the sale price when lessees sold sacred cattle. The guarantor was not to ensure that the lessee did not ask too low a price, depriving the sanctuary of revenue as he exploited sacred property to raise short-term loans, as Kent argued. Quite the opposite. The guarantor was to ensure that the lessee did not ask too high a price for the sacrificial animal being sold to a visitor. The sanctuary created the demand for sacrificial animals, but also exerted control over the lessees, through the guarantors, to prevent the lessees from exploiting what was, in effect, a monopoly given to them by the sanctuary. The sanctuary was effectively regulating the market.

The special place of cattle-raising on the estates of Delos and Rheneia does not mean that other agricultural activities were not also pursued. There are signs of terracing on the southern end of the island more consistent with agriculture than pasture, and the accounts mention large numbers of grapevines and barley. Yet Robin Osborne is surely right to suppose that the sacred land of Delos was devoted to raising cattle on a scale unparalleled on private land.[49] Not only does the *hiera syngraphe* suggest this, with its detailed regulations, but it is also consistent with the references to byres and stalls in the leases. At least fourteen estates administered by the *hieropoioi* were leased with cattle shelters, and at least eleven had sheep sheds. Widespread husbandry is also assumed by the terms of this penalty clause:

> If any of the rental is left unpaid after the (lessee's) crops have been sold, (the *hieropoioi*) are to sell for the unpaid amount his cattle and his sheep and his slaves. (*IDelos* 503.33–38)

Delos is unique in offering us a detailed account of how a Panhellenic sanctuary was organized economically, although even here, as we have seen, to get a complete picture of the cattle system centered on the sanctuary it is necessary to infer arrangements that are not spelled out explicitly. At other sanctuaries the evidence often hides as much as it reveals. At Cyrene, for example, the detailed accounts of the *demiourgoi* record the agricultural products of the sacred lands of Apollo, Artemis, and Athena, including barley, wheat, spelt, grapes, and cumin. Lessees sold these crops and paid a proportion of the proceeds to the sanctuaries. There are no provisions explicitly dealing with raising cattle, yet the maintenance (*tropha*) specified for the various officials staffing the cults, including priests and priestesses, heralds, choruses, flautists, treasurers, cooks, and guards, is measured in terms of the number of sacrificial cattle they are to share.[50] Whether these were animals raised on sacred land or simply animals brought by visitors to the sanctuary we cannot say, but the fact that sacred regulations such as these treat agricultural produce as a market commodity while simply assuming that cattle will be available for sacrifice and distribution only underscores how different cattle were from agricultural products. As long as Cyrene had citizens like Hermesandros, son of Philon, the sanctuaries could rely on piety to guarantee a rich supply of cattle. Sometime in the fourth or third century he erected a monument commemorating the fact that he had brought one hundred cattle to the sanctuary of Artemis and sacrificed them in her honor.[51]

Sacred land was also leased by the sanctuary of Artemis Elaphebolos in eastern Phokis near the borders with Lokris and Boiotia. The evidence is once again epigraphic, consisting of a single Hellenistic inscription. Despite the unpromising condition of the stone, it sheds a great deal of light on how the sacred land was exploited. After an introduction that appears to refer to the transfer of two plots and the recording of this in the shrine, the inscription records in precise detail the names of the landholders, the size of the plots, and their location arranged in sequence:

> . . . the holy and public land . . . -on son of Mynnion, Timagora the wife of Ischylos, Nik . . . above the rest, as much as for the (plot) that has been sold . . . six, Thrason son of Aristokrates being neighbor to the east . . . to the plots, one by the spring . . . Thrason son of Aristokrates neighbor to the east, to the west Thrason, son of Aristokrates. Towards the Anakieon, Amyntas, son of . . . Philon son of Kleippos neighbor to the east, to the west . . . 13 plethra, neighbor to the east . . . to the west Onymokles, son of Pasion . . . 40 plethra, neighbor to the east . . . to the west Andron son of Diomedon . . . Philomelos son Leothetes . . . 13 plethra, neighbor to the east Onymokles son of Pasion, to the west . . . son of Agrios. Another plot, 30 plethra, neighbor to the east . . . son of Agrios, to the west Polyxenos son of . . . the Hylleic estate, its neighbor to the east being the plot . . . Another plot, two plethra, another one three plethra, neighbor on the east side . . . on the west side Timandros son of Mynnion. Another plot . . . plethra . . . on the east side the neighbour being Philon son of Kleippos, to the west . . . Another, 96 plethra with . . . as neighbor on the east, on the west . . ., son of . . . plethra . . . with as neighbor on the east Thainetos son of Kallikrates, and to the west Theainetos son of Kallikrates. A plethron of wild pear,[52] with . . . -ostratos, son of Lys- . . . neighbor to the east, to the west Theopropos, son of Simmichos, the plots by those of Polyphoitas . . . 90 plethra . . . Aristos Theartios, with . . . -das son of Archedamos neighbour to the east, and to the west Aristokrates . . . 300 plethra, with as neighbor to the east Phokion son of Theodoros, and to the west along the road Philon, son of Philon. Kleon At- . . . -mon 6 and the sown(?) 110 plethra, with as neighbor to the east . . . -ikles, son of Amphidamos, to the west Kallikrates, son of Theainetos. Towards . . . plethora, with other holy(?) lands neighboring to the east, to the west Pythodoros son of Agidippos. Towards Pialeia the plot by the Hekatombion, 12 plethra, with as neighbor to the east the Hekatombion, to the west Xenokrates son of Kallimachos. (*IG* IX 1.87.20–75)

The sacred land was leased, producing revenues (*prosodous*) for Artemis and Apollo (lines 11–12). Before the leases are listed, however, the inscription begins with a reference to the transfer of two plots of land (line 5) to Artemis and Apollo, after which the plots are referred to as "the two sacred plots" (*toin hieroin chorioin*, line 13). The arrangement at Hyampolis reflects Aristotle's categories, in which sacred and public land were regarded as distinct from private holdings. The rest of the inscription gives the sizes of various holdings within the sacred estate, naming the lessee and usually naming the neighboring landholders to the west and to the east. There is no mention of either northern or southern boundaries. The description of the plot held by Philomelos, son of Leothetes, is typical. It is described as consisting of "13 plethra, with Onymokles son of Pasion as his neighbor on the east side and [name missing], son of Agrios, to the west" (lines 32–34). The absence of any northern or southern boundaries in a list of consecutive plots must surely mean that there were two natural boundaries so prominent in the landscape that there was no need to cite them. It is possible to identify this area with some confidence, since the location of both the sanctuary of Artemis Elaphebolos and the town of Hymapolis are well known. Excavations of the former show that its ancient location, Kleonai, corresponds to modern Kalapodi, while the site of Hyampolis was correctly identified by V. W. Yorke over one hundred years ago.[53] In this area, halfway between Chaironeia and Elateia, two outliers from Mount Kallidromon and Mount Akontion rise sharply from the Kephisos Valley, creating a fertile plain naturally bounded by rocky slopes on the north and south sides. Today the land is neatly divided into distinct lots and the cadastration still runs predominantly east–west. The plain extends over an area of approximately 14 square kilometers.[54] It is most probable that the land described in the Hyampolis inscription was located in this fertile plain and reached the very edge of the sanctuary. This can be inferred from the final entry, for a plot described as having "the Hekatombion neighboring on the eastern side" (lines 77–78). The Hekatombion is the place where the Phokians held their great national sacrifice, celebrating the defeat of the Thessalians, and will have been close by the sanctuary of Artemis Elaphebolos.[55]

Three notable features stand out in the arrangements at Hyampolis. The first is the dramatic range in the measurements of the plots listed, which vary from lows of 1, 2, and 3 plethra to highs of 96, 110, and 300. Such variation points to the fact that these lots did not arise from a systematic surveying of the land, in which case we would expect standardized lots such as are found in the cadastration of the Greek colonies in Magna Graecia. The Hyampolis plots were irregularly sized, and were gradual additions to the sacred land, acquired by purchase or donation, like the two plots mentioned at the beginning of the inscription. It is important to recognize that the measurements recorded in the inscription, totaling 710 plethra, do not relate to linear but to square measurements.[56] Were they linear, the total of 710 plethra would mean that the sacred land of Artemis and Apollo extended for more than 21 kilometers east–west, an impossibly high distance that would translate into an estate that engulfed all the land from Elateia to Opous. Taken as square measurements, the 710 plethra correspond approximately to 64 hectares or 158 acres. The actual area will have been bigger, since some figures are missing, but even with the addition of another 300-plethra plot the Hyampolis valley could easily accommodate the sacred land.

The second notable feature is that the leasing arrangements give no hint of duration of the lease. It is possible that a standard fixed period of perhaps ten years is intended, although if so we might expect other documents from other leasing periods to have survived, as at Delos, where one can follow the changes in leases both before the *hiera syngraphe* was drafted ca. 300 BC and over the course of the following decades. In the absence of such evidence, however, it is more likely that the arrangements spelled out in the Hyampolis inscription represent the allocation of sacred lands to lessees who expected to hold the plots in perpetuity. Such arrangements are attested elsewhere. A recently discovered inscription from Olymos in Karia reports the sale of property to Apollo and Artemis along with all vines and trees on it in return for permanent leasehold.[57] The left-hand column preserves details of the sale of the land, formerly the property of a certain Polyneikes for a total of 6,500 drachmas, paid over approximately thirty years. The right-hand column then lays out the terms for the leasing of the land. Similarly, the sacred land of Dionysos at Herakleia was also leased in perpetuity, as was the shrine of the god Hypodektes in Athens.[58] A third-century contract from Morgantina may reflect a similar arrangement. The inscription reads as follows:

> In the year when Orthon was chief priest (*hiarapolos*), in the month of Theudasios, Lyson, son of Hippias, purchases the vines and all the appurtenances that go with them, as well as two shares in the wine vat, half of the cuttings, the entire shed at the entrance, for a price of 21 (Sicilian) talents, 115 drachmai, from Theston, son of Damarchos and Satyros, son of Damon, who are guardians of orphans. Sacred properties are not included. Guarantors: Nikias, son of Kratias (followed by nine more names.)[59]

The contract exempts sacred holdings from the sale of property managed by two guardians of orphans, and the need to stipulate that the sacred land was not part of the sale may have arisen from the family's having held a long-term lease that was inalienable as long as the family honored the terms of the lease. Presumably the sacred land reverted to its divine owner upon the sale of the family estate, whereupon the sacred land was put up for lease anew. Since the Hyampolis inscription appears to record one such purchase but also to list a great many other leases as well, it may be that the acquisition of two plots by the sanctuary afforded an opportunity to regularize the claims of all the leaseholders and to produce a single, coherent, lapidary record recognizing exactly who had the rights to which plots.

Since leasing put sacred land back into the hands of individual lessees the category of sacred land can seem artificial. Moses Finley certainly asserted that it was nothing other than deme land under another name, but a good deal of sacred land both in and out of Attica was leased without the involvement of demes. In fact, as both Signe Isager and Marietta Horster have recently argued, the distinction between sacred land and other categories of both private and public land was real and important.[60] Sacred land belonged to a god. Yet there is a tension inherent in any system that assigns land to a deity who does not actively work the land and does not need most of the produce of the land. Leases are one way of resolving this dilemma, by allowing the sanctuary to reconcile the often contradictory ways in which land might be used—viticulture, oleoculture, agriculture, flock and herd management—and stipulating these in the terms of the lease. Leases in perpetuity surely arose because such arrangements, where they persisted for a long time and proved to be profitable to both parties (god and lessee), became accepted as standard and natural. The transfer of land title to a god in return for a permanent lease also looks like a shrewd strategy to forestall the seizure of private land, which, in the context of the persistent and often revolutionary calls for the redistribution of land in Hellenistic Greece, became an increasingly urgent matter.[61] It is probably not a coincidence, then, that the list of twenty-two landholders at Hyampolis reveals a cluster of names, pointing no doubt to the men of the local elite. Thrason, son of Aristokrates, possesses three separate holdings and his father one. Philon, son of Kleippos, is listed twice, and his son once. The father and son of Theainetos and Kallikrates both possess plots, as do two sons of Mynnion. Even allowing for false identifications, the clustering of names suggests that land close to the sanctuary was considered desirable and that key families controlled access to the land.

A third distinctive feature of the Hyampolis inscription is that no payment is stipulated. If land was being sold to the gods and then leased back in any regular way we would expect rental figures and payment schedules. Why do we have neither? One possibility is that terms of payment were inscribed on a completely different stele, now lost, and that the extant stone only records the dimensions of the sacred plots. Another possibility, however, is that the leases were paid in kind by a tithe fixed by tradition, so that that it did not need to be spelled out. Herakles demanded a tithe of the cattle of Augeas as the cost of cleaning his byres—a mythological reference, to be sure, but one that shows that the practice was known to the Greeks. Pausanias records that the famous bronze bull dedicated by the people of Corcyra at Delphi was a tithe from the extraordinary catch of tuna that a bull had revealed to the locals, while fourth-century accounts of sacred revenues from Cyrene refer to the "sacred tenth."[62] One-tenth was also a standard proportion of sacred revenues remitted to the royal fisc in Hellenistic Asia Minor, suggesting the practice was well established and widely known.[63] There is evidence that the land around Hyampolis was similarly tithed: Pausanias observes that the people of

Hyampolis "say that whatever beasts in the herds they name as sacred to Artemis will grow up free from disease and fatter than the others."[64] In other words, some of the cattle grazing on their land were designated as Artemis' sacred herd, although as at Delos, sacred and profane herds grazed side by side. Pausanias might have added that Artemis needed a good supply of healthy cattle since her festival, the Elaphebolia, was the greatest festival in the Phokian religious calendar. The presence of a Hekatombion by the sanctuary affirms that the celebration of Artemis involved the sacrifice of one hundred cows, payment in kind for the use of the goddess' land. If the arrangements at Hyampolis appear crude by comparison with those on Delos, the reason for this surely lies in the different scale of economic activities necessary for the running of a regional sanctuary and an international sanctuary. Yet at both we see the same general pattern. The primary need of a sanctuary is a guaranteed supply of sacrificial animals. Whether located in rich farmland or in marginal areas, sanctuaries generated a demand for what was otherwise a prestige item: beef cattle. As a result, land close to sanctuaries tended to be converted to pasture on a scale that would have been anomalous in the usual regime of mixed farming that dominated the Greek countryside. The gods required meat, and their land supplied it.

We do not possess detailed evidence regarding the exploitation of the Sacred Plain at Delphi, but there are indications that one of the clans at Delphi was also involved in the regulation of the cattle market. The evidence derives from an inscription on a pillar discovered in 1895 at Delphi, the so-called *cippus* of the Labyadai. Inscribed on all four sides, the pillar preserves regulations of a Delphic phratry that may have taken its name from the legendary founder of Apollo's temple at Delphi, the eunuch Labys. Side C contains sumptuary regulations relating to mourning and burial practices, and it is these rules that have attracted most attention.[65] Side D, however, preserves a calendar of feasts observed by the Labyadai and a curious clause that runs as follows:

> The following has been written at Phanateus on the rock inside: [Ph]a[n]otos has given the following to his daughter Boupyga: the yearling and the goat from the twelve-part sacrifice, the skins in the Pronoia sanctuary, the hides to Lykeios, and the leading calf; let anyone, whether acting privately or on behalf of a group, who enjoys the right of first sacrifice and first consultation supply what has been stipulated to the Labyadai. These are the sacrifices of the Labyadai: during the month of Apellaios, to Dionysos; at the Boukatia the first fruits to Zeus Patroos and to Apollo, and let him drink with us, the Labyadai. And let the other sacrifices be conducted in the appropriate seasons.
> (*CID* 1 1.9.31–51)[66]

Phanateus, or as Strabo knew it, Phanoteus, was a Phokian town also known as Panopeus, and today called Ayios Vlasios. It lies some thirty kilometers east of Delphi on the Sacred Way. Its presence in the inscription makes it clear that Panopeus enjoyed a special relationship with the Labyadai. In the first edition of the inscription Théophile Homolle suggested that Panopeus was the ancestral home of the Delphic phratry, but this is unlikely since local traditions at Panopeus invariably emphasized hostility to Delphi.[67] The inscription refers to the offerings from sacrificial animals given by Phanotos to Boupyga in terms that recall the perquisites offered to a priest at sacrifice: a portion of the overall meat and the hides of sacrificial animals. The myth analogizes the relationship of Phanotos (standing for Panopeus) and his daughter Boupyga to that of a person making a sacrifice and the priest who conducts it. Although we do not possess the full story, it is not unreasonable to see Boupyga as representing the Labyadai, since they too conducted sacrifices and received perquisites. In fact, the story of Phanotos and Boupyga may have been told to visitors to the oracle to explain how they were supposed to proceed, since the regulations move straight from the myth to the stipulation that the visitor was to supply the Labyadai. That is to say, the Labyadai are to receive from those enjoying the rights of *prothusia* and *promanteia* the same perquisites as Boupyga had once received from Phanotos.

Were this all the information available to us we might conclude that the regulations involved only two parties: those consulting the oracle and the Labyadai, who supervised the consultation in return for a fee paid in meat and skins. However, the discovery of a copy of this inscription at Panopeus shows that the mention of Phanoteus was more than a mythical allusion; the Phokian town also played a role in the procedure. Panopeus is too far from Delphi to have played any role other than supplying sacrificial animals, which could be driven by those proceeding to Delphi along the Sacred Way. It seems, then, that the Labyadai had reached an agreement with the people of Panopeus. Perhaps in return for the Labyadai's approval to sell animals to pilgrims, Panopeus continued to give the Labyadai the same cut as Phanoteus had once given Boupyga. Not all of the animals destined for Apollo's altar will have come along the Sacred Way, but with the Sacred Plain below Delphi providing pasture for sacrificial animals an arrangement like this with Panopeus would have resulted, as at Delos, in a constant replenishment of the sacred herds.

Exploiting Sacred Land

Generalizing about the role of sanctuaries and sacred land is a hazardous business since no two sanctuaries functioned identically and because evidence concerning them is of uneven quality. In the case of one sacred estate, around Itonion in central Greece, the presence of the sacred land is not even attested in literary or epigraphic sources and has been inferred entirely from the results of a survey in which no evidence for any other use could be found: no human habitation, no settlements in a region where they would be expected, and no signs of farming.[68] Here, for once, the absence of evidence is not evidence of absence, but of a presence harder to identify in the archaeological record: sacred land given over to stock raising. Animals can also be an opaque presence in what otherwise seem to be detailed records, as is the case of land leases from Rhamnous and Amorgos.[69] Both these leases offer very specific details of how the lessee is to plow and sow the *temenos* of the goddess (at Rhamnous) and god (on Amorgos), and how he is to manage the vines and figs. The lessee is also required to leave the estate in the same general condition in which he found it, and is not permitted to remove any dung from it. No animals are mentioned as part of the lease; in fact, the Amorgos lease specifically prohibits anyone from grazing flocks on the sacred land (line 36). *Kopros*, however, is the subject of considerable regulation. Does this refer to human excrement, to be used as fertilizer, or does it refer to animals grazing on the land, owned entirely by the lessee, and hence not covered by the terms of the lease? The former possibility is unlikely. In the case of the lease from Arkesine, the lessee was required to produce 150 measures of manure per year, at a rate of one *medimnos* and four *hemihekta* per measure (lines 20–22). This is the equivalent of 32.5 kilograms of *kopros* per measure, and approximately 4,875 kilograms per year, considerably more than the lessee and his family might be reasonably expected to produce.[70] Presumably, then, the requirement in both leases that the lessee leave *kopros* behind (150 measures at Arkesine; amount not preserved at Rhamnous) reflects the expectation that the lessee would graze his own animals on the sacred land, despite the fact that grazing in general is expressly prohibited.[71] A modern dairy cow weighing approximately 635 kilograms will produce more than 50 kilograms of manure per day.[72] Even allowing for the smaller size of ancient cattle, it would not have been difficult for the lessee who collected the manure of a single cow to have fulfilled his daily quota of approximately 13.3 kilograms. If the lessee owned a yoke of oxen, so much the better. The silence of the leases on the question of animals is instructive, and suggests that the demesmen of Rhamnous and the administrators on Amorgos were primarily concerned with the preservation of real estate, such as vines, trees, and fields.

In other cases inferring the presence of animals may help to explain otherwise inexplicable oddities in sacred regulations. In a set of late-fourth-century decisions, for example, taken by a gentilicial group on Chios called the Klytidai, a stipulation bans the use of a sacred building within the temenos and the land attached to it by any individual or by any other phratry. The Klytidai regulations consistently favor common use. After a favorable sacrifice, the sacred objects of the Klytidai are removed from private dwellings and deposited in a holy building, which the Klytidai then decide should be used in common. The purpose of this is reasonably clear: the building associated with the group's cultic activity must be left available to all members of the group. A similar decision, however, applies to the land attached to the sacred building: it too is to be used by the Klytidai in common. This would seem to preclude leasing, since the regulations specifically forbid any of the Klytidai from letting anyone else use the land privately. There are no clauses dealing with the upkeep of the land, whether as a vineyard or orchard, and no provisions for plowing the land on behalf of the Klytidai. Either it was functionally abandoned, or put to a use that was not spelled out. It is difficult to imagine what such use might be unless animals destined for sacrifice were grazed on the land. Such a practice would certainly have fulfilled the requirement of common use.[73]

The scale and type of economic activity focused on sanctuaries varied greatly, not only between Panhellenic centers and local shrines, but even between estates adjacent to each other, where one might expect a degree of similarity. At Herakleia in southern Italy, for example, the sacred lands of Dionysos and Athena covered approximately 360 hectares and 295 hectares respectively, yet despite their comparable sizes the lands were exploited very differently.[74] The estate of Dionysos, which included woodlands, vines, orchards, and land under cultivation, was divided into four large plots of approximately 90 hectares each and was leased in perpetuity, yielding revenues of 410 *medimnoi* of barley per annum. Athena's sacred land, consisting of olive trees, orchards and vineyards, was subdivided into much smaller plots of 6–8 hectares, yet produced 3,500 *medimnoi* of barley, or its equivalent in other produce, which was then sold to purchase grain. By requiring payment in *medimnoi* of barley the system at Herakleia demonstrates that it was designed to maximize the supply of grain for the population. Moreover, whatever the possible uses of the Dionysos land, rents were assessed only on its productive value as farm land. Yet the lands, despite their comparable areas, were clearly not suited to the same type of use. Athena's land was intensively cultivated, and was designed to be leased to as many as forty farmers. Even if the leasing of multiple lots by one family was permitted, the division into small plots of 6–8 hectares argues for the intention to share the land amongst a number of families. The plots are at the upper end of the range considered suitable for supporting a family.[75] The 90-hectare plots of the Dionysos estate, consisting in large part of woodland, were much less desirable as agricultural land and were also far too large to have been productively worked by a single family. The proportion of worked land to unworked on the four estates of Dionysos is as follows: Bormion's lease: 201 *skoinoi* to 646.5; Damarchos' lease: 273 to 500; Peisias' lease: 312.5 to 537.5; Philippos' lease: 308.5 to 541.5. These were mixed lots and would have been suitable for stock raising.[76] The cost of investing in herd animals made such land attractive only to the wealthy, for whom the low rental costs were an incentive to leasing land that was not good for cultivation.

A confirmation of sorts that this land was used for animals from a sacred herd comes from two apparently contradictory clauses in the leases. One clause stipulates that if anyone comes onto the land of Dionysos or grazes his animals on it the lessee is condemned to pay a penalty (line 128). This would seem to rule out stock raising. Yet the contracts also stipulate that the lessee is to build and keep in good repair a cattle byre (*boon*) measuring 22 x 18 feet (line 138). Clearly the land was intended to have cattle grazing on it, and since no one else was permitted to bring their own animals on to graze, it follows that the cattle here were those of Dionysos. In short, the division of the sacred lands at Herakleia according to such different regimes of land use—small, intensively cultivated plots and larger, woodland ranges—illustrates both an awareness of the competing alimentary demands of communities, who required both their daily bread and their feast-day meat, and the special role to be played by sanctuaries in making a place for animal husbandry in a rural economy that was largely agricultural.

Sanctuaries were by definition sacred places in the landscape, but the sanctity of sacred land also served an important economic function, marking those places where the increasing pressure to turn land over to cultivation was kept in check, thereby preserving enough land for pasture. This helps to explain why sacred land is so often closely associated with public land. At Argos, for example, the *demos* honored a local citizen for bringing back into the public domain, through persuasion and legal action, land that had been privately cultivated. He is described as having endowed Argos with "sacred and public land" (*hiera kai demosia chora*). This is the same expression as occurs in the leases from Hyampolis discussed above.[77] Such a treatment of sacred and public land as a single category was not uncommon in the Greek world, especially when contrasted with private land. Aristotle was familiar with the practice and ascribed the division of land into sacred, public, and private lots to Hippodamos of Miletos. Aristotle also advocated the division of all land in the ideal state into private holdings and common land, which he then further divided into land assigned to support religious activities and land assigned to support the cost of common meals.[78] In Boiotia, the god Dionysos and the city of Thespiai shared sacred land, no doubt with the same ends in mind.[79]

The role of sanctuaries in preserving sufficient land for herding evolved in response to increasing pressure from the eighth to fifth centuries to turn land over to cultivation, evidence for which was systematically assembled by Alfred French in 1956.[80] In an effort to explain the conditions leading to Solon's reforms, French postulated a shift from grazing to agriculture, increasing population, overcropping, soil depletion, an overreliance on imported grain, and a worsening agricultural crisis. Taken separately, each of French's arguments can be challenged. The physical profile of the Greek countryside, for example, is now better understood than when French wrote, and few scholars today would rely on Plato's *Critias* as their principal source for evidence of soil deposition or for reconstructing the pedology of Greece.[81] Furthermore, the shadows of more recent episodes in British history such as enclosure and the Highland Clearances loom over French's explanation of rural change in Archaic Greece. Nevertheless, there were deep structural changes occurring in the Archaic period, and French's reconstruction of conditions in Attica is broadly correct. The pressure to produce enough grain did not abate, and the evidence from the fourth and third centuries, which ranges from grain laws to regulate supply and demand to honors offered to donors and patrons, and illustrates an increasing reliance on their generosity, only demonstrates that the problem grew worse with time.[82] Increasing urbanism was hard to feed. Furthermore, change was not restricted to the growth of larger agglomerations that were vulnerable to food shortages and famine. Recent demographic studies paint a picture of a countryside more densely populated at this time than at any other in Greece's history, including the nineteenth century.[83] Even if, as seems likely, many people walked out from their towns and cities to farm the countryside, the proliferation of small sites—hamlets and villages—reveals a countryside dotted with agglomerations spaced often as little as two kilometers apart. The effect of this population density was to reduce the amount of land available for running large herds. The ranchland available in the Iron Age steadily diminished between 800 and 300 BC.

This shift explains a trend toward bringing previously uncultivated land into agricultural production, a process reflected, as Michael Jameson showed, in the frequent mention of *eschatiai* in the *rationes centesimarum*, accounts recording the sales of public lands in fourth-century Athens. Jameson argued convincingly that the term refers not so much to border land on the edges of the polis' territory but rather to "land that required terracing or other attentive land management."[84] Jameson plausibly suggested that as early as the Archaic period the Greeks' willingness to take part in colonial expeditions was driven by a clear preference for farming rich bottomland rather than bringing hilly country into cultivation through the laborious business of terracing. Similarly, Victor Davis

Hanson interprets the description of Laertes' estate in Book 24 of the *Odyssey* as evidence that the pressure on marginal land can be traced back to the sixth century and earlier.[85] During the Archaic and Classical periods the Greek mainland came close to reaching its carrying capacity, that is, producing as much agricultural produce as it was capable of, given the constraints of both environmental conditions and ancient technology. The process was also exacerbated by an increase in the size of the Athenian *asty* (city) at the expense of the surrounding *chora* (countryside) and the rural activities practiced there.[86]

As more marginal land was brought into production and the countryside reached the limit of agricultural capacity, cattle were propelled from the open countryside toward more areas at the extreme limits of the state's territory. Many families will have managed to keep a yoke of oxen—the poor man's slave, in Aristotle's pithy phrase—or at least aspired to, but large-scale cattle herding was an unviable economic strategy for most people in the Classical period, at least on mainland Greece, where growing communities devoted themselves to regimes of mixed agriculture.[87] The various episodes we hear of involving the movement of flocks and herds during wartime may have entailed larger aggregate numbers, but are unlikely to have included large private herds.[88] Of course herds still embodied wealth, so that slaughtering the herds of the rich was a useful demogogic gesture for would-be tyrants like Theagenes of Megara.[89] Herds would never be disassociated from wealth and status, but in the Archaic period herding moved over progressively into the religious sphere. If agriculture was the push, sanctuaries were the pull, attracting herding because of the unique needs of sanctuaries as places of sacrifice. In addition, sanctuaries, especially the largest, were often located in border lands and areas of lower population density, so that their physical location was ideal for reinforcing the association between herding and sacrifice. Sanctuaries, as sacred spaces, also had the authority to resist the pressure to turn all available land over to private cultivation. The Amphiktyons would scarcely have issued the bans of war against Amphissa had they been caught cultivating someone's private fields, rather than Apollo's Sacred Plain.

There is a danger, of course, of putting the cart before the bullock. It is possible that sanctuaries, rather than being agents in the process whereby cattle herding was pushed to the edges of cultivated territory, were themselves shaped by this shift in land use. There is a universal increase in sacrificial deposits, feasting assemblages, and bovine dedications in terracotta and bronze during the Iron Age, a powerful testament to the centrality of beef cattle to the dominant communal institutions of the time, but as David Mitten and Catherine Morgan have noted in relation to the rich deposits of bone at Isthmia, "burned bone should not be unquestioningly associated with cult activity."[90] Yet such caution may itself prevent us from recognizing fundamental changes. Morgan and Mitten's survey of bone and ash deposits from Iron Age sites underscores the ubiquity of sacrifice and feasting in Iron Age and subsequent settings. Furthermore, every site that has supplied evidence of feasting in the Iron Age, whatever its initial status, functioned as a sanctuary by the Classical period. Accordingly, the feasting attested at these sites was either conducted in a cult setting from the outset or became an activity associated with sacred space subsequently.[91] Mitten and Morgan's conclusion thus risks missing the wood for the trees: "No simple model can enable us to understand Early Iron Age cult organization; instead there seem to be almost as many forms of social and political organization as regions of Greece during this period." True, but hidden in plain view are two constants: cattle were central to the most important social institutions of the Archaic world, and by the beginning of the fifth century cattle were processed primarily through sanctuaries. Cows had become a sacred commodity.

We have already seen how large sanctuaries such as Eleusis and Delphi contributed to this process in order to meet the demand for meat focused on the sanctuary, but even modest cults could alter the pattern of land use. Near Oropos, for example, by one estimate the sanctuary of Amphiaraos came to control over 3,400 hectares, or 17 percent of the productive land in the area.[92] This was at a time when Athens took control of the Oropia and divided much of the land between two Athenian tribes. Yet even as the Athenians were satisfying a thirst for cultivatable land on their borders, a prestigious sanctuary was able to maintain control of some of the choicest land, including the fertile coastal plain. Michael Cosmopoulos argues that the *aparchai* payments from the Amphiareion to Eleusis in wheat are evidence of the land's fertility, but some of the land was also put aside to meet the sacrificial requirements of the sanctuary and its visitors. All visitors to the cult of Amphiaraos were expected to sacrifice a ram, the fleece of which remained sacred and therefore the possession of the shrine.[93] Amphiaraos was a healing god and those seeking his help slept in the sanctuary on the skin of the sacrificial animal. Unless the sanctuary continued hoarding the fleeces after incubation the administrators of the shrine will have sold them off, providing another source of revenue in addition to the fees paid by visitors.

The regulations from the Amphiareion also reveal that under normal conditions sacrificial meat was consumed within the confines of the sanctuary. The stipulation occurs as part of the procedures laid out for anyone wishing to get the god's help:

> Anyone wishing to be healed by the god is to offer in payment no less than nine obols of certified silver and should place it in the payment box in the presence of the attending priest. . . . Let him make his invoca-

> tion to the god by the sacrificial victims and, if he is present, let the priest place them on the altar, and if not, let the person who is making the sacrifice and who has invoked the god on his behalf by means of the sacrifice do it himself. The sacrificial meat is to be shared publicly. The hides of the animals sacrificed in the sanctuary are to be considered holy, and any person may sacrifice any type of animal he wishes, but it is not permitted to take any sacrificial meat outside the sanctuary. Those who make a sacrifice are to give the priest the shoulder of each sacrificial victim, except whenever there is a festival, when he [sc. the priest] may take the shoulder of each of the sacrificial victims at the public sharing of meat. (*I.Oropos* 277.25–35)[94]

These sacrifices, therefore, were regulated by a schedule of festivals in honor of Amphiaraos that resulted in banquets held in the god's honor. Those who benefited were the visitors, the administrators of the sanctuary (priest, *neokoros*), and the demesmen of Oropos.[95] The Amphiareion regulations, as with so much epigraphic evidence, often leave unstated details that would shed more light on the sanctuary's interest in animals, but, once again, it is important to notice the recurring feature of sanctuary regulations: they are usually not concerned with piety or even ritual actions, except insofar as visitors continue to provide the god with his sacrifice and his priests with their perquisites. The Amphiareion typifies the way in which the physical location of sanctuaries and the changing patterns of land use combined to put meat production into the sphere of the sacred economy and away from the secular agricultural economy, where cattle were more likely to be raised only in small numbers, primarily as draft animals.

Consumption was affected too. Homer's heroes eat gargantuan amounts of meat and little else, but the physical evidence from Classical and Hellenistic Athens points to a diet in which cereals and plant matter predominate. The most reliable way to quantify meat versus grain consumption is through trace-element analysis of zinc and strontium deposits in skeletal remains. Zinc enters the body through the consumption of meat and fish, while strontium levels serve as indicators of vegetable and plant consumption. The only study of Athenian skeletal remains to measure both zinc and strontium content finds much lower zinc levels in male skeletons of the Classical period (128.3 mg/g) compared to male bones of the Hellenistic period (186 mg/g). The bones of men from Classical Athens also had lower zinc levels than female bones from either the Classical (162.2 mg/g) or Hellenistic periods (145.6 mg/g). The evidence suggests a diet meager in meat.[96] Similarly, skeletal comparisons with the modern population and the relatively low levels of dental caries suggest a high level of consumption of pulses. Even allowing for epic exaggeration in Glaukos' boast that he and Sarpedon enjoyed the choicest meats because of their status (Homer, *Il.* 12.309–20), there is very little likelihood that any Greeks of the Classical period enjoyed a diet with even a fraction of the meat eaten in the age of heroes. Meat was increasingly a luxury, but unlike the other luxury comestible, fish, meat was consumed primarily in the context of religious festivals, and was removed from the daily diet of ordinary people.[97]

Sanctuaries evolved to play a crucial role as nodes of social interaction, exchange, competition, and negotiation in the unique culture of Archaic and Classical Greece. The processes of political centralization and growth tended to fluctuate between the city-state and larger but often more loosely organized federal leagues and *koina*. Only sanctuaries crystallized into super-regional organizations, thanks to their distinctive blending of musical performance, athletic competition, oracular consultation, sacrifice, dedication, and feasting. Their physical isolation and the distinctive rhythm of the activities taking place there also meant that sanctuaries could remain closer to an economic system that still had a place for cattle husbandry on a scale recursively linked to the growth of the sanctuaries themselves. As we shall see, however, urban sanctuaries and the societies they served were forced to handle the business of meat production very differently.

CHAPTER 8

Cities and Cattle Business

> Those who fail to pay the rents of the sanctuaries of the goddess Athena, and of the other gods and of the eponymous heroes shall lose their citizenship, themselves, their family, and their heirs, until they pay.
>
> —Demosthenes 43.58

Supplying the City

The commercial needs of cities and urban sanctuaries were different from those of Panhellenic or rural sanctuaries. Whereas the priests of Delos and Delphi may have benefited from the sanctuaries' constant stream of visitors, the interests of the local population were subordinate to the Panhellenic sanctuaries' role as international centers. City sanctuaries, on the other hand, both large and small, were closely attached to their communities. They provided the setting for religious activities at every level of Greek life from the household to the state. Whatever their city-state, most men in the Greek world were members of households (*oikoi*), brotherhoods (phratries), sacrificial groups (*orgeones*, *synthyastai*), clans (*gene*), and tribes (*phylai*). Social identity was structured on a set of overlapping and intersecting corporate groups, every one of which replicated the same broad pattern of sacrificial and feasting practices. In other words, eating food made them human, but eating meat together made them Greek.

This does not mean that there might not be doubts about some of these groups or individuals dining at public expense. Hesiod shows that the sacrificial feast could be a source of disquiet even in the Archaic period:

> When visitors flock to a public feast, do not rebuff them; since the funds come from the common stock, you get the best return of good will for the least outlay. (Hesiod, *Op*. 722–23, trans. Robertson)[1]

But the resources needed for public sacrifices and banquets were precious, and, the bigger the sacrifice (measured either by the number of animals or their size), the more irrational it became economically, since the more land given to stock raising, the less land was available for farming. Accordingly, urban sanctuaries were in the paradoxical position of driving a demand for meat in precisely the worst place to satisfy that demand.

Jennifer Neils has estimated that one-third of the Athenian year was taken up with communal sacrifices and feasting, yet of all places in Greece this was the one where environmental conditions and population density made stock raising on a large scale the least viable form of food production.[2] Of course the problem was partially ameliorated when the god being worshipped had a preference for smaller animals. The pigs sacrificed to Demeter as part of the Eleusinian cult were more easily acquired than the cows for the Panathenaia, and some cults, such as that of Aphrodite, were content with sheep, goats, and even smaller offerings such as birds and fish.[3] But Zeus, Hera, Apollo, and the other Olympian deities most strongly associated with large sanctuaries were venerated with the sacrifice of cattle. Based on the evidence from the sacred calendars of the Athenian state and the demes, Vincent Rosivach estimates that classical Athens required more than 6,500 head of cattle to satisfy its sacrificial requirements.[4] It made sense, then, for their sanctuaries to be surrounded by land set aside for their herds. It was when these deities received cult in an urban area that the realities of supply and demand became difficult to reconcile.

The problem grew worse the closer one came to the heart of Athens, where the intensity of cult grew in proportion to the unavailability of large herds of sacrificial animals. To satisfy Athena's sacrificial requirements the state resorted in part to the leasing of land located outside the city, in an area called the Nea, near Oropos.[5] The details are contained in a famous fourth-century sacred law concerning the Lesser Panathenaia:

> [De]cision of the *demos*: in addition to
> [the decision of the *boule*], let the *hieropoioi* conduct the two
> [sacrifices, both the sacrifice] to Athena Hygeia and the one to the Ar-
> [chegetes] just as it is usually done, distributing
> [to the *prytan*]*eis* five portions and to the nine ar-
> [chons three] and to the treasurers of the goddess one and to the *hie-*
> [*ropoioi* one] and to the generals and the *taxiarch-*
> [*s* three and to the] Athenians participating in the procession and to
> [*kanephoroi*] the customary portion. The rest of the meat is to be
> [distributed to the] Athenians. [. . . F]rom the 41 mnai
> [from the l]easing of the Nea, the *hieropoioi* are to buy oxen
> [with the help] of the cattle buyers and dispatch the procession
> [in honor of] the goddess and sacrifice all these cattle at the great altar
> [of Athena] . . . [Let] them distribute the meat of all the cattle
> [purchased] with the 41 mnai to the Athenian people at the

> [Kerameiko]s, just as at the other meat distributions.
> [Allocate] the portions to each deme in proportion to the number [of people accompanying the procession] from each deme . . .
> (*IG* II^2 334 B 7–27)[6]

From these details we can make some deductions regarding the amount of meat made available by a major civic sacrifice. To calculate the amount of meat yielded at the Panathenaia we need first to establish the number of animals that could be purchased for 41 mnai (4,100 drachmas). In the Thorikos calendar of ca. 430 BC two cows costing 40–50 drachmas each are listed.[7] Similarly, Athenian accounts for the years 377–373 list 109 head of cattle purchased for 1 talent, 2,419 drachmas, at an average of just over 77 drachmas per head.[8] The Tetrapolis calendar of 400–350 BC mentions five cows each at a cost of 90 drachmas. In 363/2 BC the *genos* of the Salaminioi stipulated the sacrifice of a single cow costing 70 drachmas. Michael Jameson favored the high Tetrapolis figure, 90 drachmas, in his tabulation of the cost of sacrificial animals, and has been followed by Thomas Gallant, but Marathon's reputation for good pasture land (Pausanias 1.32.7) and its association with the famous Bull of Marathon (1.27.10) probably mean that the local animals were among the prime stock in Attica.[9] It would be safer to take the average of the high figures from Marathon and the low figures from Thorikos, which happily match the numbers provided by the Salaminioi: 70 drachmas. The 41 mnai raised from the lease of the Nea would thus have purchased close to 60 head of cattle.[10] A heifer of comparable price ("no less than 50 drachmas") is attested from Kos in the fourth century.[11] Other figures from Kos are dramatically higher, ranging from 300 to 600 drachmas, but local conditions may have made cattle a much rarer sacrificial animal on Kos. Another factor influencing the total cost of a sacrificial animal may be reflected in the reference to a *bous* valued at 400 drachmas that had gilded horns (*chrysokeros*). The exorbitant price of this and other animals may, in fact, reflect the total value of the beast and the gold leaf decoration.[12]

How much meat did 60 head of cattle yield? Although Michael Jameson believed that the meat yield from the cattle slaughtered at Nichoria in the Geometric period may have been as low as 100 kilograms per animal, Eberhard Ruschenbusch has estimated the animals of the Classical period may have provided up to 220 kilograms each. Again taking an intermediate range of 150–200 kilograms each, we can safely estimate that these animals would have supplied 9,000–12,000 kilograms of meat at one sacrifice. Instructions for a fourth-century *kreanomia* survive from Koresia, on Keos, stipulating the distribution of 1.25 kilograms per man. If such a standard is applied to Athens, the festival of the Panathenaia yields enough meat to supply between 7,200 and 9,600 portions. The most reliable estimates for the number of adult male citizens in fourth-century Athens range from 21,000 to 30,000. It follows, then, that not every male Athenian citizen can have received a distribution of the Panathenaia meat, especially if, as in other cases, women and children were eligible for full or half portions as well.[13] Perhaps one in three households benefited directly. If, however, the meat were consumed at deme-based feasts open to citizen men only, 12,000 kilograms would easily have fed 30,000 men, supplying almost half a kilo per man.

The volume is impressive. The regulations make it clear that the distribution of meat was done regularly, and limiting shares to those who accompanied the procession encouraged participation. Furthermore, the meat from the Nea funds was not the only sacrifice, nor the only distribution associated with the Panathenaia. Two prior sacrifices had already resulted in at least thirteen portions shared among the major officeholders with an unspecified amount of leftover meat being made available to the *demos*. One cannot press these numbers too far, but the systematic distribution of large quantities of sacrificial meat suggests that the religious institutions of the polis were capable of adapting to the challenge of supplying the population with meat.[14] In a Hellenistic inscription from Karia, the system is laid out in explicit detail:

> Once they have completed the sacrifice and removed the customary perquisites for the priest, then let both the temple administrators and the other prescribed officials distribute the remaining meat in common to the citizens the next day in the agora around the third hour, conducting the meat distribution tribe by tribe.[15]

On Kos, the distribution of meat was linked to participation in a torch race:

> In the month of Hyakinthios, on the second day after the full moon, the great race is to be held, a torch race for boys, before the sacred procession is sent off . . . Of the sacrificial animal which the *paidonomos* sacrifices on the second day of Hyakinthios to Hermes, the rest of the meat he is to distribute among those who ran the torch race, except for the left haunch, which he is to give as a prize to the winner of the new-moon race. (*Iscr. di Cos* ED 145A.51–53, 58–62)

It is less clear whether the habit of massive distributions first arose at Athens or was perhaps borrowed from those smaller communities that benefited from the presence of an international sanctuary in their midst. At Delphi the meat from a hecatomb sponsored by the ambassadors from the Chersonese was distributed to the citizens of Delphi, prompting a decree in the ambassadors' honor.[16] The case of Eleusis illustrates how the state control of a sanctuary focused attention on managing sacrificial resources. A third-century inscription honors the *epimeletai* responsible for the Mysteries and is principally concerned with recognizing the good job they did in conducting the necessary sacrifices and providing the required animals, including the sacrificial bull, which they then distributed to the Council of 650.[17]

Ancient commentators were clearly struck by the scale of sacrifice conducted in a city the size of Athens, but characteristically interpreted the phenomenon not as a sign of the growing complexity of Athenian society but as a result of the degradation brought about by democracy. The Old Oligarch, for example, views the system of public sacrifice as an attack on the older institution of sacrifice that depended on the personal generosity of wealthy patrons:

> The Athenian populace realizes that it is impossible for each of the poor to offer sacrifices, to give lavish feasts, to set up shrines, and to manage a city that will be beautiful and great, and yet the populace has discovered how to have sacrifices, shrines, banquets, and temples. The city sacrifices at public expense many victims, but it is the people who enjoy the feasts and to whom the victims are allotted. ([Xenophon], *Ath.Pol.* 2.9, trans. Bowersock)[18]

The regulations from the Panathenaia also shed light on the logistics of meat production in the setting of a major festival. The Athenian *hieropoioi* leased the Nea, but there are no stipulations concerning a census of sacred herds as on Delos or the inspection of sacred land as at Delphi. Indeed, the leased property is not referred to as sacred land, and it is possible that the Athenians distinguished between land rendered sacred by its physical contiguity to a sanctuary and public land owned by the state and exploited to meet the state's sacrificial obligations. In the case of Athenian festivals, especially the Panathenaia, which had assumed international significance, the scale of the event exerted enormous pressure on the state. In the first place, the distance between the cult centers of the Acropolis and any ranchland suitable for grazing large herds made it impossible to employ leasing arrangements as on Delos or at Hyampolis. Large herds could not graze on or around the Acropolis. Even temporarily accommodating a sacrificial herd required the installation of cattle yards on the Acropolis, a practice unparalleled elsewhere. Little wonder that the sight of young men handling the sacrificial cattle made such an impression on the Athenians (Fig. 8.1).[19]

To compensate for the lack of a permanent sacred herd the Athenians relied on different sources: animals sent as tribute by their allies, and cattle purchased by designated cattle buyers (*boônai*).[20] In the case of the allied contributions, which consisted of a panoply and a single head of cattle,

Figure 8.1 Young cow and herdsmen from the south frieze of the Parthenon. London, British Museum. BM Images AN3279090. Copyright © The Trustees of the British Museum.

the timing of the Panathenaia, which occurred toward the end of Hekatombaion, in midsummer, meant that the festival coincided with the best time of year for culling herds. Calves born in the spring of the preceding year would now be fully grown and ready for market. Since, in the case of domesticated herds, only one bull is needed to serve the breeding stock, bull calves are regularly selected for slaughter after they have reached one year of age. It is very likely, then, that the "cow" required of each ally was not in fact a cow, but a year-old bull calf.[21] In the case of animals purchased for sacrifice, the existence of special officials to handle the purchase suggests that the animals were not grazed on the Nea. Had this been the case, the estate would certainly have been referred to as sacred land and the acquisition of sacred cattle would have been left to the officials administering the land, the *hieropoioi*. Instead the Nea seems to have been public, Athenian land whose revenues were handed over as a cash payment to the *hieropoioi*. Ironically, the remarkable scale of meat consumption at these festivals confirms Max Weber's interpretation of Athens as a consumer city, yet at the same time shows the city taking the first steps toward developing a private sector to help supply the sacrificial needs of the community.[22] Tribute and donation were not enough to feed Athens. So great was the demand that officials had to purchase animals. Even the officials' title reflects the commercial rather than religious nature of their duty. Unlike *hieropoioi,* or *hierothytai*, the title of the *boônai* contains no element deriving from holy or sacrificial associations.

The merging of sacred and profane spheres in the purchase of cattle for sacrifice suggests that the Athenians had access to private herds capable of supplying the sacrificial market. This would have been burdensome for small farms, since beef cattle are essentially unproductive until slaughtered and rely on more productive land for forage than is found in Attica.[23] So, despite the common assumption that wealthy Athenians kept large herds, the evidence for this should be treated carefully.[24] Two accounts in Aristotle suggest that private ownership of cattle was a source of tension. In the *Politics* (5.1305a) he reports that Theagenes of Megara curried favor with the populace by slaughtering the cattle of the wealthy, "which he captured grazing by the river." This odd episode makes sense only if the cattle of the wealthy were grazed together, rather than on the individual landholdings of separate farmers. How had they acquired such valuable, well-watered land? The Megarian poet Theognis alludes to the seizure of his land and the emergence of a corrupt elite who were driving Megara to ruin.[25] Behind the high moral tone of his invective may be a complaint arising from disputes over land and property in which the wealth of ranchers was both conspicuous and resented. In another episode Aristotle recounts a tale involving the Sicilian tyrant Dionysios and his tireless schemes for raising revenue:

> Because of his taxes the citizens stopped rearing sheep and cattle (*boskemata*), so he declared that he had enough money for the time being; accordingly those who now acquired any stock would not be liable to taxation. Many people immediately purchased a lot of sheep and cattle, on the understanding that they would not be liable for any taxes. But when he felt the moment was right, Dionysios had them all assessed and exacted a tax. The citizens were angry when they realized they had been tricked, and began to kill and sell off their beasts. Dionysios responded with a decree that only as many animals could be slain as were needed each day, whereupon the owners retorted by designating their animals as sacrificial victims; finally he outlawed the sacrifice of female beasts. (Aristotle, *Oec.* 2.1349b)

Allowing for the exaggeration in a story told to discredit a tyrant, we nevertheless catch a glimpse of a rudimentary system of commercial transactions in herd animals. For Aristotle's story to be plausible his audience had to accept that ordinary folks usually possessed some stock, that such animals could end up being sold, and that the commerce in animals might include their designation as sacrificial animals (*hierothyta*). Presumably most animals were not so designated in order to outwit a tyrant, but because this was the regular way animals were processed for their meat. The brinksmanship in which the people and tyrant engage is an exaggeration of a standard model, according to which a moderate surplus of animals was supplied for consumption and processed through sacrifice. That owning a large herd was unusual and faintly suspicious helps to explain Demosthenes' jabs at the Olynthian Euthykrates, who had betrayed his city to Philip. Euthykrates was known for raising many cattle without having purchased any, a sure sign that he had been corrupted by Philip.[26] Whether the charge was true is both unprovable and immaterial, but it presupposes that purchasing cattle was a possibility.

The emergence of a market in sacrificial beef may have benefited from the existence of a market for draft animals. As early as Hesiod, conventional wisdom maintained that the first steps in successful farming were straightforward: "First of all, get a house, and a woman and an ox for the plow," which makes no sense unless there were oxen to be got. Later he emends this and recommends two oxen, nine years of age, "in the prime of life," beasts for pulling the plow and not for sacrifice, although he says nothing about how the animals are to be acquired.[27] Although we have few details of the market for draft beasts, some rudimentary cattle market clearly existed. In Aristophanes' *Birds*, frightened by Peisthetairos' vision of the gods' punishments, including making their yoke oxen sick, Euelpides cries out, "Oh! don't do that! Wait first until I have sold my two young bullocks!"[28] Two inscriptions dealing with the sale of cattle survive, one from Abdera and one from Knossos. They are tentatively dated to the fourth and third centuries, and deal with irregularities arising from the sale of sick or otherwise unsuitable animals. Both decrees concern draft animals. In the Abdera regulations this can be inferred from the fact that the guarantees of good health are also required for slaves, mules, asses, horses, and hinnies, all working assets not intended for sacrifice, and the Cretan regulations exempt the purchaser from the customary daily interest to be paid when returning an animal if it has not been broken to the plow (*dedamnamenon*).[29] The Knossos decree also lays out the penalty for breaking the animal's horns. Plow animals were harnessed by a heavy yoke fixed to the base of the horns. Broken horns thus compromised the animal's ability to plow.[30]

There did exist, therefore, a market in plow oxen, so that the logistics of handling large animals in a market setting were in place in the fifth century. Cattle for sacrifice, however, were another story, and a report in Pausanias reveals that it was unusual to sacrifice draft animals.[31] The evidence for private cattle wealth comes primarily from outside Athens, and this may be another example of Athens being an exception to the rule, since it seems that not many of the cattle that were sacrificed in Athens came from Attica. Thucydides describes the evacuation of sheep (*probata*) and plow oxen (*hypozygia*) from Attica in 431 BC, but says nothing of any private herds of cattle.[32] Thucydides explains that when the Spartans fortified Dekeleia in the last phase of the Peloponnesian War, the occupation caused great hardship because the Athenians were deprived of their territory, suffered the desertion of 20,000 slaves, and lost all their sheep and yoke animals (*probata* and *hypozygia*).[33] The omission of any mention of beef cattle or sacrificial animals is notable but is consistent with the view that few Athenians had large herds of cattle. Aristotle may have asserted that wealth consisted not only of land but also of slaves, grazing animals (*boskemata*), and money, but he seems to have had sheep in mind.[34] Heroic figures such as Odysseus and Nestor had many head of cattle at their disposal, but real Athenians of status were more likely to have property like Stratokles: sixty sheep, one hundred goats, and "a fine horse he rode while serving as *phylarch*."[35]

The scarcity of cattle in the property of leading Athenians is well illustrated in the records of the *poletai* who auctioned the property of the men convicted of defacing the statues of Hermes in the Athenian agora. We only find cattle mentioned in the case of Panaitios, and the only cattle mentioned are two plow oxen, two other cattle, and four cows with an undisclosed number of calves. If Panaitios were a man of means, as is suggested by the rest of his possessions and those of the other Hermokopidai, we can infer that wealthy Athenians were not known for raising even moderately sized herds of cattle. Panaitios had eighty-four sheep and sixty-seven goats, enough to keep his household self-sufficient in wool, milk, and cheese, and even in sacrificial meat, but when it came to cattle it is unlikely that many Athenians had more than a yoke of cows and perhaps one or two more for breeding stock.[36] Xenophon mentions a miller named Nausikydes who made so much money from his mill that he could afford a large household and "many pigs and cows," but unfortunately he gives no indication of how large a herd Nausikydes had.[37] Xenophon speaks of a peaceful city attracting men rich in "grain, wine, oil, and cattle" to trade their goods, but appears to classify them as merchants rather than wealthy Athenians.[38]

The demands of a large-scale, commercialized cattle economy, even if channeled into the religious calendar of sacrifice, could not have been met by private Athenian graziers. It hardly comes as a surprise, then, that in 330/29 BC, when the Athenians were refurbishing the Panathenaic stadium, the benefactor of the work, Eudemos of Plataia, had to supply one thousand yoke of cattle to complete construction.[39] Boiotia could provide such numbers, but not Attica. Where, then, did the cattle come from to grace the altars of the gods? In the second century Polybios claimed that Pontos was more plentifully supplied than any other location with the two staples, cattle and slaves, as well as luxury items such as honey, wax, and salt fish, which were traded for Greek olive oil and wine.[40] Legal disputes over bottomry loans preserved in Demosthenes' speeches also demonstrate that Greek traders were active in the Black Sea area in the fourth century, and it is probable that the cattle trade grew in tandem with the expansion of the sacrificial calendar in Athens. Aristotle associated northern Greece with especially fine cattle, such as the cattle of Pyrrhos, and Pausanias gives a description of hunting the wild bison of Paionia with a view to capturing rather than killing them.[41] Such animals could have been conveyed down the Axios river and transported by sea to Athens.

Another possible source for cattle destined for the altars of Athens was Boiotia. Grants of *epinomia* (right of pasturage) are recorded from Akraiphia (50 head), Kopai (two grants each of 200 head) and, largest of all, a lease for four years on land rented by the city of Orchomenos to an Elatian named Eubolos for a herd of 220 large and 1000 small animals.[42] The Boiotian grants of *epinomia* reflect a particularly strong local connection with large-scale herding. Christophe Chandezon has recently suggested that the Orchomenos territories leased to Eubolos for his herds were located by Lake Kopais, an area especially favorable to the grazing of large animals.[43] Both the grant to Kallon by the city of Akraiphia and the concession to Kleuedra and Olioumpicha by the town of Kopai come from communities located on the northern shore of the lake, and each of the states appears to have exploited the region for husbandry to a higher degree than regions further south. Although grants of *epinomia* are not unusual, the Boiotian contracts are significant for two reasons. The first is that the transactions are handled by states rather than sanctuaries, and the second is that each of the grants comes from the Hellenistic period. In the Archaic period cattle had been processed primarily through sanctuaries. In the Hellenistic Age, however, we find evidence of cattle transactions conducted as purely commercial transactions. In Athens this occurred where the demand exceeded what sanctuaries could supply. Elsewhere, local conditions favored the growth of a commercial cattle industry to supply the animals sanctuaries could not. At Tauromenion in Sicily, for example, the public treasury recorded revenues of 30,000 talents, of which 12,700 talents came from the sale of cattle. Given the derivation of the city's name from Mount Tauros, the growth of a specialized cattle industry in this area should come as no surprise.[44] By the time of Plutarch, the oracle at Delphi was frequently consulted by states regarding the business of cattle breeding.[45]

Sacrificial requirements created a massive demand, but also produced vast amounts of meat. In a recent study of a bone deposit from Corinth, for example, David Reese records the bones of seven hundred goats, sixty cattle, and twelve pigs, in addition to fish and birds, from a structure smaller than 9 x 5.5 meters. Plausibly connecting this with a major festival held at the theatre, Reese notes that the meat produced would have fed 15,000 spectators at 0.5 kilograms of meat for four of five days.[46] Such sudden gluts of meat supply resulted in the treatment of some meat as a non-religious commodity. This may come as a surprise since it is often asserted that the Greeks obtained their meat from sacrifice.[47] While this is broadly true, it is not correct to suppose that all meat made its way straight from the altar to the table. In fact the enormous appetite of a city like Athens and the gluts caused by mass sacrifices there and at major sanctuaries resulted in fluctuations of supply and demand. The result was the growth of a meat market originating in the milieu of sacrifice but increasingly commercial in character. This is illustrated, for example, by a fifth-century decree of the deme Skambonidai that specifies a sacrifice for a festival, perquisites to the priest, but instead of a distribution of meat stipulates the sale of raw meat at the temple of Pythian Apollo.[48] The transformation of sacrificial victim into commercial commodity was partially resolved by allowing the same official to conduct both sacrifice and trade: the *mageiros*.[49] Although other terms, such as *hierothytes*, can also be employed for the official responsible for killing the sacrificial animal, the term *mageiros* appears frequently in inscriptions dealing with the performance of sacrifice.[50] Yet by the end of the fifth century the term has acquired a much wider semantic field, reflecting a dilution of its original religious character, and often means little more than "cook."

Aristophanes and Plato use the term in a variety of non-sacrificial contexts, dealing with the butchery of animal carcasses, the sale of meat or its preparation. In fact three of Aristophanes' heroes, Dikaiopolis, Trygaios, and Peisetairos are described either as cooks or as possessing *mageirike techne* (cooking skill). In the case of Dikaiopolis, his transformation into a chef signals his increasing urbanity, as he leaves his old rustic manners behind. Rather than becoming more priestly, Dikaiopolis looks more like a figure whose gourmet status challenges the very idea of the community. The plot of *The Acharnians* involves Dikaiopolis conducting his own private peace negotiations, and his divorce from the rest of the Athenian *demos* is signaled by the way he celebrates the Choes banquet alone, when it ought to be a communal gathering, part of the Anthesteria festival.[51]

Accordingly, by the fourth century the meat business was sometimes only tenuously connected to religion. Theophrastos' description of the shameless man, for example, carries no religious overtones:

> In marketing, again, he will remind the butcher (*kreopoles*) of any service which he may have rendered him; and, standing near the scales, will throw in some meat, if he can, or else a bone for his soup; if he gets it, it is well; if not, he will snatch up a piece of tripe from the counter, and go off laughing. (Theophrastos, *Char.* 9, trans. Jebb)[52]

Similarly, the mean man's behavior shows that even sacrificial meat could find its way to market:

> When he is celebrating his daughter's marriage, he will sell the flesh of the animal sacrificed, except the parts due to the priest. . . . Again, when he has bought provisions, he will himself carry the meat and the vegetables from the market-place in the bosom of his cloak. (Theophrastos, *Char.* 22, trans. Jebb)

There are also scattered references to the sale of meat that distinguish between commercial meat and sacrificial meat. The pseudo-Aristotelian *Mirabilia*, for example, describes the peculiar behavior of kites in Elis:

> It is said that among the people of Elis there is a type of kite that snatches meat from those carrying it across the market, but that will not touch sacrificial meat (*hierothyta*). ([Aristotle], *Mir.* 123, 842B 1)

Similar accounts of raptors seizing ordinary meat while leaving sacrificial meat untouched are reported by Theopompos (*FGrH* 115 F 76), who ties it to the Olympic festival, and Aelian, whose description may also derive from stories coming out of Olympia. He remarks,

> The kite is unparalleled as a thief. When they collect in great numbers they will fall upon meat for sale in the agora and snatch it away, although they don't touch meat that comes from sacrifices to Zeus. (Aelian, *NA* 2.47)

Even if all three accounts have a common source, the repetition of the story would make no sense unless the distinction between sacrificial meat and meat sold as a commercial commodity were generally recognized. Butchery assumed a quasi-market function precisely because the supply and demand for meat in a great city or Panhellenic sanctuary was so difficult to regulate coherently.

Religious Commerce

As the business of sacrifice grew in complexity and scale, the significance of the event as a sacred offering was overshadowed by the commercial transactions required to bring the sacrifice about. The purchasing of cattle marked the transition of cattle from the purely sacred economy into the secular sphere. Yet because cattle never lost their powerful association with religious ritual, the sacred and the profane were often intertwined. Two examples illustrate this clearly. In the mid-fourth century the citizens of Kos erected four *stelai* with some of the longest and most detailed descriptions available to us outlining sacrificial procedure.[53] In the section describing how the sacrificial animal is to be chosen we find this clause:

> When they are in the agora, the person who owns the ox or another enabler on his behalf calls out, "I am providing the ox for the Koans; let the Koans give the price to Hestia." And let the presidents take an oath immediately and make a valuation, and when a valuation has been made, let the herald announce how much the valuation was. Then they drive the ox to Hestia Hetaireia and sacrifice it. (*GHI* 62A 19–22)

The process described here is oddly complicated, but makes sense if we see the entire procedure blending the sacral business of sacrificing an animal to the state's tutelary goddess, Hestia, with a commercial transaction. The sacrificial animal, though an agent for contact with the divine realm, is also a commodity that must be purchased. The prosaic matter of negotiating a price is the clearest sign that the animal has been commoditized, yet the procedure by which the negotiation is conducted—declaration, oath taking, and proclamation—derives from the religious sphere.

As sacrifice became a commercialized activity, so too the priest's role as the official in charge of sacrifice came to resemble a paid job. At first this amounted to no more than the offering of perquisites to the priest for having arranged the sacrifice and performing such mundane duties as garlanding the altar or supplying firewood. The perquisites were simply a portion of the meat produced by the sacrifice, and often the hide of the sacrificial animal. In Olympian sacrifices, as opposed to chthonic sacrifices or holocausts, after being slaughtered the animal was gutted, skinned, and disarticulated.[54] Bones wrapped in fat, recalling Prometheus' attempt to deceive Zeus, would be burned on the altar, but a variety of products was created by the process: meat, organs, heads, hooves, and hides. Most of the meat was either consumed immediately or was distributed for consumption, but a priest's share and other perquisites were regularly kept back. In Aristophanes' caricature of a sacrifice in *Peace* the *chresmologos* Hierokles sniffs around the cooking meat trying to wheedle a portion out of Trygaios, and many inscriptions show that certain portions of meat (shoulder, flank, loin, leg, or knuckle) were set aside for the priest.[55] Other products from the sacrificial animal were subject to very different treatment, ranging from the draping of *splanchna* (intestines) across the god's statue, to the setting of a table with portions of meat for the god (*theoxenia*).[56]

Perquisites are attested in numerous inscriptions but an early-fourth-century Athenian inscription, *IG* II^2 1356, sheds particular light on the practice. The inscription, from an unidentified deme, concerns the perquisites of a number of different priestesses: the priestess of the Heroine, of Dionysos, of Hera, of Verdant Demeter, and at least two others. Each entry lists these perquisites (*hiereosyna*), before breaking them down into different categories. The entry for the priestess of Hera is almost complete and is a good example:

> The perquisites of the priestess of Hera: 5 drachmas; the hide, for each skinless (*heuston*) victim, 3 drachmas; a portion of meat (*deisia kreon*); for one-twelfth medimnos of wheat, 3 obols; for one kotyle of honey, 3 obols; for three kotylai of olive oil, 1.5 obols; for firewood, 2 obols. On the table (laid for the god), the bottom round (*kole*), the top round (*pleuron ischiou*), the half head of tripe. (*IG* II^2 1356.11–16)[57]

These *hiereosyna* amount to more than shares in the sacrificial meat. Like most of the regulations, the terms applying to the priestess of Hera stipulate a payment in money, to be paid by the sponsor of the sacrifice. In all but one of the entries, the first item after the cash payment is the skin or skins of the sacrificial victims. Next comes a money payment in lieu of hides when the animal was skinless, probably because the skin had been kept by the sponsor.[58] This is followed by the mention of a portion of meat (*deisia kreon*). Since meat was generally cut into equal portions according to weight, no further qualification was needed. The next items, however, are cash payments calculated as the price of items required for conducting sacrifice: oil, grain, honey, and firewood. The regulations therefore set an amount that the priestess is to receive, but imply that she is also to spend set amounts in order to conduct the sacrifice. The last category of perquisites comes from the choice cuts set aside for the god's table on the occasion of a *theoxenia*, a banquet symbolically laid for the gods. The same cuts are repeated in each entry, and we should suppose that once the *theoxenia* was finished the priestess was free to remove the meat and tripe for her own use. The priestly perquisite is frequently termed a *geras*, often a double share of the portions into which the sacrificial victim was cut. This has been plausibly interpreted as both remuneration and as an aristocratic privilege, but the common practices of giving the priest the skin of the victim and of linking the perquisites to the obligations of sacrifice also demonstrate the tendency for sacrifices to be regulated by state officials. In fact priests begin to resemble paid officials.[59] Take for example, these third-century regulations from Sinope:

> Good fortune. On the following conditions [the priesthood is to be arranged] of Poseidon Helikonios. The one who is appointed [shall serve as priest] for life, taking all of the allocated [money?], shall provide everything [in the sacrifices] which are public and shall take [from the sacrificial animals] that are sacrificed publicly all the hides . . . loin, tongue, while of the [private sacrifices] he will take the loin or the shoulder and will wear a crown from the 12th of Taureon until the 20th and in the month of Poseideon from the 12th to the 14th. He shall have a crown of flowers at all of the games, just like the magistrates. And he shall not be personally liable for the military tax. He shall complete that which has been arranged from the contract or he shall be liable. . . . If [anyone makes a proposal contrary to these terms,] let him be [charged] with sacrilege. (*LSAM* 1)

Bureaucracy is beginning to trump religiosity here, as minor details of obligation, payment, and performance take over. The trend is even more pronounced at Kyzikos, where a list of at least five priesthoods available for purchase contain the stipulation that the purchaser, that is, the priest, is to be paid either one or four obols from the treasury "for as long as he lives."[60]

Had sacrifice remained a private or intermittent affair, the transformation of sacrificial animals into commercially traded hides would have been on such a small scale as to qualify as no more than a type of barter, exchanging the service of the priest for a simple payment in kind. In fact this practice never died out completely, and when the Arkadian League voted divine honors to Philopoimen in 182 BC they took care to specify the distribution of edible meat (*ta krea ta brosima*), the allocation of hides from the sacrificial beasts, and the designation of the gifts (*gere*) from the sacrifice to be given to the priest.[61] The scale of polis-based sacrifice, however, meant that hides became big business. Vincent Rosivach has calculated that in the fourth century there were as many as sixteen citywide sacrifices per year, and that these were in addition to twenty deme sacrifices and between five and ten sacrifices by smaller organizations. This system produced not only forty to forty-five occasions when meat was made available in large quantities, but also a steady supply of skins. As early as 460 BC we have evidence of hides being assigned to demarchs; with the growth of citywide sacrifices the commercialization of sacrificial by-products gained momentum, culminating in the state taking control of the trade in hides. This is demonstrated by the so-called *dermatikon* accounts.[62] These are a set of records from the years 334/3–331/0 BC, during the ascendancy of Lykourgos, preserving details of the revenues accruing to the state from the sale of the hides of sacrificial animals. Lykourgos' decision to keep records of these revenues is in keeping with the spirit of his legislative and economic reforms, all of which were designed to bring order to the haphazard financial practices of the state, which were further upset by the defeat of the Greeks at Chaironeia. It is sometimes stated that Lykourgos established the *dermatikon* fund, which is strongly suggested by the fact that the first list, which is complete, only covers six months of the archonship of Ktesikles (334/3), and it is clear that under Lykourgos the Athenian state asserted control over a series of cult sacrifices and claimed the hides from the sacrificial victims as sources of state revenue.[63] Lykourgos' thrift took advantage of the fact that there was a good market for hides. In Demosthenes' speech *Against Phormio* the speaker describes a shipwreck in which a trader leaving Bosporos lost his entire cargo, including one thousand hides destined for the Athenian market.[64]

When measured by separate sacrifices, the returns were not enormous. The festival of Dionysos in the city produced 808 drachmas in 334/3 and 251 in 333/2, while the Asklepieia brought the city only 291 drachmas. Furthermore, the revenues paid into the *dermatikon* varied from year to year because some major festivals were penteteric, but the endless cycle of sacrifices and festivals meant that the *dermatikon* quickly became a major source of public funds. What Lykourgos managed to do was to create a regular income stream for the state from all major state-sponsored sacrifices. Whichever group administered a specific celebration, the expectation now was that the sacrifice had to pay back a proportion of the money expended by exploiting for profit the non-consumable portions of the sacrifice. In 333/2 BC, for example, in the archonship of Nikokrates, payments into the *dermatikon* were made by the generals, by the *hieropoioi*, by the *epimelitai*, by the *syllogeoi*, and by the *boônai*. Annual profits accruing to the city could reach close to two talents per year.[65] This was far less than the total money paid out for all animals, but at the cost of approximately 70 drachmas per head, two talents would have paid for the purchase of more than 170 head of cattle. Since they in turn produced revenue, the system demonstrates the moving of the sacrificial system into the realm of economic activity subject to scrutiny, control, and regulation. Sacrifice was becoming business.

Thanks to the economic implications of sacrifice and the value of perquisites, priesthoods themselves became the subject of commercial transaction. The evidence for this prior to 400 BC is minimal, but thereafter there are indications that a trade in priesthoods, at least in the Greek cities of Asia Minor, became more common.[66] Those wealthy enough to do so were attracted to the idea of buying priesthoods. Why? Certain priesthoods were inherited and conferred the status of tradition. Some of the oldest families in Athens prided themselves on their ancestral domination of particular cults: the Boutidai and the cult of Erechtheus, the Eumolpidai and the cult of Demeter, and the Buzygai and the Bouphonia are good examples. In some cases the sale of priesthoods may have arisen in order to satisfy the desire of those outside old families to enjoy similar honors. On Samothrace a certain Timaios, son of Straton, was prepared to pay "seven gold and sixty bronze" for the priesthood of the gods for life.[67]

In addition to conferring status, priesthoods brought economic advantages. This can be seen in a detailed set of inscriptions from Hellenistic Kos dealing with the sale of priesthoods to a number of deities, including Herakles Kallinikos (ED 238), Dionysos Thyllophoros (ED 216), and Aphrodite Pandamos and Pontia. The inscription recording the sale of the priesthood of Aphrodite is particularly revealing.[68] It is envisaged, for example, that sacrificers may substitute cash payments for perquisites: "any sacrificer who wishes may pay, in lieu of perquisites for any of those victims that it is permitted to sacrifice to the goddess, the sum specified" (lines 3–5). This clause is intended to resolve disputes "between any two persons on any grounds about the value of sacrificial victims" (lines 2–3), but also marks the transformation of the priest's role from a sacred performance to a paid service. At the same time, the regulations also stipulate a fixed rate of return to the priestess, whether sacrificers choose to pay in kind or in cash. The crews of warships, divided into eating groups by tents, are each required upon their return to offer a sacrificial victim, probably a sheep, worth at least 30 drachmas. They are then given the option of either leaving the priestess her perquisites or paying 15 drachmas. Since the usual perquisites of leg and hide would not amount to half the value of the animal, the cash equivalent appears punitive, unless the victims were worth a good deal more than the minimum price fixed by the decree.[69] If so the priestess' perquisites would also have been worth more than the low cost of the sacrificial animal would suggest. Alternatively, the stipulation may have been designed to encourage the traditional system of gifts, in which case the priestess was guaranteed a steady supply of fleeces that she was free to sell. Fixed contributions are also set at the rate of 2 drachmas per head of cattle, 1 drachma for smaller full grown animals, 3 obols for young animals, and 1 obol for birds (lines 11–12). These are described as first fruits (*aparchai*) for the treasury (*thesauros*) of Aphrodite, and it might be expected that all revenues listed belonged to the goddess. Immediately after listing these revenues, however, the *diagraphe* (contract) provides for the leasing of all these revenues by the priestess (lines 13-15). What business had the priestess leasing out revenues that were the goddess', and why would a lessee pay for a lease to collect revenues that went to the goddess? The answer is provided in line 18, in which we find that the *thesauros* was opened once per year by the *prostatai* and that half the revenues were immediately paid to the priestess.[70] It was entirely in the interests of the priestess of Aphrodite to see as many sacrifices as possible transformed into cash transactions. Ex-slaves were required to pay 5 drachmas in the year they were freed; fishermen and merchants were similarly taxed at a rate of 5 drachmas per year (lines 25–29). Half of this money went into the public bank of Kos, and was designated for the upkeep of the sanctuary (19–22). The only individual to profit from the entire scheme was the priestess, and, to a lesser extent, her lessee.

Since those purchasing priesthoods were required to begin paying in installments from the moment their purchase was approved, the system favored the wealthy who could afford to treat priesthoods as investments for themselves and their families (the priesthood of Asklepios was available to men as young as fourteen).[71] Kleoneikos, son of Eukarpos, for example, purchased the priesthood of Adrasteia and Nemesis on Kos for the extraordinary sum of 19,800 drachmas.[72] For an outlay much higher than ordinary people could afford, the purchaser essentially secured half of the cult's future revenues. There were other advantages as well, as is shown by the lengthy set of regulations for the priesthood of Hermes Enagonios:

> Let the purchaser of the priesthood of Hermes Enagonios be healthy, sound in limb, and let him be priest for life. Let the purchaser of the priesthood be exempt from military service overseas, from cavalry upkeep, from maintaining a trireme, from producing a chorus, from arranging a lamplight parade, and from every other type of liturgy. (*Iscr. di Cos* ED 145.A.5–8)

The authority granted priests also served to reinforce their social standing. The priestess of Aphrodite, for example, was empowered to exact payments "as if in fulfillment of a legal verdict" (*kathaper ek dikas*).[73] As religious activity was more fully institutionalized, social prestige and economic advantage came to outweigh piety.

As if in response to their desanctification, during the Hellenistic period priesthoods were folded into the burgeoning system of benefactions and honorifics. They remained in the control of the wealthy, but increasingly served to advertise little more than social status. At Ilion in the second century, Hermias son of Skamandrios published his generous donation of 15,000 Alexandrian drachmas to the city and referred to himself as "priest of all the gods."[74] In some cases the priesthood was part of a larger package including the very cult space itself, raising the priest to the status of founder and permanent beneficiary as well as benefactor, as a first-century BC inscription from Gythion reveals. Philemon and Theoxenos paid out of their own pockets for the restoration of a temple of Apollo in the town's agora. In return they were acknowledged for their benefaction and were declared priests of Apollo for life, while the temple was transferred to them and their descendants, to be known thereafter as "the

Apollo temple of Philemon, son of Theoxenos, and Theoxenos, son of Philemon."[75] In other cases, a priesthood was merely one of a dozen honors voted to a member of the liturgical class. Diodoros, son of Socrates, of Aphidna was honored in the first century by the members of his religious group, the *koinon* of the Soteriastai, and the stages of his career read like the *cursus honorum* of a Graeco-Roman rotarian: *archeranistes* (head of the guild), *tamias* (treasurer), and priest of Soteira.[76] He is praised in the conventional language of such inscriptions for his generosity and selflessness. His priesthood differs hardly at all from being declared a *euergetes* (benefactor).

The spheres of religion and the economy also grew closer together in the development of religious festivals as periodic fairs. Cities all possessed an agora, the central location where ordinary commercial activity took place, but certain types of commercial transaction, such as the selling of cattle and slaves, were more likely to take place in the setting of a religious festival.[77] The "profane" side of a festival is customarily termed a *panegyris*, and the association between this and market activity is a persistent one. Pythagoras, for example, famously compared life to attending a *panegyris*.[78] This is the version told by Iamblichos:

> Pythagoras said that people approached life like the crowds that gather at a festival. People come from all around, for different reasons: one is eager to sell his wares and make a profit, another to win fame from his physical strength; and there is a third kind, the best sort of free man, who comes to see places and fine craftsmanship and excellence in action and words, such as generally are on display at festivals. (Iamblichos, *de vit. Pythag.* 58, trans. Clark)[79]

Clearly this is not meant to be taken as a complete definition of a *panegyris*: the omission of any mention of a sacrifice or, in fact, of any religious activity at all confirms this. The picture Pythagoras paints is one in which the festival is as much a fair as a religious event. This has occasionally led to the mistaken belief, asserted for example by Louis Robert, that *panegyris* refers specifically to the commercial activity conducted during the festival. This has been rightly rejected by Luuk de Ligt, but the mistake is itself telling.[80] In the Hellenistic and Roman periods sacred laws and sanctuary regulations from Greece and the eastern Mediterranean frequently refer to the market activities associated with festivals. We encounter officials with titles such as *panegyriarchai* and *agoranomoi*, whose jobs are to take charge of the complicated logistics of festivals. These officials could inspect merchandise, let booths to itinerant traders, and issue instructions on such matters as prices and exchange rates.[81] So in these later periods festival and fair went hand in hand. A detailed first-century BC inscription from Andania, for example, stipulates under the heading "Agora" that the priests are to delineate an area where goods may be bought and sold, and the city's *agoranomos* is to take charge of the trade conducted in this temporary space and see to it that the same weights and measures are used as in the public market.[82] Yet in the fifth century it was customary to employ *panegyris* as a term referring to a major festival but without any particular emphasis on the festival *as* fair. Aeschylus refers to establishing common games at festivals (*koinous agonas thentes en panegyresi*), while Thucydides explains the hatred of the Corinthians for the Corcyrians as arising from their failure to offer the customary gifts at their shared festivals (*oute gar en paneguresi tais koinais didontes gera ta nomizomena*).[83] The difference between these writers and the later uses is not a matter of Classical authors avoiding the term's commercial overtones. Rather, the difference points to the fact that commercial activity had not yet come to dominate religious festivals, at least in the popular imagination. When Herodotos, for example, uses *panegyris* and its cognates to describe Egyptian festivals, he refers to specific aspects of the festival such as processions and sacrifices, but he has nothing to say about commercial transactions.[84] His contemporaries also used *panegyris* as a general designation for the festivals at Olympia, Isthmia, and the other Panhellenic games in contexts that have nothing to do with market activities.[85] It would seem that in the fifth century a *panegyris* was a public festival that included games, processions, and sacrifices. The crowds attending such festivals were there to make sacrifice and to enjoy the sights and tastes associated with a festival, and so the festivals came to take on a special economic significance, both requiring supplies of all sorts of food, and providing opportunities for all sorts of transactions, beginning with animals for sacrifice, but including an increasingly wide range of goods. Demosthenes' remark that "[h]aving a lavish supply of goods is the mark of a *panegyris* or a market, not a city" shows that by the middle of the fourth century the connection between piety and consumption was hard to miss.[86] We have seen how sacrificial practices continually reinforced the association between prestigious consumption, piety, and commensality. The commercialization of festivals in the Hellenistic period simply extended this process as the economic activity associated with festivals grew increasingly important.

Seasonal and regional fairs crystallized around gatherings that previously had had primarily a social and religious significance. The difference between the daily commerce of the agora and the economic activity associated with festivals is illustrated in three illuminating passages. In *Peace*, Aristophanes has the hero Trygaios invoke the eponymous goddess in typically exaggerated terms. He begs the goddess,

> Cause the Greeks once more to taste the pleasant beverage of friendship and temper all hearts with the gentle feeling of forgiveness. Make excellent commodities flow to our markets, fine heads of garlic, early cucumbers, apples, pomegranates and nice little cloaks for the slaves; make them bring geese, ducks, pigeons and larks from Boeotia and baskets of eels from Lake Copais; we shall all rush to buy them, disputing their possession with Morychus, Teleas, Glaucetes and every other glutton. Melanthius will arrive on the market last of all; they'll say, "no more eels, all sold!" and then he'll start groaning and exclaiming as in his monologue of Medea, "I am dying, I am dying! Alas! I have let those hidden in the beet escape me!" And won't we laugh? (Aristophanes, *Pax* 996–1016, trans. O'Neill)[87]

As a fantasy of the blessings of peace this is a singularly sober vision. One is (inevitably) tempted to ask: Where's the beef? Specialty goods from Boiotia, delicacies for Athenians as luxurious as Iranian caviar for us, are back on the menu, but steaks and spareribs, it seems, are in another category. When his fantasies turn more carnal, it is of a festival that Trygaios begins to dream. Groping the goddess' pretty attendant, Theoria, whose name is identical to that of an official embassy sent by the state to attend a Panhellenic festival, Trygaios swoons,

> Oh! fortunate Senate! Thanks to Theoria, what soups you will swallow for the space of three days! how you will devour meats and cooked tripe! Come, farewell, friend Hermes! (Aristophanes, *Pax* 715–18, trans. O'Neill)

Here then, is a distinction understandable to an Athenian audience of the late fifth century. The daily market is for profane goods. Meat remains primarily a commodity within the sacred economy. In the Hellenistic and Roman periods, festivals and the fairs connected to them would grow in number, frequency, and importance, especially as occasions for trading those goods that were more costly than the usual wares on offer in the agora. The *panegyris* would become a more prestigious market, in keeping with the greater standing of religious affairs in relation to secular. This is reflected in Pausanias' description of the spring and autumn festival to Isis at Tithorea in central Greece. His account begins with the collection of any remains from the previous season's sacrifice and continues:

> On the next day the traders make their booths out of reeds or any other material handy. On the last of the three days they hold a fair, selling slaves and cattle of every kind, and even clothes and gold and silver; then from midday they turn their attention to sacrificing. The more

> prosperous people slaughter cattle and deer, the less well off even slaughter geese and guinea hens . . . (Pausanias 10.32.9, trans. Levi)[88]

The third day at Tithorea saw the trading of slaves and cattle, the two most expensive commodities regularly sold. In the afternoon, some of the cattle ended up on the goddess' altar, and by that evening will have supplied a feast for those attending. It is because of the dramatic focus on consumption at festival time that the sanctuary of Athena Alea at Tegea relaxed its regulations on the use of sacred land for grazing at festival time. The regulations read: "During the last three days of the three major festivals they may graze wherever they wish, provided no animal is pastured in the concourse."[89]

Athens was not the only city to see sacrifice generate market activity. Others followed the same model of procession, sacrifice, and *kreanomia*. At Eretria on Euboia, the structure of polis-sponsored sacrifice imitates the Athenian process even down to the wording of the preamble:

> Exekestos, son of Diodoros, moved the motion: so that we may celebrate the Artemisia as beautifully as possible and conduct as sumptuous a sacrifice as possible, it was decided by the Council and Demos as follows . . . (*IG* XII 9.189.2–4)

In the stipulations that follow, after regulating the prizes for the various musical contests, Exekestos' bill takes up the question of sacrificial animals. As at Athens, the Eretrian supply combines animals given by subordinate communities, in this case villages outside the city, with cattle purchased from private suppliers:

> Let the villages supply select (*krita*) sacrificial animals, cattle, every year. Let the villages contribute to the purchase of select sacrificial animals as they do for the *Heraia*. Let those who provide the select sacrificial animals receive their hides. The overseers of sacred matters are to evaluate the sacrificial animals according to the law and to pay for them, should any of the villages fail to provide [their quota]. Whoever wishes to may sell in the sanctuary whatever animal he wishes to, provided he puts no imperfect animal up for sale, nor are the *hieropoioi* to make any kind of payment at all or the sellers to try to sell any animal that is not perfect. Let the demarchs arrange the procession in the agora where the sacrificial animals are sold, beginning with the public animals and the most beautiful ones, followed by the selected ones [from the villages], followed in turn by the animals offered as private sacrifices, if anyone wants to participate in the procession. The contestants in the musical games are to participate in the procession so that both the procession and the sacrifice may be as beautiful as possible. (*IG* XII 9.189.26–40)

It is noteworthy that the law calls for the villages to supply Eretria, assumes that a coherent set of obligations has already been spelled out for another festival, the *Heraia*, yet also acknowledges that villages may (and therefore probably did) fail to meet their obligations.[90] To that end, the officials in charge of sacrifice were empowered to resort to private sellers, provided that no animal that failed to meet the stringent test of perfection was purchased.[91] As at Athens, the commercialization of the meat supply arose in response to pressure on the sacred economy to keep up with demand.

Similarly, a set of regulations from the sanctuary of Artemis Kindyas from Bargylia in Asia Minor is concerned with the dedication of a silver deer to the goddess and distribution of meat for a public feast in her honor. The inscription begins with stipulations, unfortunately badly mutilated, that deal with the breeding of cattle for sacrifice.[92] The cattle were presented publicly to the demos for approval in April after they had been fattened for sacrifice over the course of the preceding months. The *boutrophoi* (the officials responsible for the animals' care) are recompensed for taking care of the cattle, but their payment is never referred to as a sale. It is difficult to know whether this points to the existence of a sacred estate belonging to Artemis, in which case the *boutrophoi* were like professional cowboys tending the cattle of the goddess, or whether the payment of the *boutrophoi* was simply the purchase of their cattle rendered in terms that recast a financial transaction as a sacred one.[93] In either case, the involvement of the entire community in regulating how the goddess was to receive her sacrifice (and the community its meat) resembles the Athenian practice of treating stock raising as a practice that straddled private, public, and religious spheres.

CHAPTER 9

Sacred Law

> These are the four laws of Erin: Patrick's Law, not to kill clerics; and Adamnan's law, not to kill women; Dáire's law, not to kill cattle, and the law of Sunday, not to transgress at all.
>
> —*Martyrology of Oengus the Culdee*[1]

Custom and Law

In a culture whose political development saw the polis emerge as the dominant form of state organization, sanctuaries played a central role in state formation. Within the city and its territory they provided locations where social relations were manifested through sacrifice and feasting. The feast, in particular, emphasized group solidarity and inclusion. Other institutions, such as the sacrifice, drew attention to hierarchies of wealth and prestige, in which benefaction and reciprocity were channeled through pious actions to the credit of the donor and the benefit of the community. Panhellenic sanctuaries were the settings for similar performances, often on a vastly greater scale, as when Jason of Pherai ordered the transportation of thousands of sacrificial animals from Thessaly to Delphi in order to celebrate the Pythian Games.[2] The athletic contests that took place at sanctuaries in the Archaic period allowed a distinctive pattern of elite competition to take shape before becoming the vehicle for assertions of community pride. Only Croton could boast of a wrestler like Milo, who, after carrying an ox up and down the stadium at Delphi, proceeded to eat the entire beast.

Yet, as we shall see in this chapter, the role of sanctuaries went further. In the Archaic state in particular, authority was a deeply contested issue. I refer not only to political power and the question of elites and masses, but the more intangible matters of value, meaning, and influence. Sanctuaries profited from a unique position as places vested with authority because of the presence of the gods. That status allowed the distinctive institutions of the sanctuary to become the dominant institutions of the polis: sacrifice, feasting, and dedication constituted a nexus of practices that led to prescriptions on ritual behavior, which in turn served as a model for lawgiving. The cattle complex, we shall see, was not a relic of a distant

Greek pastoral past, but a formulative feature of the mentalité of the Greeks that continued to shape their development.

We can begin to chart the importance of sanctuary-driven institutions by comparing the structure and organization of sanctuaries with that of poleis. From the point of view of the activities that took place there, sanctuaries were clearly special locations. It was here that humans performed the business of sacrifice that forged a connection with the divine, and the special status of sanctuaries derived not only from their function as places of competition and mediation, but also from the fact that the economic activity focused on sanctuaries is distinctly different from what took place in the private or secular economic sphere. They possessed some of the best land in Greece, and their sacrificial requirements skewed the normal relationship between land, cultivation, and diet, placing much greater emphasis on herding and meat production. As a result of dedications they often possessed enormous wealth, yet were much more vulnerable than towns equipped with armies and walls; they regulated the use of land through leasing arrangements that could be carefully overseen by administrators, often drawn from families who passed priesthoods down from generation to generation. The administration of sanctuaries provided a model that was copied even in the most humble sacrificial groups. Yet, in virtually every way, the sanctuary operated according to a set of social relations at odds with those of the polis. Where the Greek city so often succumbed to *stasis*, the sanctuary was rarely riven by internal discord.[3] Where cities had contradictory and shifting answers to the question of who is to be included in the body politic and to what degree, sanctuaries operated according to clear rules: they belong to the god (or hero).[4] Contrast this with the situation at Athens, where women could not be full citizens, but where after Perikles' law of 451/0 BC a citizen's mother had to be a freeborn Athenian for the child to inherit citizenship; or the situation at Gortyn, where the woman could fully inherit property and dispose of it as she wished, even marrying a slave and bearing free children. Civil society tussled with the question of eligibility for high office, but in sanctuaries priests were either hereditary or appointed. Smaller cult associations were either closely affiliated to community groups such as demes and *phylai*, and were run by officers such as demarchs and phylarchs, or were constituted from fee-paying members and appointed their own *archeranistai* or presiding magistrates. Sanctuaries and cult associations were free of the dilemmas that fueled democratic and oligarchic revolutions, which so often concerned restrictions, or the lifting of restrictions, on the sheer number of participating citizens. In smaller cultic associations either blood ties or elective affinities defined inclusion. At Delphi, potentially anyone could make a dedication and consult the oracle, an egalitarianism reflected in the questions so often posed to the oracle in Plutarch's day:

not weighty state issues such as "Should we fight the Persians?" but private and individual matters such as "Should I get married?" or "Should I lend money?"[5] At Dodona, the questions posed to Zeus were often as mundane as "Where is my missing cup?" and as personal as "Is my son safe?"[6] So, while the democratic assembly opened with the words, "Who wishes to speak?" but really favored those with professional oratorical skills, it was actually consultation of the gods that was truly open to all. In fact, while a democratic body such as the *ekklesia* was open only to adult male citizens, participation at a major cult spot such as Eleusis was unrestricted: even women and slaves could be initiated.

Paradoxically, sanctuaries were also unambiguously authoritarian. Consider, for example, a sixth-century sacred law from Megara Hyblaia: "For all the sacred law of the god as follows: whoever makes a sacrifice contrary to the wishes of the *archaomos*, let him pay back an eighth."[7] The law does not need to cite the will of the people or the council. It is the god's law, even though it is only concerned with procedural matters rather than any general principle of piety or holiness. The official presiding over the sacrifice, the *archaomos*, has authority to determine whether the sacrifice is correct or not, and if the range of his power is not exactly equivalent to that of a tyrant, in its own field of operation it is nevertheless absolute. One could appeal to the god for a more favorable oracle, but his decision was not subject to debate by assembly or judgment by jury. Accordingly, the god's servants possessed an authority that was based not on compulsion, but on the visitor's willingness to accept that the priest was backed by the god's sanction. At Olympia a set of judges had the final say in determining whether one was Greek (and might participate) or was not (and could not). Compared to the relative clarity of participating within a sacrificial group or attending a sanctuary, citizenship was a much more complex matter, and legal cases from the fifth and fourth centuries reveal a widespread anxiety about who could legitimately claim to be a member of the citizen body. In the case of Neaira's daughter and her marriage to the Archon Basileus, the sources of outrage were that she was not, in fact, the daughter of an Athenian, that her marriage was the result of a deception, and that she had profaned holy rites by performing actions that were supposed to be done by the legitimate wife of the Basileus.[8] The underlying problem was not impiety in a strictly religious sense, so much as the fraudulent claim of citizenship. Establishing citizenship required that one was presented as an infant by one's father to his demesmen, and reintroduced as a young man.[9] Accordingly, a regular tactic in legal cases was to cast doubt on the citizenship status of the opponent, while suggestions that a woman had lied about her citizenship status were a reliable way for an orator to prejudice a jury.[10] For the Greeks, "bastard" was a label with stronger political than moral connotations.[11]

The distinction between divine affairs, in which the lines of authority are clear, and secular matters, where authority is contested, is critical to the shape of polis culture. When an early law such as Drakon's law on homicide allows the members of the victim's phratry to pardon a killer who has committed murder unintentionally, the law derives its authority not from the state, which in the seventh century was still weak and had no formal procedure for prosecuting a murderer, but from the power of traditional kinship structures and religious ties.[12] This is only one instance of an underlying feature of early Greek law: it is sacred law, supported by belief in the gods shared by groups bound together as sacrificial partners, and promulgated by religious officials, that provides the basis for the development of legal procedures covering the community at large. The laws of Archaic Greece, in fact, emerge from the milieu of religion. This is borne out by the treatment of homicide in the Athenian courts—once again our best-documented cases—in a manner more concerned with ritual than forensics.[13] Oaths were more important than investigation, and were frequently sworn on the entrails of a sacrificial victim, as if the fear of perjury were a more reliable test of innocence than any dispositive evidence.[14] Susan Guettel Cole has shown that the most suitable animals for oath sacrifice were adult male victims, usually of three different species, and that the concept of an oath victim goes back to Homer. She notes that "[t]he ritual required a correlation between adult male animals and the males who swore because the destruction of the animal represented the enormity of the conditional curse implied in the oath."[15] Appeals were made to written legal texts for the weight of precedent they brought to a case rather than for their probative value, recalling the weight placed on the authority of *exegetai* and *hieromnemones*, religious officials charged with expounding sacred matters such as the proper conduct of a funeral.[16]

The distinctions made between different types of homicide also point to a religious sensibility keenly attuned to the horror of pollution, and it is from this religious outlook that a legal sensibility emerges. This is demonstrated by the treatment of cases of intentional homicide involving those who had gone into exile or had received pardons for earlier offenses. Such cases were overseen by the *ephetai* and held near the seashore at Phreatto. In an odd but revealing twist to these cases, the accused was required to attend offshore, in a boat.[17] Raphael Sealey has plausibly argued that the *ephetai* continued to have jurisdiction over what must have been a relatively small number of such cases after the principal instances of intentional homicide had been transferred to the Areopagos.[18] In other words, the odd procedure at Phreatto, as a relic of earlier practices, can serve as an indicator of the mindset surrounding the original judicial proceedings for intentional homicide. The concern for keeping the accused isolated reveals that the underlying desire was to protect the community's land from pollution.[19]

Even though the *ephetai* retained competence in all non-Areopagos trials, that is, in all cases not involving premeditated murder by an Athenian citizen, they did not hear all these trials in the same place. The different locations of these ephetic courts also shed some light on their religious associations. Trials involving the deaths of metics and slaves, as well as involuntary manslaughter, were held at the Palladion. Cases of justifiable homicide were heard at the Delphinion, while animals and inanimate objects were tried at the Prytaneion. Evidently the Athenians attached significance to the location of the trial and saw it as reflecting some important aspect of the crime. In the case of the Palladion, its ambiguity is appropriate. On the one hand, it houses a cult of the city's tutelary deity, Pallas Athena, yet the temple itself lies outside the walls of the city, making it a fitting setting for trials involving victims, metics, and slaves, whose status also puts them outside the full Athenian community.[20] The decision to put trials for involuntary manslaughter here, reflected in the aetiological story of Demophon, who inadvertently killed another Athenian, also reflects the ambiguity of the charge: the defendant is potentially guilty of wrongdoing, but not of the intention of doing wrong.[21] At the Delphinion those claiming justifiable homicide were tried in a setting that recalled Apollo of Delphi, who had justifiably slain the monstrous Python.[22] The Prytaneion is, at first sight, the one location that seems to have no religious association, since it is the building that housed the serving *prytaneis* of the *boule*, but here too, a cultic connection fits the class of homicide. Just as animals and weapons, themselves incapable of intention but capable of causing a death, were tried here, so too the ax used by the first bull slayer in the cult of Zeus Polias was tried here.[23] The religious origins of Athenian law may also help explain the peculiarity of Athens' having had two lawgivers: Drakon and Solon. Drakon's name, as Karl Beloch noted nearly one hundred years ago, means "the Serpent," and recalls the many associations between Athena and the sacred snake worshipped on the Acropolis, itself an avatar of Kekrops and Erichthonios.[24] It is entirely plausible that the deep sanctity associated with the snake was applied to traditions surrounding Athens' first lawgiver, and that the earliest Athenian law known to us, itself an object of veneration to the Athenians, was originally a sacred law of homicide.[25]

The thesis that sacred law provides the foundation for civil law in the early Greek states was first argued eighty years ago by Kurt Latte, who believed that the Panhellenic oracles habituated the Greeks to a fundamental notion of law and order.[26] More recently, Adalberto Giovannini has described piety (*eusebeia*) as the "primordial virtue" of the Greeks, and has argued that the Greeks saw civilized life, governed by agreed customs and laws, as arising from a fear of the gods and their power to punish wrongdoers.[27] In a set of regulations from Cyrene in the fourth century, for example, the preamble reflects that the very detailed rules by which the community was governed originated with Apollo:

> Apollo declared by oracle: dwell in the land of Libya forever, performing purifications, abstinences, and paying tithes. (*SEG* IX 72)

The regulations concern pollution, marriage, participation in Cyrenaian cults, tithes, and intestate inheritance law. In short, the entire fabric of social, religious, and legal life was codified in accordance with a mandate from Apollo. The idea that state law ultimately derived from the gods was also expressed by Demosthenes in his prosecution of Meidias for assaulting him at the Dionysia. He appeals to the jury:

> You surely realize that all your choruses and hymns to the god are sanctioned, not only by the regulations of the Dionysia, but also by the oracles, in all of which, whether given at Delphi or at Dodona, you will find a solemn injunction to the State to set up dances after the ancestral custom, to fill the streets with the savour of sacrifice, and to wear garlands. (Demosthenes 21.51, trans. Murray)[28]

Latte attributed the influence of sacred law to the weakness of state authority in the early Greek state, placing special emphasis on the state's inability to enforce decisions. Sacred laws, by contrast, reinforced what was often customary ritual practice, emphasizing the procedures that were necessary for maintaining the goodwill of the gods. They were efficacious because they were bolstered by a greater authority. Oaths sworn to the gods, for example, were more powerful than a judge's decision, since those who perjured themselves were guilty, in Xenophon's words, of waging war on the gods.[29] In other words, religion provided what the early state lacked: a model for judicial action that went beyond the principle of self-help. Sacred law was sanctioned by authority and defined by procedure. The influence of ritual and sacred law remained powerful. Even into the Classical period, Greek law was always more concerned with procedure than with underlying legal principles.[30]

The importance of religious procedure to the emergence of law is reflected in Hesiod's admonitions in the *Works and Days*, where we encounter a concern for actions that are proper both in the sense of being ethically right and ritually correct. Some of these dictums take a religious event such as a sacrifice as the occasion for making a point about piety, such as the notion that one should not cavil at seeing good meat go up in flames on the god's altar:

> If you are present when offerings are ablaze on an altar, do not carp at what is consumed; this too the god rather resents. (Hesiod, *W&D* 755–56, trans. Robertson)[31]

Here the moral of the injunction is primarily concerned with the individual's personal response to the sacrifice. Do not be niggardly about sacrifices, says the moralist, because what the flames prevent you from consuming the gods will get, and they don't appreciate your parsimony. But in other cases the underlying moral is less important than the external behavior that marks the ritual, as for example in the injunction to cleanliness:

> Never pour a libation of sparkling wine to Zeus after dawn with unwashed hands, nor to others of the deathless gods; else they do not hear your prayers but spit them back. (Hesiod, *Op.* 724–26, trans. Evelyn-White)[32]

In fact, what Noel Robertson calls the "counsels of prudence" in Hesiod are as much about correct ritual behavior as they are about morality, and include a list of taboos devoid of moral significance: do not urinate facing the sun; do not have intercourse following a funeral; do not cross a river without washing and praying; do not cut the withered from the quick on a bough with five branches; do not use water in which a woman has washed if you are a man. What unites the injunctions is an underlying concern for prescribing the behaviors that constitute right action and lead to prosperity.

The rituals prescribed in the *Works and Days* that work between men and gods are also a model for the relations between humans:

> Be careful to avoid the anger of the deathless gods. Do not make a friend equal to a brother; but if you do, do not wrong him first, and do not lie to please the tongue. But if he wrongs you first, offending either in word or in deed, remember to repay him double; but if he ask you to be his friend again and be ready to give you satisfaction, welcome him. He is a worthless man who makes now one and now another his friend; but as for you, do not let your face put your heart to shame. (Hesiod, *W&D* 706–14, trans. Evelyn-White)

In ethical terms, this can be read as a Greek version of the universal principle "Do unto others," although the culture that recognized two kinds of *eris* may be said to have inflected this commandment uniquely: "Do unto others, *precisely as* they do unto you." Yet Hesiod's advice is notable not just for its validation of vengeance ("remember to repay him double"), but also for the fact that his advice reflects piety interpreted by a legal consciousness: the injunction to repay the offender double is also a statement of the legal principle of self-help, while the advice to welcome the offender if he is willing to give satisfaction is equally an endorsement of compromise and settlement. Hesiod's poetry reflects a society taking the first steps toward developing a structure for settling disputes in an orderly fashion, but at a time when central authority was more personal than institutional it was the realm of the gods that offered guidance as to how human justice could be accomplished.

It is this combination of divine and human concerns that Hesiod explores in his direct advice to the *dorophagoi basileis* (bribe-eating princes). He bids them to render straight judgements, not only because that is the right thing in itself, but because the gods are at hand, watching:

> You princes, mark well this punishment you also; for the deathless gods are near among men and mark all those who oppress their fellows with crooked judgements, and reck not the anger of the gods. For upon the bounteous earth Zeus has thrice ten thousand spirits, watchers of mortal men, and these keep watch on judgements and deeds of wrong as they roam, clothed in mist, all over the earth. And there is virgin Justice, the daughter of Zeus, who is honoured and reverenced among the gods who dwell on Olympus, and whenever anyone hurts her with lying slander, she sits beside her father, Zeus the son of Cronos, and tells him of men's wicked heart, until the people pay for the mad folly of their princes who, evilly minded, pervert judgement and give sentence crookedly. Keep watch against this, you princes, and make straight your judgements, you who devour bribes; put crooked judgements altogether from your thoughts. (Hesiod, *W&D* 706–14, trans. Evelyn-White)

More than a reflection of archaic piety, this passage shows a Greek world on the cusp of the momentous shift from a prelegal age, in which decisions were uncodified expressions of power coming from the elite, to a time when a formal apparatus would allow the orderly promulgation of laws. A clearly defined procedure for handling disputes is about to emerge.[33] Yet that human justice evolves not separately from the justice of Zeus, but with divine justice as its model and the source of its authority. The thousands of spirits acting as divine enforcers to whom Hesiod makes reference are there both to witness any evildoing, but also any crooked judgements. The spirits of vengeance are keeping an eye on both wrongdoers and judges. This suspicion of those with the power to render judgement has led some scholars to argue that the codification of laws in Archaic Greece arose from the tension between elites and non-elites. But more likely than an incipient class tension in the eighth and seventh centuries is a more diffuse unease occasioned by the growth of stratified societies.[34] Early Greek lawcodes, though often associated with the name of a single lawgiver, are not constitutions laying out a coherent set of procedures and reflecting a common set of legal principles. Most Archaic laws were limited in scope and specific in application.[35] So it is a mistake to treat the codification of Archaic law as equivalent to the drafting of a constitution or to imagine the Archaic Greek communities as characterized by rigid legalism.[36] The process is, however, important because of the cumulative effect of such lawmaking. It habituated Archaic Greek society to the rule of written law.[37] Even if most early laws were about specific matters such as the amount one could spend on a funeral, and were more often concerned with procedures rather than underlying principles, the practice of regulating so many areas of life by passing laws inculcated the habit of law. It was only later, in retrospect, that Solon, Lykourgos, Zaleukos, and the other semimythical lawgivers were elevated to the status of constitutional founding fathers.[38]

For the rule of law to take hold a further step is necessary: the orderliness it provides is only possible if there is no party within society above its reach. This is explored by Hesiod in both the *Theogony* and *Works and Days*, in which the triumph of Zeus and the emergence of the Olympian order suggest a model of the social contract. Just as each of the gods has their *timai* (honors) and attributes, so too society is articulated into different trades and roles. Humans each have their *kleros* (lot). All in heaven and on earth are subject to the will of Zeus, who himself is constrained to recognize fate and destiny. All is orderly, at least potentially so, as long as those in authority recognize that their power is not unlimited. It is this dilemma, the tension between power and responsibility, that makes the death of Sarpedon in Book 16 of the *Iliad* not only poignant but a crucial test of Zeus. He has the power to save Sarpedon, but he must not, if he is to distinguish himself from a tyrant. That Zeus *chooses* to recognize the disastrous consequences should he intervene to save Sarpedon serves as a model for princes: even supreme power must not act arbitrarily. Hesiod develops the notion of constraint further: there is a correct way of doing things, which through repetition becomes customary and is an antidote to the danger of arbitrary authority.[39] *Themis* gives rise to *nomos* (in its Herodotean sense of habitual practice). In turn, *nomos* in the sense of "law" is fundamentally rooted in the earlier sense of *nomos* as custom.[40] For that reason, Greek law arises in a religious setting and can never be divorced from that setting.

The continuing importance of conventional piety in shaping law can still be seen in a legal and political dispute three hundred years after Hesiod, by which time laws have been written down and procedures codified. According to the speaker of Lysias' *Against Nichomachos*,

> But of course, gentlemen of the jury, we are not to be instructed in piety by Nicomachus, but are rather to be guided by the ways of the past. Now our ancestors, by sacrificing in accordance with the tablets, have handed down to us a city superior in greatness and prosperity to any other in Greece; so that it behooves us to perform the same sacrifices as they did, if for no other reason than that of the success which has resulted from those rites. How could a man show more piety than mine when I demand, above all, that our sacrifices be performed in accordance with ancestral custom. . . . (Lysias 30.18–19, trans. Lamb)[41]

Even after decrees have been passed and laws written down, what the speaker appeals to is not the law *as written*, but rather the Hesiodic notion that good fortune arises from right action. The written law is evidence of that piety, but the speaker recognizes that writing allows for debate and the possibility that Nichomachos will act as an exegete, interpreting the law to his own ends. Trumping such chicanery is ancestral custom. Understood this way, written laws are an instrument for regulating right action and thereby guaranteeing prosperity, but are less important than the procedures themselves, in this case sacrifices, that the laws record. The law only points to what is already customary. In one sense, Lysias is juxtaposing laws with custom, seeing the former as powerful yet historically contingent (based on decisions made by the people and written at one particular time), while seeing in custom a power that transcends history, since it is capable of connecting the present Athenians to their ancestors through the repetition of the same actions. One might be tempted to dismiss Lysias' words as empty rhetoric, but it is worth remembering that if all legal speeches of the Classical age are to some degree tendentious, this is also counterbalanced by a powerful pressure to be normative; a speech could not convince a mass jury if it did not express conventional sentiments. Lysias' audience took for granted that their laws arose from their religion, or, as Herakleitos put it, "All human laws are nurtured by one divine law."[42] In a similar vein, but less generously, Plato compared the efforts of human lawgivers to legislate a just society to beheading the Hydra. Compared to their well-intentioned but feeble efforts, it was the authority of religious law exemplified by Apollo's oracle at Delphi that constituted "the greatest, the finest, and the primary instance of lawgiving." Only Apollo was capable of legislating concerning the most serious matters: "the founding of temples and sacrifices, and the rest of the services appropriate to the gods, heroes, and spirits."[43] Those who championed the authority of the state and those who dismissed it could at least agree that the gods' laws came first, and that the first obligation of the individual or the state was to act in accordance with divine law.

The connection between piety and law was strengthened by the fact that the Greek construction of piety was based on performance rather than feeling. Convention was therefore not the dress in which belief was wrapped. It was religion itself. And because divine law was the foundation for human law, it followed that the procedures for the formal operation of law were themselves quasi-religious conventions, to solemnize and formalize the business of justice. In practice, law is a kind of theater. Participation involves acting certain roles, such as the secretary, the archon, or the jury, and has a script involving generic routines such as *anakrisis* (preliminary investigation) or *ephesis* (transfer of case to a different court), routines that give legal operations their dramatic shape and their distinctive form—but the use of wigs and archaic garments, or the elevation of the judge on a dais (in modern versions of legal theater), or the solemn swearing of an oath and the making of a sacrifice to Zeus Horkios serve a different function. They separate legal proceedings from the mundane, and infuse the proceedings with dignity and importance. Theatricality and piety, then, are both ways of giving legal proceedings authority and force. In the case of the Greeks, the origins of law in the sphere of religious ritual and prohibition meant that sacred law provided a model for how both public and private law might develop within the state. Accordingly, sanctuaries played a central role in the development of law since they were places that were defined by human recognition of authority. Sanctuaries are set apart; hymns tell the story of the divine presence; but it is humans who build the *temenos* wall and the altar, humans who bring sacrifice, and humans whose performances prove and attest the god's presence. Sanctuaries offer a space and place for ritual, the repetitive performance whose formulaic character and solemnity suggests the power behind it. All these are features required by law.

Leges Sacrae

While all laws retained an air of holiness, sacred laws nevertheless constitute a distinct group within the entire corpus of Greek laws, at least in terms of the classification of inscriptions within the major corpora. As a result it has become normal to speak of sacred laws as if they constituted a coherent body of law, but recent studies have repeatedly expressed dissatisfaction with this assumption. The editor of one volume of sacred laws, advocating a comprehensive understanding of the class, nevertheless remarks that the evidence for sacred laws defies clear-cut classification.[44] Another recent attempt to create a typology identifies two ideal types: regulations concerning holy things, and regulations concerning procedures.[45] A third approach has been to classify together laws and decrees, the prime purpose of which "is to regulate religious matters" but to exclude such religious texts as sacred calendars.[46] Yet once one begins classifying sacred laws according to their content one risks creating a set of categories without seeing how they relate to each other. The subjects covered by sacred laws are very broad-ranging: the calendars of sacrifice for demes, *orgeones*, and the city; regulations for the use of sacred land, including leasing, inspection, and agistment; terms of the appointment of priests and the sale of priesthoods, with particular attention to priestly duties and perquisites; and regulations for conducting festivals, usually involving some combination of procession, musical and athletic contests, sacrifice, and feast. Furthermore, sanctuaries often arbitrated land disputes and the decisions of boundary commissioners were frequently erected in sanctuaries, making them in one sense another category of sacred law, as were the manumission records which served a very real legal purpose—testifying to the individual's legal status—and which were also commonly erected in sanctuaries.

If, then, sacred laws operated across a wide-ranging field, and all laws intersected at some point with the sacred, a typological approach may not be the most productive way of understanding a phenomenon that is amorphous and heterogeneous. Since sacred laws reflect social practices, another approach to understanding their significance may lie in identifying the assumptions that underlie such laws and asking whose authority is invoked, and to what end. As Sally Humphreys has observed, "laws and forensic argument [are] forms of discourse about the state of society."[47] Consider, for example, one of the longest inscriptions from the early fifth century, the so-called *stèle du port* from Thasos:

> From the road of the dyke . . . of the shrine of Herakles. From the road of the shrine of the Charites, in this road let him not make a threshold nor draw water for . . . nor establish a well nor place . . . nor make. . . . Whosoever does any of these things contrary to what is written, let him owe a hundred staters to Apollo Pythios and a hundred to the polis. The magistrates in whose term he does this are to exact [the fine]. If not, they themselves shall owe the fine double to the god and to the polis. And if the fine has not been written for the offender, let him not pay it. Who has not done this, . . . let him show . . . let him be. But when he has put it in order (?) then let him use the building.
>
> Let each occupier keep the road clean against his own property. If there is no occupier, the owner of the building [shall do it]; the *epistatai* shall clean it themselves each month. And if anything falls, let them do. . . . From the shrine of Herakles as far as the sea, let the *epistatai* clean this road. Let him take up what comes from the *katoikiai* (?) and what is on the road, when the magistrates request it. Whosoever does not do

> any of these things according to what is written, let him owe a *hekte* for each day to the polis; the *epistatai* shall exact it and keep half themselves. Let no one of those in the brothels in this street go up for the sake of being seen nor shall a woman look from the windows. For whatever of these things he does, let the occupier owe to the polis a stater for each offense; the *epistatai* shall exact it and keep half themselves. From the balcony that protrudes into the road, let him not pour water. If he does, let him owe a *hemiekton* for each day, half to the polis, half to the *epistatai*; let the *epistatai* exact it. From the shrine of the Charites up to the buildings where are the money exchange and the symposion and along the road by the *prytaneion*, between these let no one throw out dung nor solicit as a prostitute. Whosoever does any of these things, let him owe a *hemiekton* to the city on each occasion that he does so; the *epistatai* shall exact it and keep half themselves; if not let them owe double to Artemis Hekate.[48]

The inscription deals with public space and represents an assertion of control over that space by the state. The identity of the body making the decision is not preserved, since the preamble is missing, but the mention of *archoi* (magistrates) and *epistatai* (supervisors) suggests a civic provenance, as does the fact that the fines for all the infractions cited are payable to the polis. At the same time, the religious aspects of the regulations are prominent. Apollo is to receive half of the considerable fine of 100 staters if anyone tries to sink a well or build beside the road, possibly one of the roads leading toward the agora and the public and religious buildings clustered in that area. Religious landmarks such as the shrine of the Charites, tentatively identified with the Passage of the Theoroi, provide important reference points for the topography of the city, as do major public buildings such as the *prytaneion* and symposion. In fact, the juxtaposition of religious buildings with key public buildings serves to invest the latter with the same status as the former. In terms of public space, this Archaic community had begun by setting aside areas to the gods—sacred space became the first community space. In one sense, the community must be a religious body before the state can emerge. Religious action required religious space, which was part of the articulation of public space. Similarly, the ritual cleanliness one would expect in a religious setting was also extended into the public setting, so that the Thasians' conception of themselves as a polis is an imaginative act predicated on their understanding of themselves as a sacrificial community.[49] Accordingly, just as prostitution would only be considered acceptable within the religious sphere in the setting of the cult of Aphrodite, so too the Thasos regulation seeks to control its presence, not banning it outright, still less imposing a public morality, but asserting an authority over where and how it is to be on view. The law is not concerned with the morality of prostitution, but only with its visibility in the public sphere.

Another Thasian law, usually seen as dealing with the quality of Thasian wine, has recently been interpreted by James Davidson as a politically motivated ban on drinking in public bars, an activity associated with the *demos*.[50] It may be useful to read the *stèle du port* in a similar light. In the early fifth century, at the time the regulations regarding public cleanliness and prostitution were promulgated, Thasos was ruled by an oligarchy of 300. If the *stèle du port* is the work of an oligarchic administration it would be an early, and certainly not the last, example of a ruling elite attempting to clean up town, both figuratively and literally. But even if the regulations are not political in the narrow sense of oligarchic, the law does exemplify an underlying feature of all laws that touch upon sacred matters: the inclusion of gods and sanctuaries gives the regulations legitimacy. This is the source of their efficacy.

The close association between divine law and the emerging legal system of the polis also underscores the weakness of the latter. The gods are invoked precisely when the body claiming the authority to make law, here the polis and possibly an oligarchic cadre, cannot guarantee that its authority will be otherwise recognized. The sacred laws of the Greeks are fundamentally ideological because they arise from a weakness in the state. As societies move from pre- and proto-states, which are often face-to-face communities with mechanisms of settling disputes that rely on familial and tribal structures of authority, and begin developing greater degrees of complexity, the emergence of the state as an entity separate from any one "big man," or even from a ruling elite of elders or princes, renders the judicial power of the elite problematic. Within the tribe the individual is tied to the elder by connections of blood, marriage, clan, and moiety. Authority is embedded in a web of relationships. As the polis takes shape, not only as an urban agglomeration but as a community, these ties become diffuse and are weakened by the very complexity of the polis: thus, *xenia*, a system of personal ties guiding the treatment of individual strangers, is not a sufficiently elaborate mechanism for treating an entire class of foreigners, metics; and customary law may shape inheritance procedures within the family, but is a crude tool for handling multigenerational and blended family disputes of the sort we encounter in the fourth century, for example in the dispute over the estate of Apollodoros.[51] Achilles and Agamemnon fall out over a single mismanaged instance of gift giving, yet by the fifth century the Athenians were expected to resolve disputes over bottomry loans and shipwrecks that had occurred hundred of miles away.[52] In scale, complexity and nuance, the growth of the city-state required the articulation of laws.

This process, by which the state essentially asserted itself, was modeled on those practices that marked the community as a uniform, cohesive body bound by sacrifice, so that when, in the Hellenistic period, separate communities sought to amalgamate through *sympoliteia* agreements and form a new entity, their first priority was not to establish boundaries or prerogatives but to assert the unity of the new community through a sacrifice. Thus, when Latmos and Pidasa forged their union, the ritual marking their *sympoliteia* was laid out in the agreement as follows:

> Let the officeholders lead forth a sacrifice so that the city may be of one mind; and let them establish a new tribe, named Asandris, in addition to the existing ones . . . [53]

Other stipulations follow, and the new constitutional arrangement is solemnized by a joint sacrifice:

> Let 100 Pidasan men, whom the Latmians select, and 200 Latmians, whom the Pidasans select, swear to abide by the decree and this constitution, sacrificing over a bull and a boar in the agora.

This and other instances of *sympoliteia* confirm that the integrity of the polis was rooted in its cohesion as a community bound by sacrificial action and by law. Yet the Greek *poleis* did not maintain the fiction that their laws had been handed down directly from god, and even if lawmakers were accorded heroic status, it was equally clear that communities were capable of passing laws. But if legislation lies within the realm of human activity, that very fact poses inevitable questions: Who makes the law? By what authority? With what sanction? Aristotle recognized as much:

> For the law has no power to compel obedience beside the force of custom, and custom only grows up in long lapse of time, so that lightly to change from the existing laws to other new laws is to weaken the power of the law. Again, even if alteration of the laws is proper, are all the laws to be open to alteration, and in every form of constitution, or not? (Aristotle, *Pol.* 2.1269a, trans. Rackham)

Such dilemmas arise precisely as deep-seated social change manifests itself. Such changes, in turn, can be understood as arising in the transformation of Iron Age societies from a "big man" model to a more complex, hierarchically organized society. The former is characterized by dissipated power, in part because it relies on the giving away of wealth, which accords well with Lefkandi. What will follow will be secondary states, of different sorts: patrimonial proto-states such as sixth-century Athens under Peisistratos, to use Walter Runciman's categories, and citizen states, of which fifth-century Athens would be an example.[54] It is conventional for historians to observe the changes between these stages in broadly descriptive terms: temples get bigger, walls embrace larger areas, coinage is introduced, but often it is only the most obvious of these changes that excites analysis. Coinage, for example, raises issues of fiduciarity and the transformation of barter and exchange to a monetized economy. But law, too, is both a marker of change and an explanandum. We ought to ask not only what sacred laws were, but why the Greeks continued to rely on them.

Any answers to the questions of legal legitimacy, authority, and sanction that invoke human agency are subject to contestation. If it is the *demos* (people) that exercises authority, the elite will respond that nobility is under siege. This is illustrated in the poems of Theognis:

> Among rams and asses and horses, Kyrnos, we look for those
> Of noble breeding, and a man wants them to mate
> From worthy stock. Yet a noble man does not mind marrying
> A base woman of base birth if she brings him money in abundance,
> Nor does a woman shrink from becoming the wife of a base man
> With wealth; she prefers a rich husband to a worthy one.
> Money is what they honor; the noble weds a base man's daughter,
> The base a worthy man's: wealth mixes stock.
> Thus do not be amazed, son of Polypaos, that the citizens' stock
> Is growing feeble, for what is noble is being mixed with what is base.
> (Theognis 183–92, trans. Miller)[55]

Theognis' abiding contempt for the people, masquerading as eugenics, is nothing more than contempt of the ruling class for the *demos*, a prejudice that feeds directly into the antidemocratic tradition.[56] The Thasian regulations cited above are intriguing for the absence of the term *demos* and the repeated references to the polis, a more neutral term than *demos*. In avoiding populist language the decree may be inadvertently advertising its true political provenance. Yet any elite claim to authority must rest on claims of moral superiority: being fit to rule because they are better, as conveyed in the charged descriptor *kalos k'agathos* ("fine and good"). In other words, the pressure on the emerging state to enact laws, which were made necessary by the greater complexity of civic life, foregrounds a political debate about who is fit to make those laws. Under such conditions, the flourishing of sacred laws—both regulations concerning sacred matters and laws that invoke the sacred—is not simply a matter of customary piety, nor even of sacred regulations providing a procedural model for civil law. The Greeks turn repeatedly to the realm of the sacred both for guidance and as a way of finding the kind of efficacious authority lacking in the embryonic state.

The corollary to this is a trend in the Archaic period for the distribution of power to undergo an important shift. Claims to authority based entirely on blood or personal worth, *arete*, might be secure for the princes in the *Iliad* but are beginning to look shaky for Hesiod's *basileis*. Increasingly the claims of the *basileis* gave way to a different model, according to which power was invested in magistrates, whether a single *aisymnetes* such as Pittakos of Mytilene, or boards of officials such as the *ephetai* in Drakon's law or the *epistatai* in the Thasian laws. Vestiges of the earlier practice remained, as in the election of an *archon basileus* at Athens, whose authority was primarily religious. The steps required in this shift are conceptual and procedural: the community must be aware that it is an entity capable of asserting and delegating power, and it must develop mechanisms for doing so: magistracies, elections, assemblies, debates, votes, all of which reinforce the rule of law. Accordingly, the structural changes in Archaic society continue to cluster around the need for laws that are equipped with the power to win consent, since Archaic Greek society had so few mechanisms to coerce consent. Armies were not used to police internal order; police forces were nonexistent. Given the underdeveloped mechanisms of civic order in the early Greek state, sacred law was the catalyst necessary to allow proto-states to transform themselves into full, complex states.

This process was also expressed in concrete ways. Take, for example, the development of the demes, the regional divisions of the Athenian state and the Attic countryside. It is customary to interpret demes from the point of view of the Kleisthenic reforms in the late sixth century and to regard them as the building blocks of the Athenian democracy. Regional organizations of varying size, reflected by their bouleutic quotas, demes were the first affiliation of the Athenian citizen. He introduced his son to his fellow demesmen to ensure his acceptance into the brotherhood of citizens, and the larger units of social and civic affiliation were constituted out of clusters of demes: *trittyes*, *phylai*, and ultimately the assembly and council. Yet demes were also religious bodies that relied on their control of sanctuaries to reinforce their cohesion. This is illustrated in a fourth-century inscription dealing with the leasing of property by the deme of the Peiraieis. The preamble reads:

> In the archonship of Archippos (321/0 BC), when Phynion was demarch, the Peiraieis lease Paralia and Almyris and the Theseion and all the other sanctuaries (*temene*) on the following conditions . . . (*IG* II^2 2498.1–3)

The terms of the lease reveal that the deme was the corporate unit that owned a number of sanctuaries. Nothing is said regarding the formal cult practices associated with them, and the inscription breaks off just as it begins to deal with the "building in Almyris" (*oikia en Almyris*), which probably refers to a small structure housing the cult. It is very likely that the cults controlled by the Peiraieis operated in a way comparable to the cult of Egretes, and that the land was leased on the condition that it was periodically made available to the deme for religious observances. We have already seen private groups of *orgeones* operated as both sacrificial groups and landowning corporate entities, and the Athenian state functioned similarly—the Panathenaia culminated in a sacrifice celebrating Athena and all Athens, and at the same time provided a set of ritual performances around which an Athenian identity coalesced. So too demes relied on religious authority to provide cohesion. No doubt neighborliness and blood connections of birth and marriage also contributed to making the deme into a community, but for the Athenians what defined the deme as a municipality was not contiguity or regionalism, but shared cults. The former were real but incidental; the latter were both real and constitutive. And the deme's religious identity brought tangible benefits to the demesmen. The deme's coffers were filled twice a year with lease payments, and windfall profits accruing from the sowing of more land than stipulated by the lease accrued to the demesmen (lines 22–23). The inscription also reveals that demes exploited the different kinds of land and the types of economic activity that were associated with them. The lease distinguishes between Paralia, Almyris, and the Theseion on the one hand, sanctuary lands that were primarily under cultivation, and the lands designated "the Thesmophorion, the Schoinous territory, and the other pasture-lands."[57] Such pastures were carefully regulated, no doubt because overgrazing is a more serious problem than overcultivation.[58]

If sacred laws are critical to the transformation of incipient states, their widespread existence outside of sanctuaries also points to a remarkable feature of the Greek polis. Although modern scholarship debates the precise meaning of the term *polis,* it is apparent that in practice the peoples of the Greek city-states thought of their communities in a number of different registers: one was as physical territories, another was as aggregations of kinship groups, a third was as communities of enfranchised citizens, and a fourth was as religious communities.[59] As a result, not only were the regulations devised by a strictly religious entity, such as the cult of Bendis, sacred laws, but so too were polis laws and legal institutions, even leases negotiated by demes. Every institution was relentlessly attracted to the sacred in order to embed its authority more firmly in the fabric of Athenian life, as is especially reflected in the foundation myth of the Areopagos as homicide court. The story as told in Aeschylus' *Eumenides* draws upon the well established cycle of tales concerning the house of Atreus, and deals specifically with aftermath of Orestes' murder of his mother, Klytaimnestra. As with most tragedies, the audience was well aware of the story's outline, and knew that this was only one in a string of murders that had included Klytaimnestra's murder of Agamemnon and Agamemnon's killing of Iphigenia. Each episode permitted the playwright to explore issues of guilt and conflicting loyalties in his own particular way. What makes the *Eumenides* unique is that it tries to reconcile older patterns of blood vengeance with the concept of a court comprised of ordinary men. It can be classified as an aitiological myth, using the intervention of Athens' own goddess at the climax to give a divine imprimatur to the play's resolution. Dramatically this has sometimes left modern audiences dissatisfied, but to Aeschylus' contemporaries the message was powerful: justice will now be institutionalized, controlled by the state, but under the eye of Athena. This is not to deny that Aeschylus' work can and should be read in the context of a vigorous debate over the powers of the Areopagos in the mid-fifth century—the reforms of Ephialtes occurred no more than three years before the play was first produced in 458 BC—but regardless of whether Aeschylus was writing a conservative defense of the Areopagos or vindicating democratic restrictions on the court's powers, the fact that he could plausibly weave the ancient legend of the cursed family of Atreus and a contemporary court into a single narrative is powerful testimony to the Athenians' willingness to accept a divine pedigree for their institutions.[60]

In the case of laws passed by the assembly and concerning state cults, such as the reception of Asklepios into Athens or the Panathenaia, the overlap between religious law and the legislative competence of the polis is hardly remarkable. It is not surprising that the traders from Kition sought permission from the assembly when they wanted to set up a cult of Aphrodite, but even laws that were not about sacred matters were infused with a religious sensibility.[61] This bifurcation is evident in the public imprecations erected at Teos in the early fifth century, dealing with a range of threats to the community. Poisoning, blocking the wheat supply, revolting, betraying the city, or otherwise harming the community all call down on the guilty party a curse: may he perish, both himself and his descendants. The text continues:

> Those magistrates who have not pronounced this curse during the festivals of the Anthesteria, the Herakleia, and the Dieia shall be liable to the curse.[62]

It was understood, then, that magistrates were supposed to pronounce the curse in order to protect the city, as it were, preemptively. Moreover, the magistrates were also expected to take an oath that invoked a curse on their own heads should they mistreat, persecute, arrest, or execute any citizen without authorization from the assembly.[63] The power of the magistrates, in short, is quasi-sacerdotal; at religious festivals they summon the gods to protect the city and punish its enemies. And at the same time, the same religious festivals are the setting for the oaths whereby the power of the magistrate is, at least notionally, constrained. By constantly bringing the gods into civic affairs, through public oaths, prayers, sacrifices, and festivals, the Greeks created a discourse that offered an alternative to the rancorous and unstable discourse of the political community. In the face of antagonisms structured around noble versus base (*chrestoi* versus *poneroi*), the leaders of the people versus the few (*hoi prostatai tou demou* versus the *oligoi*), and all the other overdetermined polarities that fed *stasis* for three hundred years, the Greeks still managed to create a regular, stable cultural existence, because the rule of law and the underlying security of the community acknowledged a universal, uncontested, apolitical authority: the gods. It is this that led Demosthenes to describe the gift of Athenian citizenship as an "honorable and sacred thing."[64]

Accordingly, when conflicts between ordinary men occurred in a religious context, the affront was twofold, both against the human victim and against the community bound by their devotion to the gods. Thus, the speaker of Lysias 30, *Against Nichomachos* claims,

> when he taxes me with impiety for saying that we ought to perform the sacrifices named in the tablets and pillars as directed in the regulations, he is accusing the city as well: for they are what you have decreed. (Lysias 30.17, trans. Lamb)

This is a motif running through Demosthenes' conflict with Meidias (Speech 21). The prosecution was for assault, but in emphasizing that the assault occurred at the Dionysia and was only one act of violence among many that occurred during his *choregia*, Demosthenes effectively makes the attack on him equivalent to an attack on the city and its holy rites. When Meidias smashes the crowns and apparel being prepared for Demosthenes' chorus, it is an attack on sacred objects, because as Demosthenes says, "all apparel provided for use at a festival I regard as being sacred until after it has been used" (21.16). Based on this and the actual assault, Demosthenes brought the charge against Meidias not as a private complaint but as a charge of having profaned the festival (21.28). Not only did the law call for complaints to be lodged about any misconduct during the festival, but as a *choregos* Demosthenes argued that there had effectively been two assaults: one on the individual named Demosthenes, and one on the incumbent *choregos*, thus constituting an attack on the office and by extension on the community whose laws created and regulated that office. He offers the analogy of attacks on a serving judge or archon and then applies the principle to his own case:

> For the chorus master was insulted as well as Demosthenes, and that is a concern of the State, as well as the fact that this occurred on the very days on which the laws expressly forbid it. When you are framing your laws, you must scrutinize their purport; but when you have passed them, you must uphold them and put them in force, for that is required by your oath and by justice as well. (Demosthenes 21.33, trans. Murray)

By returning time and again to the theme of his *hybris*, Demosthenes portrays Meidias as not just his personal enemy, but an enemy to the rule of law, religion, and the entire Athenian community.[65] The democratic state was built on the foundation, whether fictive or real, of the community as a sacrificial community.

CHAPTER 10

Authority and Value

> Il y a quelque commerce entre les bêtes et nous, et quelque obligation mutuelle.
>
> —Montaigne, *Essais* 2.11, "De la cruauté"

Sanctuaries, Law, and Authority

If sacred law was the necessary precursor to the emergence of secular law codes in the early Greek state, it is also true that cattle wealth spurred the growth of a monetized economy by combining wealth, value, and exchange into a single institution. As we shall see, it was the central position of the sanctuary in Greek life that would make this possible by giving form and expression to notions of authority and value. Before investigating the role of religious action in the origins of coinage, however, we should look further at the importance of the sanctuary as a locus of authority in the Greek world. On the one hand, the city, as an imagined community, was a society defined by its religious observances; on the other hand, the international sanctuary was its opposite: a religious locale that gradually assumed civic and juridical functions.

Various factors have been adduced to explain the growth of sanctuaries beginning in the Archaic period. In a famous aphorism, W. G. Forrest once noted that colonization did more for Delphi than Delphi did for colonization, while the phenomenal growth in the number of sanctuary dedications attests to the role played by sanctuaries as outlets for status display and competition among the aristocratic elites of Archaic Greece.[1] At the same time, the continuing tendency of Greek communities to frame laws as sacred matters also enhanced the status of sanctuaries. This complemented the internal workings of states, as when Delphi was called upon to authorize the use of the sacred land at Eleusis or the introduction of new cults to Athens, and also explains why in a time of crisis, such as Xerxes' invasion, the Athenians might resort to the Delphic oracle for guidance—but it also resulted in sanctuaries playing a unique role in areas of legislation that were at the limits of the competence of the polis, particularly in interstate relations. The sanctuary of Artemis Heleia at Alorion, for example, played an important role in shaping relations between Triphylia, in the southwestern Peloponnese, and Arkadia, further inland.

The Arkadians provided the priests to the cult and continued to do so even after Triphylia had come under the control of Elis, thereby allowing Triphylia to assert an identity distinct from the rest of Elis.[2] Supraregional sanctuaries not only mediated between states but could also serve as agents in helping a regional identity to coalesce, as happened among the Pisatans, whose claim to the *prostasia* (guardianship) of Olympia under the heroes Pisos and Pelops was a critical element in the articulation of Pisatan ethnicity.[3] The repeated attempts by the people of Phokis to take control of Delphi, most notably in the Third Sacred War (355–346 BC) also served to cement a Phokian identity.

A more concrete role played by extraurban sanctuaries was the definition of borders. This was accomplished in two ways, either by authorizing commissions to survey the borderland and delineate it clearly, or by rendering border lands neutral. The latter is demonstrated by the case of the Messenians and Spartans who shared the cult of Artemis Limnatis, which was located on the edge of their respective territories. Similarly, when the Aitolians crossed the Corinthian Gulf in the 240s, the sanctuary of Artemis located midway between Kleitor and Kynaitha offered asylum to the people from both towns.[4] Significantly, those who sought refuge were herdsmen with their animals. Angelos Chaniotis has noted that it was common for Cretan sanctuaries to be located in the *eschatiai*, the farthest reaches of a state's territory, where transhumant herders from various states mingled.[5] In such areas a neutral border zone rather than a fixed border line will have been more practical since, as we have seen, herders were increasingly forced into areas removed from cities. They moved across borderlands that were less easy to police than the city center, in areas that were often feared as wilderness, and were by their very nature open to contestation.[6] Grazing therefore added an economic complication to the already complex legal status of borderlands, and as we noted earlier the Boiotians and Athenians found it easier to share the Skorta plain as common pastureland rather than to contest its ownership.[7]

A revealing glimpse of the isolation of the hinterland and its desirability is offered by a remarkable inscription from Thessaly. The decree records the details of an investigation by a group of judges called upon to decide the fate of an upland region whose ownership was contested by two towns, Kondaia and possibly Gonnoi. The judges collected evidence from various witnesses, including a shepherd, Ladikos son of Harmodios, who claimed on oath that one of the parties in dispute, the Kondaieis, had asserted control of the territory by collecting the transit tax (*paragogion*) in the district. The shepherd goes on to assert that he knew the territory in question well, having grazed the uplands for so long.[8] A second witness, Pantaios, son of Kleoboulos, testified that he had seen men from Kondaia farming and grazing their flocks in the area. The judges' reliance on testimony about how the land was used points to the difficulty in determining the ownership of a territory lacking any permanent settlement. Because the area was subject to transient occupation by grazing flocks, it was difficult for a bureaucracy based on settled communities to operate here. In fact, the testimony of Ladikos and Pantaios accords well with the picture of isolated uplands painted by Dio Chrysostom in his Euboian Discourse, in which the entire apparatus of law, debate, and punishment is also imagined as a characteristic of urban existence.[9] The Kondaieis could show that they had cut a track into the backcountry and collected taxes from the herders using the pastureland, thereby asserting control over the land's use and justifying their territorial claim. The decision of the judges was recorded on a stele erected in the temple of Apollo Kerdoios in Larisa. Grazing herds complicated the problem of borderland, requiring a complex legal apparatus and the sanction of religious authority. The very fact that this was not settled land made it difficult to apply conventional law.

In some cases a sanctuary itself would claim borderland pastures, as in the case of the sanctuaries of Telephos and Pan on Mount Parthenios, close to the boundary separating Tegea from Argos.[10] We have already seen that sanctuaries were hungry for sacrificial animals, and that this led to the annexation of land, but large international sanctuaries were also uniquely located to assert neutrality in border areas. Eleusis lies almost exactly halfway between Athens and Megara, while Delphi and Olympia are situated in the least densely urbanized regions of Greece, making it possible for the sanctuaries to retain territorial independence, despite the attempts by their neighbors in the fourth century to wrest control. Because of their need for a plentiful supply of meat, sanctuaries were keen to have their territory clearly delineated. One of the fullest descriptions of a border, in fact, comes from Delphi. An inscription from the first or second century BC records the border between Delphi and the Phokian towns of Phlygonion and Ambryssos. The text reads as follows:

> God. Good fortune!
> . . . the so-called De . . . iss . . . is to belong to the . . . while the territory on the left, as the water flows, is to be the property of Delphi as far as Aigoneia. From Aigoneia, as the water flows, [the border shall run] through the ravine. The hill called Kerdon [constitutes the border] as far as the road bearing towards the holm oak. The territory on the righthand side belongs to the Phlygonians and Ambryssians, while the territory to the left is the property of Delphi. [The border shall run] from the road to the holm oak that stands on Lookout Point. From Lookout Point [it continues] straight on along the ridge to the sharp rock. From the rock straight on to Petrachos. From Petrachos straight on to Parnassos. Everything that is towards the east is the property

of the Phlygonians and the Ambryssians, while everything to the west belongs to Delphi. The water that flows past Aigoneia is to be common to all. If there are any sanctuaries in these areas, let them continue to function according to customary usage. (*FD* III 2.136.18–33)[11]

The boundary described here is an imaginary line joining conspicuous landmarks such as ravines and hills whose names were familiar to the local population (Kerdon Hill, Lookout Point, the Oak). It runs broadly north–south up the slopes of Parnassos, permitting the territory to be assigned to the left and right, that is, to the west and east, of the border. A recently identified border inscription from the eastern side of Phokis operates similarly, using toponyms such as the *diodios* (passage), the *lakkos* (reservoir), and the *skopia* (lookout).[12] The Delphi border inscription also asserts the neutrality of sanctuaries in borderlands. Here they are described as operating "according to customary usage," effectively divorcing the question of their control from the wider issue of the border.

Sanctuaries were concerned not only with their own borders but were well suited, because of their prestige, to play an important role in settling boundary disputes between other states, either supplying the commissioners who determined the border, or applying the sanction of the god's approval to make the boundary line binding. Boundaries were commonly determined by boards of judges, known as *geodikai*, *dikastai*, or *ekdikoi*, who surveyed the land before rendering a decision, which often used prominent topographical markers (as in the Delphi boundary inscription) such as ridges, rocks, roads, and trees to demarcate the boundary.[13] The decisions of such boundary commissioners remained in force only as long as the states in dispute submitted willingly to their authority. To that end, commissioners were often chosen from a neutral state, or states. The arbitration of a boundary dispute between Hermione and Epidauros, for example, was left to commissioners from Miletos and Rhodes.[14] In other cases, states affiliated religiously to the contestants in the dispute were called upon to arbitrate. This occurred in the dispute between Miletos and Myus in the early fourth century. In this instance the case was submitted to a jury of at least twenty-five commissioners drawn from the cities that worshipped at the Panionion. The final decision was ratified by Strouses, the Persian satrap, who presumably could have issued a decision on his own authority.[15] The decision to operate through the channels familiar to the Greeks reflects a sensitivity on the part of the Persian authorities toward the Greeks of Asia Minor and also shows how common was the practice of relying on religiously sanctioned arbitration.

The boundaries thus determined were usually published on *stelai* erected in sanctuaries belonging to the contesting parties, or at a major international shrines such as Delphi (or both), thereby receiving the imprimatur of the gods. The boundaries established by these covenants thus became sacred.[16] It is particularly important to keep in mind the role played by sanctuaries in fostering the orderly settlement of boundary disputes since there continues to be a tendency to see in the polis an evolutionary dead end, as if the Greeks were incapable of solving the problems of interstate relations because of their attachment to their separate states.[17] In fact, even before the fourth century and the various experiments with common peace treaties, the Greeks had regularly experimented with federalism, and would continue to explore other forms of super-polis relations, such as *isopoliteia* and *sympoliteia*, all of which were predicated on the sharing of cults and religious obligations.[18] Sanctuaries were the locales in which a form of Greek identity that transcended the polis could express itself. This is especially borne out by the most common form of interstate diplomacy in the Hellenistic age: the recognition of a sanctuary's right of inviolability (*asylia*), a privilege highly prized and yet utterly dependent on the goodwill of those acknowledging the status. Sanctuaries provided a network through which Greek culture and identity expressed itself even as new political forms, in the shape of the Hellenistic kingdoms, imposed themselves on the eastern Mediterranean.[19]

If borderlands were largely uncultivated, they were certainly not unoccupied. Sanctuaries were ideally suited to play a role here, either as owners of contested land or as mediators in disputes over land use. In fact, agreements on how to exploit borderlands often reflect the fact that pastoralism and border territories went hand in hand. Drawing on earlier studies that found a clear distinction between ownership and possession of land, Angelos Chaniotis has recently observed that international agreements in the Classical and Hellenistic periods use possessive nouns ands adjectives to signal legal ownership, but employ verbs such *echo*, *katechô*, and *kektêmai* to signal possession. Two verbs that frequently occur in this context are *nemomai* and *karpizomai*, which Chaniotis glosses as "hold, inhabit," and "exploit," but which also reflect the way land was used: "grazed and planted."[20] Frequently, both sides in a dispute agreed to share water in the borderlands, as in the Delphic inscriptions quoted above: "The water that flows past Aigoneia is to be common to all." Water was especially important to shepherds moving across the landscape, and the right to graze border lands, *epinomia*, was a fraught issue because of the potential for conflict.[21] Thucydides, for example, reports that Argos administered the sanctuary of Pythian Apollo (probably in the territory of Asine) and leased its lands to Epidauros for pasture. When the Epidaurians failed to send the required sacrificial animals in payment, the Argives, prompted by Alkibiades, used this as an excuse to declare war.[22] Provisions regarding pasturage were frequently stipulated in border agreements, as is shown in the case of Messenia and Sparta.[23] One of the agreements between them maintains that water is to be shared, an understandable clause in the context of flocks moving through a border zone.

The advantage of marginal lands to grazing was that such exploitation was by its very nature transient and could therefore be managed without making a permanent claim of ownership. Thus, in the arbitration of the dispute between Hermione and Epidauros over their borderland, the commissioners canceled any complaints arising from earlier incidents over grazing in the common land.[24] Such disputes were, of course, unavoidable—the Oxyrhynchos historian claimed that the Phokians and Lokrians had gone to war repeatedly over such incidents—but by declaring the pastures common land and refusing to investigate earlier infractions, the commissioners were effectively creating a neutral zone in which one could graze animals without having to assert ownership of the land itself.[25] In the case of a *sympoliteia* between Ilion and Skamandroi, Francis Piejko has also suggested that an agreement to share pastureland was one of the terms of the accord.[26] Another way of sharing border land was to limit access to pasture, so that herders from both sides of the border could exploit it. For example, when the Lokrian towns of Myania and Hypnia agreed to a *sympoliteia* in the second century, they settled the question of the border zone by simply guaranteeing that no one could stay on it long enough to claim it:

> If any of the shepherds who has pastured his sheep in the area in the past brings his sheep before they are washed, let him take them away once he has washed them. But all the shepherds who have not previously used the land as pasture, while they are washing their sheep and putting them to pasture here, may remain and fold their sheep here for ten days.[27]

Questions of ownership aside, the desire to keep herds moving made sense because of the danger of overgrazing. Angelos Chaniotis cites four Cretan treaties dealing with common pasture that stipulate penalties for overpasturing. A second-century treaty from Priasos, for example, states: "If anybody grazes [his flock], he shall be exempt from charges; but if he does damage he shall pay the fine according to the laws of each city."[28] In cases where the borderland had been assigned to a sanctuary, sacred laws addressed the problem of overgrazing. This can be seen in the well-known regulations from Arkesine on Amorgos that stipulate that "[n]o one shall be allowed to bring flocks into the sanctuary; if anyone does bring them in, the flocks are to be sacred to Zeus Temenites. Anyone who wishes can make an indication to the council and be rewarded with half."[29] Priests at the temple of Athena Alea in Tegea also tried to address overgrazing by limiting the amount of time sheep and pigs could be grazed on sacred land without penalty.[30] The point of such regulations was to control an activity that was troublesome and potentially damaging to the natural environment. The effect, however, was a bonus: a small herd of animals for sacrifice was on hand for sale to visitors, just as at major sanctuaries where a much larger supply was needed. The Tegean inscription refers specifically to the sacred herd attached to the sanctuary: "The priest may pasture twenty-five sheep, a yoke of cattle, and a goat. . . . The chief sacrificial officer may pasture any animal that is without blemish" (lines 1–3).

Herding favored the intermittent usage of pastureland, but seasonal fluctuation and the transient movement of herders from different states did not fit well with a polis-based legal system that was increasingly taken up with land claims and leases, both of which implied permanent or at least long-term occupation.[31] Accordingly, sanctuaries and the apparatus of sacred law filled a void that the laws of the polis were ill-equipped to manage. The sole instance in which *epinomia* was treated as an internal matter is a Lokrian law from the sixth century that deals with the distribution of land recently annexed. The new settlement treated grazing as a matter of private use of public land, and, unusually, made the right a matter of inheritance. The relevant clauses read as follows:

> This law concerning the land shall be in force for the partition of the plain of Hyla and Liskara, both the separate lots and the public. The right of pasturage shall belong to parents and to the son; if there is no son, to the daughter, if there is no daughter, to the brother; if there is no brother by relationship, let a man pasture according to the law . . . (ML 131–6)

The language of the law is elliptical, and it is only the references to successive degrees of familial relationship (parents, son, daughter, brother) that allow us to infer that the law is concerned with the transference of the right of pasturage by inheritance. What is clear is that the provisions were made necessary by the annexation of a piece of territory that was then divided into allotments (line 18) between two waves of settlers (16–17), and that some of the land was intended for cultivation (6–7). The apparent contradiction between a regular *anadasmos* (land distribution) into family plots and the inheritance of pasture rights has led to some confusion in the analysis of the Lokrian law, but the solution is straightforward. The land that had been annexed, including the rich bottomland, was surveyed by the *demiourgoi* and assigned to the settlers, who were free for a period to exchange their lots (line 18 and Text C). In addition to the private lots, the cadastration put aside two other kinds of land: *ta apotoma* and *ta damosia*. These surely correspond to land for the gods (*apotoma* is from the same root as *temenos*) and land for communal use.

The new Lokrian land division corresponds precisely to the categories recognized by Aristotle, who believed that the division of land into sacred (*hiera*), public (*demosia*), and private (*idia*) portions was a characteristic of Greek city planning that went back to Hippodamos of Miletos. Aristotle proposed dividing all land into two categories, common land and private lots, and further subdivided common land into two portions: land given over to religious observances, and the land assigned to support the expenses of common meals.[32] The same arrangement applies in the Lokrian distribution of the land in the plain of Hyla and Liskara. In such a land division there would be no need to spell out the conditions of inheritance of the private lots, since these would be subject to whatever inheritance law already pertained in the community. Public land, however, if primarily given over to pasture would require different treatment, since the land itself was not liable to inheritance. Only its usufruct as pasture could be inherited by family members. We do not know how long this system of common pasture remained in effect. An increase in population may eventually have led to the land being enclosed and turned over to cultivation, or the annexation of the land by an emerging power such as the Aitolians of the third century may have caused the arrangement to lapse. In this as in many other areas the inscription remains opaque, but it is clear on one matter: the law designed to settle these complex issues required the assistance of the gods to be effective: "This law (*tethmos*) shall be sacred to Pythian Apollo and the gods who dwell with him" (lines 14–15).[33] It is probable that a copy of the law was erected in the temple of Apollo, possibly the Apollonion at Naupaktos. As in so many other areas, the sanctuary imbued what would seem to us a civil matter with divine significance.

Historians' fascination with the polis, inherited in part from Aristotle, has meant that Archaic and Classical Greece are often viewed largely from the point of view of the polis and its institutions, as if this encompassed all sanctuaries. Yet the Greeks conceived of sanctuaries as somehow distinct from the polis. Strabo, for example, describes early Ilion as a *katoikia* (colony) and *hieron* (temple), but not yet a polis.[34] In the case of large sanctuaries, such as Olympia, Delos, Delphi, and Eleusis, it was their function as religious centers that led them to develop as urban agglomerations, rather their growth as cities that enhanced their religious function. The development of Greek society thus followed at least two distinctive tracks: one focused on the *asty* (town), and the other on the *temenos* (sanctuary). These parallel lines of development led to two distinct forms that recursively influenced each other. Without sanctuaries it is difficult to see how Greek law in the setting of the state could have evolved as it did. Similarly, the economic system of cattle husbandry, sacrifice, and feasting that was so crucial to the shape of Greek society appears to have been generated by, and focused on, sanctuaries before crossing over to the city. At times the fusion of polis and sanctuary institutions is so complete that one is tempted to dismiss the distinction: the polis without religion is unthinkable, to be sure, but the hierarchy of sanctuaries, from small cults to communal urban altars, to extramural *temene*, to Panhellenic sanctuaries exhibits a growth and development that is not subordinate to the growth of the polis. Often the transformation of sanctuaries is parallel to the growth of *poleis*, or even independent of it. Sanctuaries were frequently the engine of change in the Greek world. This was certainly the case in the development of Greek law, and, as we shall see, it was also true of the development of a monetized economy.

Money

An important example of how sanctuaries could shape the growth of *poleis* lies in the complex issue of the origins of coinage and the evolution of a monetized economy. Here too the persistent influence of the bovine *habitus* reveals itself in the unexpected origin of coinage in the milieu of sanctuary-based sacrifices and dedications. Based on evidence such as the foundation deposits from the Archaic Artemision at Ephesos, it is usually argued that coins began to circulate in the Greek world shortly before the middle of the sixth century.[35] The practice of minting coins reached Greece from the ancient Near East, where previously transactions had been understood as a straightforward exchange of goods. As the Sumerian proverb expressed it, "No one will give away a cow for nothing."[36] But barter can be cumbersome. Consider, for example, the story of Jacob's wages in the Book of Genesis. Negotiating his release from Laban, Jacob explains how his pay may be computed:

> Today I will go through all your flocks. Take out of it every black animal among the sheep, and every speckled or spotted one among the goats. Such shall be my wages, and my honesty will answer for me later: when you come to check my wages, every goat that I have that is not speckled or spotted and every sheep that is not black shall rank as stolen property in my possession. (Genesis 30:32–34)

This is an especially interesting passage since it connects domestic animals and wages in such a way that the use value of the animal is exchanged for the labor of the hireling.[37] In time such exchanges would be commuted from barter to payment in coin, the introduction of which would lead to immediate and striking consequences. Direct barter would give way to a monetized economy, in which value was assessed and transformed into a medium, coins, that increased the flow of goods and facilitated the expan-

Figure 10.1 Stone money (rai), Yap, Micronesia.

sion of trade. A fragment from Herakleitos illustrates how the Greeks quickly came to understand such exchange: "All good things are an equal exchange for fire and fire for all things, as goods are for gold and gold for goods."[38]

The entry of coinage, however, into the Greek world and the mental steps that permitted the transition from barter to a monetized economy are more complicated than the simple adoption of an existing Near Eastern practice. There is evidence, for example, that both gold and silver were used in unminted forms, as bullion, for trade, and that the standardization of weights was a necessary precursor to the appearance of a monetized economy.[39] Furthermore, the denominations of coins—obols, or "spits," and drachmas, or "handfuls"—points to an intermediate stage between barter and money. Yet it is a stage that requires explanation. The "spits" were iron bars used for spitting roasted animals and a "handful," or drachma, is an understandable way of referring to a bunch of them, but as proto-money the spits seem almost as unwieldy as the stone wheels laboriously cut by the Micronesian islanders of Yap (Fig. 10.1).

Plainly, the most important impetus in the shift from barter to money, if there was an intermediate stage during which spits were used, was not utility. Instead, what obols and drachmas reveal is that money could only be adopted by a society that had already developed a sophisticated understanding of value that went beyond barter.[40] As Aristotle understood, a first step was the recognition of an object's exchange value, which was quite distinct from its use value. The use of an object gives it its direct value: a knife, a plow, a horse, each has value residing in its use to cut, to plow, to carry. Based on its use value, an object can be traded or bartered for something else, giving rise to its exchange value.[41] The well-known story of Archilochos' meeting the Muses while on his way to sell his cow at market, and exchanging the cow for a lyre, is an example of a direct barter.[42] Even at this point, however, the exchange value of a shoe or knife is closely connected to its use value since the exchange is generally predicated on the assumption that the exchange is designed to supply the object so that it can be used. Its use value is implicit in the exchange. But there is a further distinction to be made, when an object becomes the measure of the value of an exchange regardless of whether the object providing the standard of value is itself exchanged. These different conceptions of value can be seen in Homer. For example, Iphidamas offers one hundred cattle as a bride price, and we are to imagine that the animals will be given, that the bride's father is satisfied, and that the bride will be handed over in exchange.[43] Already such negotiations involve use and exchange value. However, when we are told that the tassels on Athena's *aegis* are each worth a hundred cattle, there is no expectation that Athena paid for these, or that an actual exchange can or will take place.[44] Even when Glaukos and Diomedes famously exchange armor, and the poet notes that Glaukos must have been crazy since he exchanged golden armor for bronze, he computes the disparity of value in terms of a third item: cattle. The golden armor was worth one hundred head, the bronze only nine.[45] In fact, throughout the Homeric poems, value is expressed in terms of cattle without cattle actually being exchanged.[46] The cattle, therefore, are the measure of an object's value. Richard Seaford has recently explored the different ways that value is expressed in the Homeric poems and concludes that cattle provide the only measure of value in Homer.[47] In fact, cattle are instrumental in stimulating the very conception of property. This is exemplified by the habit of branding cattle, attested by seals incised on the flank, going back at least to the middle of the Bronze Age.[48]

This is not altogether surprising, since the associations of cattle with pastoral wealth and the broader scheme of values that were clustered around cattle make them an obvious measure of value. At the same time, however, one might have imagined gold or silver serving a similar function, especially since by the time of the Iron Age, precious metals—either as manufactured objects or bullion—could convey status as readily as cattle and would be more easily manipulated and stored. In fact, it is possible that there *was* a measure of wealth that derived from gold: the talent. Hélène Nicolet-Pierre has recently drawn attention to the derivation of the term "talent" from the word for a balance, and has noted the appearance of gold disks in Tomb 3 from grave Circle A at Mycenae, some of which were found attached to the crossbars of balances. Relating these to scenes in Homer in which Zeus weighs the destiny of a hero in a balance, Nicolet-Pierre raises the possibility that Homer's "talents" retain a memory of Mycenaean objects of extraordinary value—objects that, by association with balances, also had an exchange value. Yet, tellingly, when the poet ascribes an exchange value to such talents, he reverts to the bovine rate: a fat ox is worth more than half a talent.[49]

Whether the talent originated in a memory of Mycenaean gold disks and balances, it is worth exploring further the role of spits and handfuls in crystallizing Greek notions of value, since this once again illustrates the central place of cattle in the mentality of the Greeks. Ingrid Strøm has noted two features in relation to the iron *obeloi* found in Late Geometric and Archaic sanctuaries that are especially revealing. The first is that, Zeus sanctuaries aside, the deposits of iron *obeloi* tend to occur in clusters of six at the sanctuaries of Apollo, Athena, Artemis, and Hera.[50] These clusters of six obeloi, constituting a drachma or handful, are frequently found with *lebetes* (cauldrons), most of which are of North Syrian provenance. Strøm concludes that the sanctuaries where the obeloi were found witnessed the introduction of a Near Eastern banquet tradition in the late eighth and early seventh centuries. Strøm infers that "the Greek sanctuaries in question [where such dedications were made] had an independent economy and a high level of organization in the late eighth and early seventh century BC."[51] Given the importance of the iron spits to the adoption of coinage, one might go further and argue that the Greeks' religious and social commitment to sacrifice, feasting, and dedication was responsible for establishing a concept of value centered on the supreme sacrificial animal.[52] At the same time, as Sitta von Reden has noted, a number of early laws from Crete stipulate penalties calculated in terms 5 or 100 *lebetes*.[53] Vessels connected with sacrifice and feasting thus from an early date conveyed a conception of value both from their association with religious practice and legal sanction.

This nexus of cattle and value was also strengthened by the fact that cattle have a disproportionate value as a food source, that is, after sacrifice, when compared with other animals. Consider, for example, the data from Kassope in the fourth century. Excavations at the site recovered over 22,000 bones of cattle, ovicaprids, and swine. Of these, cattle bones comprised 25 percent of the total number of bones, but this amounted to over 257 kilos, or more than 55 percent by weight.[54] Even with the addition of wild animals to the diet, domesticated cattle continued to overshadow other animals. Cattle thus became the measure of wealth, and the metal accoutrements associated with feasting on them become the first objects exchanged not for their own use value, or even for their exchange value, but because, as happens in a monetized economy, the objects that measure wealth become wealth. This is why using a five hundred–dollar note to light a cigar makes us cringe, despite the fact that the ink and paper wasted are worth no more than a few cents. The derivation of *pecunia* (money) from *pecus* (cattle) reflects the same transference of value from the original object to the medium of exchange.[55]

The process by which value moved from cattle to objects associated with them may also have been helped if, as seems likely, Bronze Age copper ingots were cast in the shape of oxhides.[56] This view, once vigorously proposed by Charles Seltman, has been attacked recently by David Schaps, who argues that the similarity between the ingots and animal skins is purely coincidental: the ingots were cast with protuberances at each corner to serve as handles, creating a shape that suggested an animal's hide.[57] But there are reasons to connect cattle and copper. Enkomi, on Cyprus, was the center of a vigorous copper industry between 1600 and 1300 BC and the development of this industry was evidently associated with a so-called Ingot God, a forerunner of the Minotaur, depicted as a man with horns and associated with bars of the precious metal.[58] Comparanda from African pastoral societies also support the view that cattle can combine exchange value and symbolic importance. While the Tshidi of southern Africa raise cattle and measure wealth in terms of the size of a man's herd, some important transactions such as fines, loans, payments to healers, and marriage contract payments are conducted using tokens known as "cattle without legs" (*dikgomo tse di tlokang maoto*), whose value is pegged to the market value of the real animals.[59] In Lesotho there exists a "bovine mystique" according to which cattle are treated as a commodity unlike any other, set apart and not subject to conversion into cash.[60] In part this special status arises from the fact that cattle are living creatures, so that even if they have value as a commodity they are also connected to us by the kind of affective bonds that can only exist between living creatures.

This reconstruction is supported by the few literary accounts from ancient sources dealing with both spits and money, and by the archaeological record as well. Thus, the *Etymologicum Magnum*, probably drawing on Ephoros, asserts that

> Pheidon of Argos was the first of all people who coined money, in Aigina; and giving the coins and taking spits in return, he dedicated them to Hera of Argos. (*Et. Mag.*, s.v. *obeliskos*)

Pheidon's career was the subject of imaginative storytelling, and there are plenty of reasons to be skeptical of this account. Why Pheidon should have chosen to begin minting on Aigina rather than at Argos is not clear, even with Strabo's claim that he did so because Aigina was an emporion.[61] Furthermore, the notion that Pheidon took all the spits out of circulation and replaced them with coins is difficult to understand. Did this not amount to emptying his own treasury and putting his wealth into circulation beyond his control?[62] Nevertheless, the discovery of nearly one hundred iron spits at the Argive Heraion seems to offer dramatic support for the general claim that in the sixth century *obeloi* and *drachmai* originally referred to handfuls of iron spits, used first for roasting sacrificial animals and subsequently dedicated as valuable objects by virtue of their close association with the high-status business of sacrifice and feasting.[63] The value placed on sacrificial animals was first transferred to the objects by which they were processed, in turn establishing a concept of value based neither in use nor in exchange. Once value had been thus rendered abstract, it was possible for it to be expressed symbolically by objects that retained only a slight metonymic relationship with the original measure—in this case, the names of the denominations. Instead of cows, wealth could be expressed and measured by coins that depicted cows (Fig. 10.2). The image thus served as an advertisement of the community's wealth, with the cow, like an ear of wheat, suggesting the territory's fertility, even as the association of the image with the concept of value reinforced (and was reinforced by) the value of the coin's bullion.[64] Later, once coinage was well established, it was an easy mistake to imagine, as Plutarch did, that the process had worked in the opposite direction, moving from coinage to value. According to the *Life of Theseus* 25, the Athenian king "minted coins and engraved them with cattle either commemorating the bull of Marathon, the general of Minos, or to encourage the citizens to farmwork. From this come the Athenian expressions 'worth a hundred oxen' and 'worth ten oxen.' " But in fact it is much more likely that the expressions came before the coins, and that like other pastoralists, the Greeks had measured value first in terms of oxen and only later in terms of money.

This, however, is still not the complete story. As a number of modern commentators have recognized, the shift to a monetized economy has enormous consequences for the development of the state, since money's success does not depend solely on such characteristics as its portability, durability, and availability. The effective circulation of money also relies on fiduciarity, the intangible element of trust that allows it to function as a medium of exchange and as a substitute for an entire range of commodities.[65] This is the fundamental difference between a US dollar and a Zimbabwean dollar: in the one case it is guaranteed by a powerful state whose economic authority is recognized, and in the other it is not. How did the emerging states of the sixth century accomplish fiduciarity? Some clues are suggested by the iconography of Greek coins, which make claims in

Figure 10.2 Fifth-century nomos, Thourioi (Lucania). American Numismatic Society acc. no. 1997.9.172.

at least three categories. The first is that the states are independent (conveyed by symbols specific to one state or another: the rose for Rhodes, Arethusa for Syracuse, owls for Athens, and so forth). Euripides retrojects the association between regions and particular symbols to the heroic age when the chorus in *Iphigeneia in Aulis* speaks rapturously of the statues of nereids, dragons, and bulls that adorned the Greek ships at Aulis, signaling the regions from which they boats came. The bulls, for example, are said to represent the Alpheios River "that flows by Nestor's country."[66] Second, the coins frequently depict bulls, horses, ears of wheat, and other simple assertions of the state's wealth. Finally, the gods are frequently shown, particularly Zeus, Athena, and Poseidon, to convey the unmistakable message that the community enjoys the favor of heaven. Other elements play a part as well, such the formulation of standardized weights and measures. Together these allowed the states of the sixth century to develop trade zones in which states subscribing to one or other set of standards—Aiginetan, Corinthian, Athenian—were able to participate in steadily more complex trade networks. The adoption of coinage by all the Greek states suggests a successful instance of peer polity emulation.

Leslie Kurke has recently argued for a more political explanation of these changes, suggesting that the adoption of a silver standard for early coinage represents a democratic tilt, the triumph of a middle-class ideology over an aristocratic preference for gold.[67] The rise of a monetized economy, however, should first be seen against the earlier notions of value and worth located in the context of ritual and sacrifice. Critical to fiduciarity and the emergence of a monetized economy are elements such as regularity, reliability, and repetition. That is to say, once money is in circulation it is not difficult to understand why it continues to circulate. As Richard Seaford has noted, the successful circulation of money creates the very confidence that sustains its circulation, or, as an earlier economic theorist put it, "[money's] wide acceptability, paradoxically, depends upon its wide acceptability."[68] The question is, where does this confidence come from? Seaford draws attention to the significant difference between the instability in exchanges based on gift giving, distribution of booty, retribution, and the transient type of sacrificial distribution of meat described by Homer and the formal, standardized pattern of sacrificial distribution characteristic of sanctuaries in the emerging *poleis* of the Archaic period. Whereas booty is haphazard and unritualized, sanctuary-sponsored sacrifices proceed according to repeated, and therefore predictable, patterns. Seaford's analysis is neatly expressed:

> [T]he prestige objects of gift-exchange are also the objects of that divisive distribution of plunder that contrasts in both epics with the ordered communality of traditional sacrificial distribution. The development of the polis, in Homer generally neglected, will transcend these contradic-

> tions by the synthesis of sacrificial communality with the permanence of prestige objects, and with their typicality, mass production, and substitutability (by mere symbol) . . . a synthesis favourable to the genesis of coinage.[69]

As sanctuaries became the focal point for a ritual formerly managed by princes and "big men," sacrifice took on increasingly commercial overtones: animals could be purchased, their value fixed by the authority of the priests whose business it was to conduct the entire affair. As Plutarch noted, "At Delphi, he who makes sacrifice buys his own meat."[70] And as the Archaic period witnessed the growth of monumental temples and more elaborate sanctuaries, these sacred spots became the locations where two parallel forms of economic activity took place. On the one hand, the treasuries that adorn the sanctuaries point to the state's adopting the same practice of wealthy dedication practiced by high-status individuals. The states are increasingly identified with the athletic prowess and piety of their greatest men. But at the same time, the development of a regular calendar of games and *theoria* (consultation) commercialized the entire business of sacrifice, a process that we have seen transferred even to the smallest sacrificial groups who repeated the same practices of land leasing, purchase of sacrificial animals, and calendrically regulated sacrifice.

If sacred law was the necessary precursor to the emergence of secular lawcodes in the early Greek state, it is also true that cattle wealth spurred the growth of a monetized economy by combining wealth, value, and exchange into a single institution. Only one state resisted the trend: Sparta. There the regular banquets and meat distributions associated in other states with sacrifice were replaced with a communal eating system, the *syssition*. Divorced from the milieu of the sacred, Spartan communal dining relied on contributions of barley flour, wine, cheese, and figs, supplemented by hunting (boar) and by fishing and netting (dove and quail). Without any sacrificial association, beef played only a minimal role in the Spartan diet.[71] Consistent with their idiosyncratic dietary practices, the Spartans also rejected the monetized economy, preferring to retain iron spits. As Plutarch noted, iron currency could not be exported anywhere outside of Sparta and had no value anywhere else.[72] Not for the only time, Sparta was the exception that proved the rule.

Repetition and Aggregation

As production, sacrifice, distribution, and consumption merged into a single coherent system, a tendency toward standardization emerged. This is reflected by the practice of parading animals before sacrifice, a practical demonstration of the power and wealth of the body sponsoring the sacrifice, and as we have seen, a performance widely repeated. The Athenians required their allies to send a "cow" (but probably in fact a bull calf) to the Panathenaia, the Thebans had the rest of the Boiotians send a cow to the Ptoia, the Eretrians required subordinate communities to stock the Artemisia similarly, and the hamlets of Oinoanda were required to do the same thing in the first century AD.[73] These processions preceded the sacrifice and often took place along processional routes that monumentalized the route between sanctuary and city center.[74] The banquet that followed was also a reaffirmation of the importance of the sponsor and host.[75] It was a set of practices that came to be widely repeated in the Hellenistic period. For example, when Attalos II paid for a new festival at Delphi, the Attaleia, he required the sacrifice of three suitable cattle to Apollo, Leto, and Artemis, the meat being consumed at a public banquet.[76] It was also a practice that came to be highly regulated, as is attested in the lengthy inscriptions typical of the period. In the decree from Ilion recording the benefaction of Hermias, for example, details concerning the procession, sacrifice, and distribution constitute approximately half of the inscription.[77]

The repetition of procession, sacrifice, and feast helped a standardized system to emerge. As a result, Greek communities experienced an astronomical year based on the movement of sun and moon, a civil year based on the appointment of officeholders, and an alimentary year based on the schedule of meat consumption. This is well illustrated by the various sacred calendars known to us from Attica, specifically from the demes of Thorikos, Erchia, and the Marathonian Tetrapolis, and from the *genos* of the Salaminioi, as well as sacred laws issued by a variety of corporate bodies, from small sacrificial associations up to the state.[78] The evidence from the sacred calendars is especially detailed. We learn, for example, that the inland deme of Erchia did not schedule any sacrifices involving cattle. Over the course of a year the deme spent 421 drachmas to pay for fifty-six sacrifices, the most common of which were sheep (thirty animals). The Salaminioi, a *genos* rather than a deme, spent a comparable amount, 429 drachmas, on twenty-two sacrifices, of which six were sheep and one a cow. The calendar from Thorikos, located close to the Laurion silver mines in southern Attica, records few details concerning prices, but does reveal that the deme scheduled fifty-four sacrifices, of which twenty-eight involve sheep and only two cows. The anomaly in the Attic calendars is Marathon, where in alternating years the deme spent 668 drachmas and 732 drachmas to pay for thirty-nine and forty-six sacrifices, respectively. Although the number of sacrifices is not excessive, and is, in fact, exceeded by both Erchia and Thorikos, the expenditure is noteworthy, and includes the cost of five cows.

The data allow different interpretations. One way would be to emphasize the lack of any consistency between demes and sacrificial groups, either in terms of expenditure, choice of animal, or preferred divinities (or heroes). Certainly local conditions influenced the choices made by demesmen in relation to their sacrificial obligations. But more significant by far is that fact that the same underlying pattern exists, according to which the demes imitated each other and replicated the same basic practices. Being in a deme was not merely a matter of registration in a place to which one returned to vote on local issues. It entailed participation within a sacrificial community that on a regularly scheduled basis met to sacrifice to a variety of gods and heroes. Some of these were purely local, such as the eponymous hero Thorikos; others were local versions of well-known gods, such as Athena Hellotis, and others still were gods whose worship tied the local association to the larger region, as in the case of Zeus Polieus, who received a sacrificial pig from the deme of Erchia both "on the Acropolis in the city" and in the deme as well.[79] To honor this diverse set of divinities, from entirely local heroes to Olympian deities, the demes employed the same set of sacrificial practices, a template that fashioned the demes into larger versions of households and smaller versions of the city.

Demes conducted a schedule of sacrifices according to a calendar devoted to the gods and heroes specific to their community, defined by locality. The same was done on a smaller scale and with less frequency, but still according to a schedule, by those sacrificial groups that existed below the level of the deme. Corporate groups of all types, in fact, possessed sacred land and exploited it to support their commensality. On Mount Helikon, for example, a group calling themselves "the Hesiodic *synthutai* of the Muses" possessed land sacred to the Muses.[80] Groups such as the *genos* of the Salaminioi also owned sacred land in common, which was leased to generate revenue for sacrifices. In the case of the latter, a shrine of Herakles was rented out by the *genos* to raise the 530 drachmas needed to meet the demands of their sacrificial calendar.[81] There are at least forty-seven known Athenian *gene*, many of which claimed a deep ancestry and exerted control over some of the central religious rites at the heart of civic life. The Bouzygai, for example, conducted sacred plowings on Skiros, at Eleusis, and beneath the Acropolis to ensure the fertility of Attic soil, while the Eteoboutadai ("real sons of the Herdsman"), a distinguished clan, provided the priestess of Athena Polias and the priest of Poseidon Erechtheus, two of the most important cults of the city.[82] Aischines considered it a mark of prestige that his phratry shared altars with the Eteoboutadai (2.147). Even if the antiquity of some of these groups was fabricated to enhance their status, their continued existence reflects a basic feature of life in Classical Athens: the individual was defined by participation in an array of collective units that were variously constituted, by political function, through blood ties or according to locality, but which were all sacrificial groups. Sacrifice was the glue of the community.

Among the sacrificial groups that bridged the gap between individual household and deme were the *orgeones*, whom Pollux defines as "those in the various demes (*kata demous*) who conduct particular sacrifices on set days."[83] The records left by these sacrificial associations demonstrate how they were structured along the same lines as the larger corporations of Athenian life. One of the best examples is that of the *orgeones* of Egretes, a minor Attic hero. Like the *hieropoioi* serving the major gods of the state, the *orgeones* leased the sacred property of the hero, appointing a treasurer (*tamias*) to collect the semiannual payments. In the case of these modest leases, however, there was no question of dividing large swaths of land connected to the sanctuary into smaller plots. The entire *hieron* was leased to one individual, Diognetos of Melite, for ten years at 200 drachmas *per annum* on the condition that he kept the shrine in good repair.[84] The *hieron* consisted of a building, whose timbers and rooftiles Diognetos had to supply, a requirement also found in state regulations concerning the cult of Athena Nike.[85] The terms of the lease are particularly concerned with the upkeep of the land. Diognetos was expected to tend the trees, replacing any that died during his tenancy. Of particular interest is the clause requiring the lessee to make the building available to the *orgeones* "when they sacrificed to the hero during Boedromion." This stipulation is accompanied by regulations requiring Diognetos to make the kitchen and dining room with couches available to the *orgeones* at the time of the sacrifice in honour of Egretes. The fact that the only sacrifice specified is during Boedromion, presumably on the hero's birthday, supports the view that the shrine of Egretes served as a farm for most of the year. Yet the transaction that made possible the agricultural exploitation of the land was designed not to generate profit to the owners. They had not invested in land in order to increase their wealth. Rather, in exactly the same manner as a much larger sanctuary, they leased the land to guarantee revenue and resources for their sacrificial needs. The driving forces at work here were not investment, improvement, return, and profit, but stability, availability, and the guarantee of a successful sacrifice (and distribution of meat).

Even sacrifices sponsored by individuals replicated the patterns of sacrifice practiced by every group in Greek society, exploiting piety to assert status. As Theophrastos said of Olympias, "Whereas others sacrifice tens and hundreds of animals, Olympias sacrifices them by the thousand or ten thousand."[86] Theophrastos is especially illuminating on the significance of individual sacrifice. He illustrates *mikrophilotimia* (pettiness), for example, as follows:

> When he has sacrificed an ox, he will nail up the skin of the forehead, wreathed with large garlands, opposite the entrance, in order that those who come in may see that he has sacrificed an ox. (Theophrastos, *Char.* 21, trans. Jebb)

Private sacrifices, however, were more likely to be modest, as a mime by Herondas shows:

> Graciously accept
> this cock I sacrifice . . .
> Our well is neither ready nor deep or else
> we'd have offered an ox or a sow of crackling skin
> and not a cock for the healing of illnesses
> that you wiped away, O Lord, with the stretching forth
> of your gentle hands.
> (Herondas, *Mime* 4, trans. Fowler)[87]

Private sacrifices were not necessarily tied to a calendar and might take place for just about any conceivable reason. Theophrastos' superstitious man runs off to make a sacrifice in order to expiate the most insignificant omen, such as a mouse gnawing through his flour sack.[88] Similarly, in Menander's *Dyskolos*, the old woman is prompted to make a sacrifice to Pan and the Nymphs because of a bad dream.[89] The endless repetition of sacrifice made it so habitual that inevitably it could lose its significance for some. Theophrastos describes the shameless man who, after a sacrifice, will salt the meat, store it away, and then go to another man's house for dinner.[90] On the other hand there are sacrifices that reflect the popular feeling that the gods were actively present, and that any stroke of luck deserved thanks. Three soldiers from Kassope survived a battle and sacrificed to Herakles for their deliverance.[91] The most developed form of private devotion was the foundation of a cult by a single individual, as in the case of Xenophon, who founded a cult to Ephesian Artemis at Skillos. His description shows that it imitated larger cults, with land set aside for grazing animals and sacred laws requiring the user of the land to pay a tithe to the goddess.[92]

The seamless transition from individual to group identity made possible by the repetition of sacrificial rites is demonstrated in a *lex sacra* from Selinous.[93] The law outlines the obligations of a gentilician group, the *homosepuoi,* and sets a timetable for sacrifices to be conducted before the Kotytia festival. It is comparable to the *cippus* of the Labyadai at Delphi or the regulations of the Klytidai on Keos. However, the regulations shift between the collective and the individual. Thus, "Let them sacrifice a full-grown sheep to the pure Tritopatores" (line 15) refers to the entire group, while "Let him take out the public sacred objects (*hiara*) and put out a table before them, and burn a thigh and the offerings from the table and the bones" (line A 18) appears to concern an individual.[94] The individual in question may be the principal officeholder acting on behalf of the collective, but the subsequent clauses deal with an individual conducting a sacrifice for himself: "Let him invite whomever he wishes. And let it be possible to sacrifice after a year, at home" (lines A 20–21). How these individual sacrifices fit into the sacrificial program of the *homosepuoi* remains unclear. Even if the *oikos* ("home") in line 20 refers, in fact, to a sacred shrine within a *temenos*, as Kevin Clinton has suggested, the law envisions an individual conducting sacrifices individually.[95] This is borne out by the second column of the inscription. It begins,

> [If a . . .] man [wishes] to be purified from *elasteroi*, having made a proclamation from wherever he wishes and whenever in the year he wishes and in whatever [month] he wishes and on whatever day he wishes, having made the proclamation whithersoever he wishes, let him purify himself. (B 1–3)

What follows are guidelines for the purification of an individual under the malignant influence of an *elasteros*.[96] The stipulations show how the individual can reintegrate himself into the group by first confronting the *elasteros*, appeasing it, and then sacrificing a sheep on the public altar. Since there is no mention of a holocaust it is not unreasonable to conclude that the sheep was available for consumption after the sacrifice. The *homosepuoi* thus affirmed the reentry of the individual into the group by sharing the sacrificial feast. The Selinous *lex sacra* exhibits a certain elasticity. The rules derive from a decision of the group, but are sufficiently loose to authorize private sacrifices. A similar fluidity can be seen in the regulations for the cult of Herakles on Chios, where the sacred law defines the perquisites owed to the priest on the occasion of sacrifice but also informs us that sacrifices could be conducted either by the *genos* ("when the *genos* conducts a sacrifice . . . ") or by an individual ("whenever an *idiotes* sacrifices . . . ").[97]

The blending of individual action and group sponsorship in the law from Selinous reflects a dominant feature of Greek society, the persistent tendency toward aggregation. The consumption of meat obtained from large animals heavily favored the meeting of men in (literally) consumer groups. The consumption of meat was for the Greeks, therefore, both a social event and a religious occasion. Furthermore, unlike grain, which could be stored and apportioned as needed, sacrificial animals always favored aggregation in large groups: as *orgeones*, *genetai*, and *demotai*. As Yulia Ustinova has recently demonstrated in a study of the *orgeones* of Bendis in the Piraeus, small and large sacrificial groups overlapped.[98] The different definitions of these groups, some, such as *gene*, by descent (real or fictive), some, such as demes, by locale, point to their different purposes, but they all serve the same function: to aggregate Greek men into sacrificial and commensal bands. The larger the aggregation, the more complex the internal ordering of the group, requiring laws, office-holders and the control of resources such as lands and herds, but it is the constant repetition of the same pattern, fundamentally predicated on the killing and sharing of a large animal, that keeps the cattle complex at the heart of the Greek *habitus*, the web of practices whose repetition rendered them so natural as to be unremarkable. Accordingly, there is an underlying unity beneath the thousands of different cults with their different regulations.[99] Though there innumerable ways of deciding which animal the god wanted, what parts could be eaten, which parts were to be burned to a crisp, and which parts could be taken away or left in the lap of the statue, these are no more than endless variations on a theme.[100] The repetition of these practices constituted being Greek. The cattle experience continued to exert a powerful influence over life in Greece, shaping religious practice and economic practice in ways that the Greeks simply took for granted. Indeed, so important were cattle to the Greeks that, even though sheep and pigs far outnumbered cattle in the frequency with which they were sacrificed (according to Attic sacrificial calendars), cattle occur with almost twice the frequency of sheep and pigs combined in depictions of sacrifice on Attic black- and red-figure vases.[101] This is not to dismiss the specific restrictions and requirements of different cults. It would be valuable to know what conditions, for example, led Asklepios to accept goats at Cyrene but to refuse them at Epidauros.[102] One would like to know why the cult of Thasian Herakles banned the sacrifice of pigs and goats, a restriction confirmed from the bones excavated at the Herakleion.[103] But whatever the reasons for the ban, it is just as important to see the very existence of sacrificial regulations of this sort as one of the constituent features of being Greek. Calendars of sacrifice and feasting provide a paradigm of Greek behavior, repeated by each corporate group in Greek society, in an endless fugue of variations permitting the expression of a host of group identities operating under the overarching identity of being Greek.

The ubiquity of sacrifice among the Greeks may perhaps prevent us from recognizing how unusual it is that the same set of practices should be repeated at each level of Greek society, from the individual to the household, from the brotherhood to the deme, and from the tribe to the state. At every level of Greek society, notions of *themis*, customary "rightness," could not be divorced from ritual action involving sacrifice and feasting. This relentless repetition prompted Greeks institutions to evolve by peer emulation in ways comparable to the effects of peer polity on state formation.[104] What we find in the case of the Greeks as sacrificial groups is that the concept of peer emulation can be mapped on a vertical as well as horizontal axis. Not only is one polis like another, but so too do the smaller sacrificial groups within a community resemble the larger ones (and vice versa). One deme has a calendar of sacrifice and leases sacred land; other demes copy these arrangements, altering them according to their needs and capacities. Similarly, demes emulate the city, whose meat distributions are copied by the *phylai*, who honor their heroes as do the *gene*, who lease their property to members from within their ranks, in exactly the same way as *orgeones*, *synthiastai*, and *homogalaktoi*. All meet regularly to make sacrifice and share meat. Greek society operates like a set of Matruschka dolls, in which the size of the sacrificial group varies, but the shape remains the same. Accordingly, the economic activity of the Greeks cannot be understood solely in terms of primitive versus modernist models, since neither approach recognizes that a sacred economy also existed, which was designed to meet the sacrificial needs of the Greeks, and operated according to a completely different set of assumptions, values, and pressures. We catch a glimpse of this in the *Suda*'s definition of the expression "the seventh cow" to mean "without sensation": "When paupers did not have anything living to sacrifice they molded a cow out of dough. When the six living animals were being sacrificed—sheep, pig, ram, ox, bird, goose—a seventh, the one made from dough, was sacrificed, and since it was lifeless, it was adapted into a proverb." Better to sacrifice a cow made of flour and water than no cow at all.[105]

CHAPTER 11

Conclusions

> Compute the number of the cattle of the Sun, O stranger, and if you are wise apply your wisdom and tell me how many once grazed on the plains of the island of Sicilian Thrinakia . . .
>
> —Archimedes' *Cattle Problem*, lines 1–4[1]

Archimedes' Cattle

The mathematical problem known as Archimedes' Cattle Problem, odd as it may seem, is a good example of how the Greeks had moved from a pastoral society operating in the bovine idiom to what we might call a post-pastoral society. By this I mean a society no longer directly dependent on cattle production for all aspects of the community's well-being, as might be said (though with increasing inaccuracy) of the Maasai, Dafla, or Bahima people who have provided comparanda for this study of early Greece. Instead, the Greeks had become a society characterized by large settlements, international sanctuaries, nuanced social hierarchies, literacy, a fully articulated notion of statehood, networks of trade, and other similar indicators of social complexity. Yet, the inheritance of pastoralism remained powerful, like a marker in the cultural DNA of Greek society, capable of influencing action, judgments, and ways of thinking in spheres quite unrelated to actual cattle business.

Archimedes, a third-century BC mathematician, challenges his audience to determine the size of the herd of the cattle of the Sun, providing us only with proportions of bulls to cows in four herds arranged according to the color of their hides—one milk white (*to men leukoio galaktos*), the second sleek and dark-skinned (*kuaneo chromati*), the third tawny-colored (*xanthon*), and the fourth dappled (*poikilon*).[2] The designation of the animals as the cattle of the Sun serves, then, as a convenient narrative package in which to express a mathematical problem, where we might employ an abstract formula. The conceit, however, is a nice one, since it locates the mathematical problem in a broader cultural matrix: the cattle of the Sun appear in the *Odyssey*, as any of Archimedes' readers would know, where they are described not as a single herd, but as seven herds comprised of fifty cattle each. This total of 350 may allude to the approximate length of the solar year, appropriate since these are the cattle of the Sun, after all, so that by alluding to Homer's cattle of the Sun—note Archimedes' use of "Thrinakia," the epic name for Sicily—the Hellenistic scientist presents his mathematical work in a form that evokes the great accumulation of epic storytelling and myth that were the foundation of Greek culture.

At the same time, the emphasis on the color distinctions within the herds suggests that Archimedes was drawing on other traditions and associations, some connected to epic, others not, but all certainly pastoral in origin. For example, it is a little hard to imagine Archimedes creating the four color categories if colored herds were unfamiliar to the Greeks. It is more likely that by Archimedes' day, in the late third century BC, the Greeks were familiar with herds, not simply single animals, that had been bred for color. How does this concern pastoral practices? Breeding for color need not be merely an aesthetic choice. Pastoralists recognize that dark- and light-colored cattle have different responses to heat absorption and stress, as well as different rates of growth. Dark cattle are much more susceptible to heat stress, but this is a problem encountered less often at higher altitudes, where the dark cattle quickly absorb warmth from the rising sun. Lighter cattle, by contrast, are hardier and more heat resistant. Accordingly, breeding herds of different colors suggests an awareness of environmental variables such as elevation, daytime heat and nighttime cold. Archimedes' colored herds may reflect an understanding of the challenges of adapting cattle to the natural environment.[3]

Environmental factors aside, breeding for color is also connected to the symbolic value of cattle in certain pastoral societies. The Ankole cattle of Uganda, for example, range from light brown (*siina*) to dark brown (*mbindi*), but the preferred color is a medium reddish-brown (*bihogo* for cows, *ruhogo* for bulls). Herds of a single color are especially prized as an indication of the skill and power of the herder to control breeding from only the most desirable bloodlines. Not surprisingly, in the past single-color herds were associated with kingship, since only the king, or *omugabe*, had the means to amass such distinctive wealth. In Greek culture, there is evidence for particular herds reserved for kings, such as the four hundred cattle belonging to the royal family of Epiros.[4] There is also a marked preference for individual sacrificial animals to be of a single color, usually white, such as the bull Athena gives to Glaukos in order to win Poseidon over, but there are also references to herds of a single color.[5] Euboia, for example, was known for its white cattle, and white cows were often associated with Hera and Io.[6] In the *Argonautica*, Apollonius refers to the cattle of the Sun as "milk white and rejoicing in golden horns."[7] His description may reflect both Greek and Egyptian notions of purity, since both cultures emphasized the importance of sacrificial animals being unblemished (*kritos*).[8] In *Homeric Hymn III to Apollo*, for example, the god prophesies that he will receive unblemished hecatombs, while Herodotos describes the Epaphos bull thus: "If they see even one black hair on them, the bull is considered impure."[9] The Greeks were probably also familiar with the famously aggressive Aithiopian red bulls, which were reputed to be flesh-eating.[10]

It is also possible that, just as Ankole royal herds conveyed status in their uniform coloring, the cattle of the Sun reflect a time when the colors of the herds of the Greeks functioned as the basis of a wider code of social and economic practices. Andreas Eckl has demonstrated that among the Herero people of Namibia the color and hue of cattle are part of a complex set of taboos, prescriptions, and proscriptions, specifying which cattle could be sacrificed and consumed by different individuals and groups. One result of this seemingly arbitrary system is that in times of drought or sickness, provided that not all of the animals have perished, there will always be some cattle preserved from sacrifice and left to form the nucleus of a new herd.[11] The Greeks were similarly aware that herds had to be managed. According to Androtion, an ancient law at Athens forbade killing sheep that were either unshorn or had not been bred, so as to guarantee further generations of animals.[12] It must be emphasized that I am not arguing for a reductive view of Greek culture, according to which every significant cultural gesture, value, or idea arises from the fact that once upon a time the Greeks bred cattle. Rather, I am suggesting that by recognizing pastoralism as a powerful, affective experience that continues to shape a culture long after it has been superseded as the dominant economic practice one may understand more fully the distinctive practices of Greek culture.

Pastoralism, in fact, lies at the very heart of an entire epistemology, as is demonstrated when one considers the clash of pastoral and non-pastoral cultures. Utterly different mindsets are revealed, for example, in the confrontation between property-owning Europeans and indigenous hunter-gatherers in eighteenth-century Australia. The differences between them were not solely a matter of technology but of values, conceptions, and ideation. One group saw the land as property belonging to no one because it was not exploited, giving rise to the legal justification for its seizure as *terra nullius*.[13] The other saw the land as the creation of ancestors in the Dreamtime, recreated by corroborrees and songlines, and temporarily occupied by humans. In technological terms, in AD 1770 the indigenous people of Australia were Paleolithic hunter-gatherers. More importantly, in epistemological terms their experience of the world was that of people who did not live in a domesticated cosmos: no herds, and no sacrifice.

Once domestication of large animals had become an intrinsic part of many human societies, it initiated deep-seated cultural changes not only in lived experience but in the realm of ideas and symbols that have left their mark in the material record. Evidence of cattle burials, for example, either in association with or even separate from human burials can be traced back in central Europe as early as 3500 BC. Attempts to explain these burials have included the cosmological (the animals are sacrificed to appease the gods and reassert cosmic order); eschatological (the animals are companions for the human dead in the next life); sociological (the animals are evidence of the individual's wealth). A convincing interpretation has been hard to agree upon, in part because there is no consistent pattern in how the burials relate to humans. The cattle represent only 25 percent of animal bones found in human burials, yet represent 62 percent of animals buried in contexts not associated with a human burial.[14] The burials, then, do not display a compelling pattern of deposition vis-à-vis humans. They do, however, occur at a time when there is evidence for the emergence of a distinctive sun symbolism on incised amber and Neolithic pottery, in which sun disks, wagons, and wagon-wheel motifs occur together. This combination of motifs has been plausibly interpreted as evidence for a late Neolithic belief system in which the sun was imagined to be conveyed across the heavens in a wagon pulled by cattle.[15] According to this scenario, the buried cattle will have been prized for their power as draft animals, perhaps imitating divine creatures.[16] Their use as draft animals may also have been connected to both long-distance trade and elaborate funeral ritual, since the existence of wagons suggests the building of roads, while the earliest wagon ruts in Europe occur at the edge of a long barrow.[17] Many of the best examples of sun-wagon decoration occur on vessels associated with the Globular Amphorae Culture, and appear to coincide with a marked decline in the consumption of cattle compared to other sources of animal protein: at the Neolithic sites in the Kuiavia region of Poland, for example, cattle amount to 65–87 percent of the domesticated animal consumption in most periods, but at the time of the Globular Amphorae Culture this drops to 19 percent.[18] The correlation is suggestive: as the cosmic significance of the cattle became more firmly fixed, actual consumption decreased. It is difficult to escape the conclusion that the animal had become an object of sacred significance. It is entirely possible, then, that the original cattle of the Sun were oxen that dragged the sun's chariot across the sky.[19]

Epic and Pastoral *Habitus*

Heirs to the pastoralists more than to the hunter-gatherers who came before them, the Greeks inhabited a dichotomous world: on the one hand were the conditions of their own day, and on the other the imagined world of heroes. Fundamental to this, we have seen, is Homer. The notion that epic articulated concepts such as *arete* and *time* ("excellence" and "worth"), notions that remained central to Greek culture, is hardly a remarkable observation, but the degree to which these values are themselves part of Greece's pastoral inheritance has not been explored fully. The *Odyssey* is concerned with the survival of the *oikos* and the preservation of its wealth, an anxiety shared by wealthy and poor alike. Cattle serve both as a symbol and species of that wealth. They provide a focal point for the narrative not only because cattle are the measure of value in a premonetized economy, but also because arrayed around cattle is a constellation of institutions central to Iron Age Greek society. *Xenia* (guest-friendship), sacrifice, and feasting are the dominant institutions of Homer's world and together constitute the normative practices of civilized life. Each is informed by the habits of mind arising from herding. It is the possession of cattle that marks a man as prosperous. It is their sacrifice that shows that a hero is pious. It is the consumption of meat that reinforces social bonds, and it is the invitation to a stranger to share this that connotes proper behavior.

Nor are cattle integral to these institutions merely as commodities. It is the balance between breeding and the accumulation of wealth, on the one hand, and sacrifice, consumption, and sharing of wealth, on the other, that foregrounds moderation. To withhold animals for sacrifice would starve the community and be impious. At the same time, the indiscriminate killing of animals, gluttonous feasting, and waste violate the delicate balance underlying the cattle system and are punishable. In fact, it is around these two axes that the *Odyssey* revolves, and inasmuch as the *Odyssey* constantly returns to the question of which behaviors make social life possible and which threaten it, one might say that for Homeric society the cattle idiom is paradigmatic.[20]

The continuing hold of these values and practices on the Greeks can be seen by comparing the Homeric system of sacrifice and feasting with the observations of Athenaios of Naukratis, the rhetorician who flourished at the end of the second century AD, close to one thousand years after Homer. When Athenaios wants to capture the liberality of a man like Konon, he asserts that rich men show their generosity with a sacrifice and a great feast. The mark of a truly liberal man is that he will butcher an ox for an uninvited guest.[21] No one would claim that in Athenaios' time the cultivated elite of the Graeco-Roman world relied on cattle the way that the pastoral Bahima do, drawing off small amounts of blood to take nourishment directly from their long-horned Ankole cattle. Yet the institutions grounded in pastoral economics—gift giving, raiding, patronage, sacrifice, and feasting—continued to function as the fundamental institutions of ancient Mediterranean culture. Pastoralism established the *habitus* within which Greek culture continued to change.

Perhaps the most distinctive aspect of social life that reveals this is magic, since it is here that the human and animal spheres are closest. Take, for example, this old Macedonian spell against the evil eye:

> The cow gave birth to a calf.
> Its mother, licking it, undid the evil spell with her spittle.
> And I, the mother, have undone the evil spell by licking.
> May the child live and prosper and not suffer any harm!
>
> Dappled cow
> Dappled calf
> Went to market and was not evil-eyed.
> Thus may the child not be evil-eyed.[22]

Sympathetic magic like this operates precisely because the bovine and the human are felt to be interchangeable categories. Such identification is not the result of mere familiarity, otherwise one could just as easily invoke a turnip. The magic relies on a deep-seated notion, rooted in the experience of watching, molding, and making life-and-death decisions about cow and calf. Herds must be nurtured. In Greek folklore, herds were thought to be threatened by an evil spirit known as a *smerdaki,* whose baleful influence caused animals to abort. The herdsman could sanctify his pastures with holy water and minimize the *smerdaki's* influence by heeding the advice of the priest: "he must not shed blood in his herd, for this act [would] make things worse."[23] The domesticated herd shapes the human consciousness of society, even if the identification is as crude as referring to unmarried girls as "unbulled" or "unyoked."[24]

Across a range of values and behaviors cattle are a recurring motif in the mythology of early Greek society. They are the measure of wealth and the medium of exchange. How we treat them marks us as either pious or not. In fact, so tight is the hold of the cattle idiom on the Greek imagination that cattle-themed myths provide the outlet for expressing in narrative form an experience as utterly unrelated to stock raising as the Greeks' first contact with Spain, Italy, and Sicily. They help crystallize notions of reciprocity between humans and gods, as objects of sacrifice, and normative social behavior within human communities, through feasting. They also help to define new ideas of the hero, since as the object of raids they provide the opportunity for heroic performance in ways analogous to warfare. Homer's heroes, models for the big men of the Iron Age, are cattlemen, huge of appetite, generous but not wasteful with their meat, scornful of trade. They are defined through their cattle habits, measuring wealth in terms of cattle, appeasing the gods with sacrifice, and fashioning social relations around the sharing of meat and the bartering, or stealing, of cattle. They are not, however, the mere relics of an earlier age. In the hands of poets and storytellers such as Homer and Stesichoros the various stories employing cattle motifs proved infinitely flexible, operating as allegories for the confrontation with death or giving expression to the anxieties of an age of increasing cultural contact. The hero, as a focus for storytelling and ritual, bridged the gap between the golden age and the present. In the Archaic period, as communities at home grew larger and those abroad established footholds around the Mediterranean, heroes assumed a central position in the imagination of the Greeks. The tombs of heroes provided a venue for veneration, forging a connection between the present inhabitants of a region and their ancestors.[25] Heroic genealogy allowed communities to assert relationships based on the fictional ties of blood. Together heroic poetry and hero cult provided the basis for a symbolic system that would underpin Hellenic identity and the state formation of the Archaic period. In its fully developed form, this system embraced all who could show a blood affiliation to a descendant of Hellen through one of three strands: Aiolos, Doros, or Xouthos (from whom was descended Ion).[26] This Hellenic genealogy is not identical to the network of Homeric heroes, but it is complementary. Stories of Homeric heroes and local heroes could overlap, so that Achilles was identified with Phthia in Thessaly, yet was also acknowledged as the best of the Achaians. Sometimes the connection between the catalogues of Homeric heroes and local heroes was more tenuous. Schedios, for example, rates a mention in the Catalogue of Ships as a captain of the Phokian contingent, but accomplishes nothing in the *Iliad*, other than to die twice at the hands of Hector. He was, however, honored with a tomb in Phokis.[27] Hellenic identity, in fact, was shaped by exactly this tension, between the pull of the local and the push to be part of something larger.[28]

The bovine idiom, that legacy of the Greeks' familiarity with pastoralism from the Neolithic onwards, was especially effective in helping the Greeks navigate the tumult of the Archaic peiod. This was a period when the balance tipped from epichoric identities in favor of a Panhellenic identity (although it would be a mistake to think that regionalism was ever entirely supplanted). These were the centuries of the Olympic Games, of the writing down of Homeric poetry, of Hesiod and the earliest of the Homeric Hymns, and of the emergence of Delphi, Olympia, Nemea, and Isthmia as sanctuaries serving more than their local populations. In terms of institutions this was a period of intense "intersocial organization," as Greg Nagy puts it.[29] And as the characteristic practices of Classical Greek culture took shape, this process often occured in the shadow of their cattle experience. "Whence did the graces of Dionysos appear with the ox-driving dithyramb?" Pindar asks, setting the origins of the genre in a cultic procession.[30] In terms of the Greek *imaginaire*, it is the age in which the Olympian order was established.

Like their heroes, the gods of the Greeks reflected the values and mindset of a stock-breeding society: Zeus who transforms himself into a bull, cow-eyed Hera, Apollo the cowherd, and Hermes the cattle rustler. But there are contradictions at the heart of the bovine idiom as well: the gods are tauriform, yet invoked through sacrifice, while the society in which they were worshipped looks to sacrifice for meat, yet is moving toward an increasing reliance on mixed farming dominated not by pastoralism but by agriculture. Herding is pushed to the margins or reduced to the ownership of only one or two plow oxen, while the institutions of sacrifice and feasting demand an increasing supply of sacrificial animals. The place where these transformations were effected and contradictions resolved is also the place where the pastoral economy and religion intersected: the sanctuary. As we have seen, the hold of the pastoral legacy on the Greeks is modulated through a set of religious institutions that exist both within and without the polis. The sanctuary, with its smoking altar, lowing cattle, and bloody knife, is ubiquitous throughout the polis, and yet the very most important sanctuaries exist quite independently of the polis. This is the distinctive and peculiar institution of Greek life, shaped by and in turn shaping the transformation of meat from a physical source of nutrition into a medium of contact with the divine. And it is that recursive quality that reveals the pastoral *habitus* of the Greeks, an inevitable tendency to evaluate behavior through the lens of cattle business.

The traditions that form a part of a society's *habitus* are capable of functioning at different levels of consciousness. The pinstripe knickerbockers of baseball players are a conscious connection with the past, deliberately evoking the game's century-long continuity. On the other hand, a tradition such as Spanish bullfighting seems to be a practice of deep antiquity, but continues to exist in part precisely because its origins are, in fact, murky, allowing it to serve a multitude of purposes within a modern setting: a reassuring affirmation of Spanish identity in a global culture, a gesture to conservatism at a time of cultural change, even, as one critic expressed it, as "the prime target of the libidinal hermeneutics of the early 20th century."[31] It is, however, the notion that a practice is simply innate to a society that shows it has become embedded within its *habitus*. Thus, while baseball is played from Japan to Venezuela, its uniform marks it as an American institution, rooted in the vacant lots of nineteenth-century New York. Similarly, *corridas* can found from the high plateau of Madrid to dusty villages in Mexico, affirming cultural affiliations through an assertion of continuous practice. When the customary is equated with the natural it has fixed its roots deep.

In the Greek setting, the practices that most fully connoted Greekness were the clearest evidence of Greece's bovine legacy. This can be seen from Greek descriptions of foreign practices. Describing the treatment of the Epaphos bull in Egypt, for example, Herodotos writes:

> [The priests] lead the animal marked for sacrifice to the altar where they are going to sacrifice it. Next they kindle a fire, and after that, they pour libations of wine over the altar so that it flows onto the victim. Then, after calling upon the god, they cut the animal's throat, after which they cut the bull's head completely off. They next flay the animal's body before calling down all sorts of curses on its head. In places where there is a market with Greek traders, they will take the bull's head to the marketplace and sell it; in communities where there are no Greeks, they hurl the bull's head into the river. This is the curse they pronounce on the heads: they pray that if there is any evil about to befall either those making the sacrifice or Egypt generally, may it fall upon the head instead. Now with respect to the heads of sacrificial animals and the wine libations poured over them, all Egyptians conduct all their sacrifices according to the same set of customs; and because of this custom no Egyptian will eat of the head either of this or of any other kind of animal. On the other hand, they have different practices when it comes to the way of disembowelling the victims and of burning them . . . (Herodotos 2.39)

The passage fluctuates between the familiar and the bizarre. As in Greek practice, the sacrifice is set at an altar, requires fire, and begins with a procession, a libation, and a prayer. The preliminary stages thus contain no surprises. Two common practices are omitted, sprinkling the sacrificial beast with grain and gaining its assent by inducing it to nod, but Herodotos does not discuss the omission. Then, after the critical moment of cutting the animal's throat, the sacrifice begins to depart from the Greek model. The animal is beheaded and flayed. In Greece head and hide constitute two of the most common perquisites given to the priests officiating at the sacrifice. Here, however, the head is treated completely differently. So great are its magical properties that it becomes a vessel into which malign power can be poured. Yet selling the head in a Greek market reflects its ambiguity: its power remains efficacious to the Egyptian, so much so that if no market is available it must be hurled into the Nile, but to the Greeks it is no more than a commodity. In other words, the bull's head is as only as significant as the two communities allow it to be. Herodotos goes on to describe how dead cows are thrown into the river, while bulls are buried, and their bones collected and transferred to the sanctuary of Atarbekhis.

In his description of the Skythians the Greek historian also reveals a fascination with the sacrificial habits of non-Greeks. Take, for example, his account of the Taurians:

> It is the custom of the Taurians to sacrifice to the Virgin Goddess all shipwrecked sailors and any Greeks whom they have captured having come by sea. This is what they do: after performing a preliminary rite, they hit the victim on the head with a club. There are some who say that they toss the body over the edge of the cliff, for their temple is built on a promontory, and impale the head on a stake. Others agree about the head, but maintain that the body is not tossed off the cliff but is buried. The Tauri themselves maintain that the spirit to whom they make this sacrifice is Iphigenia, the daughter of Agamemnon. When they get their hands on an enemy this is what they do: after they have cut off the enemy's head, they take it home. Sticking it on a long pole they then set it up over the house, usually near the fireplace. They say the heads hanging over it serve to protect the whole house. They live by war and plunder. (Herodotos 4.103)

Herodotos' entire treatment of other cultures, whether Egyptian, Persian, or Skythian has long been recognized as reflecting the ethnographer's fascination with otherness, the mirror in which we see ourselves more clearly.[32] In the account of the Taurians he stresses the alien and the exotic. Unlike the Greeks they sacrifice humans, and unlike the Greeks they kill the very people who most deserve the formal treatment of *xenia*, or hospitality: those who have been shipwrecked and those who have come over the water and are therefore vulnerable to the elements and chance. Furthermore these anti-Greeks treat the bodies of the dead in a manner that is all the more despicable for being a parody of a religious ritual. Instead of washing and cremating or burying the body, they dismember it and then treat it as a hideous apotropaic trophy. Herodotos' treatment of Egyptian and Skythian religion confirms not only his belief that custom is king, but also explores where the limits lie—this side of which one is acting Greek and beyond which one has become something else. That border is defined by sacrifice, a clearly defined cultural system capable of replication from the household hearth to the Altis at Olympia. How appropriate that the Greeks themselves recognized that their relationship with cattle was definitive, as is reflected in Anaxandrides' remark to an Egyptian, "Neither our practices nor our customs are in accord, but are completely opposite. You worship cattle; we sacrifice them to the gods."[33] Butchering a large animal is the gesture that most forcefully conveys a distinctively Greek identity. Little wonder, then, that the founding heroes of the Athenian democracy, Harmodios and Aristogeiton, should be portrayed in the very act of hacking the tyrant to pieces, or that ever after, when the Athenians announced honors to those whose actions had strengthened the democracy, they did so immediately before a festival sacrifice, in which cattle would similarly shed their blood for the benefit of the community.[34] In this respect, democracy was the final gift of the Greeks' pastoral legacy.

Notes

Chapter 1
Cattle Habits

1. Pollan 2006.
2. See Adams 2000 on meat eating and spousal abuse. For a more historically nuanced reading, see Donovan 1990.
3. Singer 1975; Regan 1983.
4. Dombrowski 1984.
5. Porphyry, *Abst.* 3; Aristotle, *Rh.* 1.13 1373b (Empedokles); Aristotle, *Pol.* 1.3.7 1256b. For discussion, see Sorabji 1993, 156–57.
6. Aristotle, *The Nichomachean Ethics*, trans. H. Rackham (Cambridge, Mass., 1982).
7. Heath 2005.
8. Strabo 5.2.7.
9. Aristotle, *HA* 3.21 on milk production; 8.7 on feed and increasing size; 5.31 on ticks and lice; 8.23 on foot disease (probably caused by *Fusobacterium necrophorum* or *Bacteroides melaninogenicus*) and lung disease (possibly pneumonia); 9.50 on castration; 8.7 on Epirote cattle; and 9.45 on Paionian bison. See also Arrian 2.16 on the rich pasture and fine cattle of Epiros.
10. Georgoudis and Baltas 1998, 28.
11. Carlson 2001, 88–90.
12. Homer, *Odyssey,* trans. S. Lombardo (Indianapolis, 2000).
13. Zeus' favor: Homer, *Il.* 4.45–50; Poseidon's anger: Homer, *Il.* 7.446–53. See also Adkins 1972 and Heiden 1997.
14. On sacrifice and communication, see Prescendi 2007.
15. On the growth of the brain, see Krantz 1968; Fialkowski 1986; and Stanford and Bunn 2001.
16. Evans-Pritchard 1940, 19.
17. Lincoln 1981, 16.
18. The centrality of the cattle experience to any group need not be equated with the mistaken view that their ethnic identity is natural, fixed, and unchanging. For an excellent appreciation of the importance of cattle to a pastoral society undergoing rapid change, see Oppong 2002.
19. Bourdieu 1967. Tolstoy anticipated the notion one hundred years earlier: "There are two sides of the life of every man, his individual life which is the more free the more abstract its interests, and his elemental swarm-life in which he inevitably obeys laws laid down for him" (*War and Peace,* trans. L. Maude and A. Maude [London, 1922–23], 9.1).
20. Rigby 1969, 24.
21. For a recent treatment of the notion of culture in the work of ancient historians, see Hall 2004.
22. See Evans-Pritchard 1953; Adebayo 1991, 2.
23. Runia 2006 (quotation from p. 1).

24. Benton 1961. The close association of birds with bulls on Late Bronze Age vases was noted by Furumark (*CMP*, 434), and is especially prominent in Cypriot art; see Karageorghis 1956. Hesiod may also be familiar with a connection between water birds and cattle; at *Op*. 448–50 he interprets the cry of the crane as a signal to start plowing.

25. Tartaron 2001, 22–24.

26. Milichos River: Pausanias 7.22.11; Sybaris River: [Aristotle], *Mir.* 169; Kerkyra bull: Pausanias 10.9.3–4.

27. Derks 1995.

28. Gallant 1991. For more recent studies of risk management in strategies of husbandry and agriculture, see Halstead and O'Shea 1989, 3–4; and Hesse 1995, 212.

29. Xenophon, *HG* 6.4.29.

30. Bourdieu 1980. See also Narotzky 2007, 7.

31. On the *boule*, see Rhodes 1972, 132. On the iconography of sacrifice, see Van Straten 1988, 51–68.

32. García Soler 2001.

33. *Suda*, s.v. *boos*.

34. Aristophanes, *Eq*. 1113–30. See Wilkins 1997, 2000; and Zumbrunnen 2004, 670.

35. Aristophanes, *The Knights*, trans. R. A. Neil (Hildesheim, 1966).

36. On the exchange of insults, see most recently Rosen 2007, 148–49.

37. Galinksy 1972, 81–100 (the comic hero) and 294 (on the multiplicity of roles). The voracity of Herakles is the theme of Book 10 of the *Deipnosophistai*, announced by Athenaios at the close of Book 9.

38. Athenaios 10.1 411c (= Pindar frag. 168b).

39. Atheniaos 10.2.

40. Loraux 1990; for references to Herakles' gluttony, see 30 n. 39.

41. A similar trajectory can be seen as Arnold Schwarzenegger has moved from the ludicrously hypermasculine beefcake, the aptly named Mr. Olympia, of *Pumping Iron* and *Hercules in New York* to *Junior*, posters for which show him heavily pregnant. Even his elevation to the governorship of California represents a modern apotheosis (of sorts).

42. Graber and Richter 1987.

43. Theophrastos, *Char.* 5.

44. Theodoros of Hierapolis, *On Athletic Contests*, in Athenaios, 10.412e.

45. *Suda*, s.v. *Polydamas*.

46. Euripides, *Autolykos* (= Athenaios 10.5).

47. Sophocles, *Aj*. 20–35, 65–70.

48. Plato, *Lg*. 782c.

49. Athenaios 2 Epitome 21.

50. Henrichs 2003, 212.

51. Bremmer 1992.

52. For a diametrically opposite movement, the adoption of meat eating in a broadly vegetarian culture, see Khare 1966.

53. Diogenes Laertius 8.36.

54. Plutarch, *Mor.* 792F, trans. E. L. Minar Jr., in Plutarch, *Moralia*, vol. 9 (Cambridge, Mass., 1969).

55. Athenaios 3e.

56. Diogenes Laertius, *De Pythag.* 12.

57. Athenaios 8e–9f, 25d; Schol. ad Homer, *Il.* 16.747. See Dalby 1996.

58. Hippocrates, *On Ancient Medicine*, trans. W.H.S. Jones (Cambridge, Mass., 1923).

59. On maenadism and *omophagia*, see Rapp 1872 and Henrichs 1978. On the ecstasy of Dionysos' followers, see Portefaix 1982.

60. Lonsdale 1979.

61. For treatments of the *boukoloi* as rebels, see Alston 1999. For the *boukoloi* in the Greek novels, see Winkler 1980; Rutherford 2000a. Shaw 1984 remains indispensable. On *boukoleo* in Attic Greek and Kratinos' *Boukoloi*, see Rosen 2003.

62. For the *boukoloi* as priests of Dionysos, see C. Alexander 1932; Cumont 1933. On the *boukoleion*, see [Xenophon], *Ath. Pol.* 3.5 and Chap. 5 below.

63. For the Cypriot evidence, see Karageorghis 1991, 139–48, 180–81, pl. C–CIII ("sanctuary" models, bull figurines and protomes) and Karageorghis 1993, 118–22, pl. LXVII–LXX.

64. Demosthenes, *De cor.* 218.

65. Theokritos, *Id.* 17.127–28.

66. Sutton 1993; Spear 1993.

Chapter 2
The Paradoxes of Pastoralism

1. Baudrillard 1994, 129.

2. Troy et al. 2001; Bruford et al. 2003, 906.

3. Cymbron et al. 2005; Pellecchia et al. 2007; Achilli et al. 2007.

4. On domestication, see Helmer 1992, 92–95. For Argissa Magoula, see Boessneck 1962. On chronology, see Gautier 1990, 145, and esp. Perlès 2001, 41–42.

5. Meadow 1989; Vigne et al. 2005; 12.

6. Arbuckle 2005; 18–19.

7. Downs 1960.

8. Marciniak 2005, 40. For interpretations of meat eating according to a cost/benefit analysis, see M. Harris 1985.

9. Ingold 1974.

10. Hamilton 1989, 129.

11. Longevity: Finch and Stanford 2004; brain development: Stanford and Bunn 2001.

12. R. Foley 2001, 306.

13. Cordain 1999, 23; C. S. Larsen 1995.

14. Stanley 1995. See also Allen 1994.

15. Rubin 2005, 39.

16. Lepetz 1996.

17. Boessneck 1962; Gejvall 1969, 33–34.

18. Mirov 1945.

19. Lorenz 1971. Gould (1979) uses neoteny to explain the changes in the drawing of Mickey Mouse. As the size of the mouse's head increases in proportion to body the animal becomes more appealing and less ratlike. On neoteny in domestication, see Wayne 2001. On human neoteny, see Bogin 1999 and Shaner et al. 1989, 73–78. Aristotle (*HA* 2.1) comments on the heavier appearance of wild cattle and wild boars compared to their domesticated relatives.

20. On symbiosis, see Budiansky 1992.

21. Hecker 1982.

22. Gautier 1990, 10.

23. Cauvin 1978, 141; Helmer 1992, 155.

24. Cauvin 1994. On animal symbolism, see Helmer et al. 2004.

25. Hodder 1990.

26. Ducos 1978; Helmer 1992, 81–82; S.M.J. Davis 1995, 148.

27. Davis and Valla 1978; 608–10.

28. Ehret 2001, 228.

29. Reinold 2005. On the Kerma bucrania, see Chaix and Hansen 2003, 119. The association of cattle with elite status remained embedded in Sudanic society. See Strabo 17.2 on Meroe.

30. Patou-Mathis 1997; Trinkhaus 1985.

31. Wendorf 1968; Camps 1992; Guilaine and Zammit 2001, 103–9.

32. Wahl and König 1987.

33. Kikawada and Quinn 1985.

34. Ohrmazd Yasht 7; Khwarshed Niyayesh 6.

35. De Bary et al. 2001, 24.

36. Strabo 5.2.7.

37. Spenser 1934, 217. In Book 2 of *The Faerie Queene* Spenser describes the dispossession of Albion by Brute, who finds the land a "saluage wildernesse, unpeopled, unmanurd, unprou'd, unpraysed," but possessed by indigenous giants, the offspring of Diocletian's fifty daughters (!) (2.10.5).

38. Chang 1993, 691.

39. Ibn Khaldun, *Al-Muqadimah* 54–55.

40. Gellner 2000.

41. For the collapse of the tribal economy of the Dafla, see Sikdar 1982.

42. Simoons 1968. For recent mitochondrial analysis of the breeds, see Bruford et al. 2003.

43. Baker and Manwell (1991, 235) note that the mithan's "local socioeconomic importance is out of all proportion to their relatively small numbers."

44. Simoons 1968, 55.

45. Simoons 1968, 174.

46. On the danger of using the study of cattle systems, particularly African examples, to create a false universal type, see Boyce 1987.

47. Evans-Pritchard 1946.

48. Wilson et al. 2003, 106.

49. Lévi-Strauss 1963, 89 and 101. See also Mullin 1999, 208.

50. Infield 2003; see also Evans-Pritchard 1934. On the impact of color-related taboos on feasting, see Crandall 2000, 123.

51. Schapera 1965; H. F. Morris 1964; Coupez and Kamanzi 1970. Among the Bahima these songs are sung by herders, while in neighboring Rwanda similar songs are performed only by a bardic class.

52. Plato, *R.* 459a. The notion predates Plato and can be found in Theognis 7.183–192.

53. Aristotle, *Pol.* 1254b.

54. Euripides, *Med.* 230–34.

55. Lienhardt 1961, 21.

56. Philostratos, *VS* 501. In Philostratus, *Lives of the Sophists*, trans. W. C. Wright (Cambridge, Mass., 1921).

57. Kruijt 1906, 155.

58. Junod 1912, 376, cited by Lévy-Bruhl 1996, 44.

59. *The Táin: From the Irish Epic Táin Bó Cúailnge,* trans. T. Kinsella (Philadelphia, 1985).

60. *Suda*, s.v. *agelarchês*; Plutarch, *Rom.* 6.

61. 1 Enoch 85–90. See Halperin and Newby 1982.

62. The next stage of this manipulation is upon us: the fusing of human and bovine DNA to create a chimera for medical research (see http://news.bbc.co.uk/2/hi/health/6121280.stm).

63. Peires 1989, 180.

64. For a description of a sacrifice illustrating all these elements in combination, see Siimets 2007.

65. Hahn 1896. See also Isaac 1962.

66. Detienne 1989, 20.

67. On hunting and sacrifice, see Meuli 1946; Burkert 1972, 1987, Vernant 1981. The foundation for all later studies of sacrifice is Hubert and Mauss 1899.

68. On domestication and sacrifice, see J. Z. Smith 1987. For differences between hunting and sacrifice, see Jensen 1963.

69. N. Marinatos 1986.

70. *Suda*, s.v. *apoboukolisas* and *boukolêsas.*

71. Plutarch, *Quaest. conv.* 8.8.729. On the differences between the hunter's response to killing and the herder's, see M. Lambert 1993.

72. For examples, see Cumont 1937.

73. Plato, *Euthphr.* 12e–13c.

74. Aristotle, *MA*. On the implications of animal motion, see Laks and Rashed 2004.

75. On expiation and cleansing in Israelite and Mesopotamian sacrifices, see Milgrom 1981.

76. Dietrich 1988, 37.

77. On Fontbregoua, see Villa et al. 1986. For the status of humans and animals, see Boulestin and Gomez de Soto 1995.

78. Runnels et al. 2009.

79. Bégouën and Clottes 1987. For cultic interpretations of Le Tuc d'Audoubert, see Bossinger 2005, 39. On recent approaches to European stone-age rock art, including discussion of Chauvet, see Clottes 1998.

80. Reinhardt et al. 1994.

81. Mussi 2001, 351, 364–365; Bachechi and Martini 2002.

82. On sacrifice as a communal activity, see Rudhardt 1992.
83. Stanner (1963) attempted to interpret the Murinbata *punj* ceremony (an adolescent initiation) as sacrificial. For a decisive rebuttal, see Keen 2005.
84. On the bull in Neolithic religious systems in Europe, see Hayden 1985; also Schwabe 1994. Stories of flesh-eating horses revolve around a similar tension between domestication and the threat of wildness; see Detienne 1971.
85. On Gilgamesh as a meditation on the socialization of the individual, see Abusch 2001. For an overview of earlier readings, see Vulpe 1994. The theme of nature versus culture is noted by Kirk (1970, 142–143).
86. A. R. George, *The Babylonian Gilgamesh Epic: Introduction, Critical Edition and Cuneiform Texts,* 2 vols. (Oxford, 2003).
87. *Epic of Gilgamesh*, LB 1.351–60. See Lugalbanda in the Mountain Cave in Oxford's Electronic Text Corpus of Sumerian Literature (http://www-etcsl.orient.ox.ac.uk/).
88. Sallaberger 2008, 26–29.
89. On the ideological significance of distance, see Helms 1988. On the Humbaba episode, see Blenkinsopp 2004; J. P. Brown 2001, 131.
90. *The Epic of Gilgamesh: A Myth Revisited,* trans. D. P. Jackson (Jerusalem, 2001).
91. Hiltebeitel 1980.
92. Ornan 2003.
93. Michalowski 2003, 112.
94. Willis and Curry 2004, 46.
95. Willis and Curry 2004, 46.
96. M. S. Smith 2001, 32.
97. M. S. Smith 2001, 32.
98. Smith 2001, 32.
99. Green 1984.
100. Wengrow 2001.
101. Herodotos 2.38, 3.28. See also Cicero, *Rep.* 3.9.14.
102. Plutarch, *De Is. et Os.* 29 and 43; Diod. Sic. 1.85.4.
103. Porphyry, *On Images,* frag. 10.
104. Pyramid Texts, Utterance 307. In *The Ancient Egyptian Pyramid Texts,* trans. R. O. Faulkner (Oxford, 1969).
105. R. L. Alexander 1993.
106. For the orthostats from Alaca Höyük, see Bittel 1976; for Murshili's seal, see Neve 1993.
107. Bonatz 2007.

Chapter 3
Cattle Systems in Bronze Age Greece

1. For Argissa Magoula, see Boessneck 1962. On breeds, see Mason 1996.
2. Theopompos, *FGrH* 115 F 38.
3. On cattle origins, I have benefited from the generosity of Georgios Arsenos, Aristotle University of Thessaloniki, who shared with me his unpublished paper

"Animal Husbandry during the Archaic and Classical Periods in Greece with Particular Reference to Cattle Production." On Celtic origins, see Alderson 1992. For an overview of Neolithic and Bronze Age cattle, see Payne 1985.

4. Cymbron et al. 2005.

5. For climatic conditions in Greece, see Boyazoglu and Flamant 1990. For the plow in the Neolithic Aegean, see Crouwel 1981; Pullen 1992. On cattle breeding in the mountains, see Zervas 1998. For evidence of Bronze Age cattle raised for traction rather than meat, see Greenfield and Fowler 2005, 111–112, and for the significance of traction in general, see Sherratt 1983.

6. Preziosi and Hitchcock 1999, 64–65.

7. Rehak and Younger 1998, 165.

8. Hadjisavvas 2003, 22.

9. Killen 1964; Halstead 1981a, 1981b.

10. Bennet 1985, 237.

11. Schoep 1999, 212. Regional environmental variations are also emphasized by Halstead 1981b.

12. Watrous 1984.

13. Palaima 1989, 98.

14. Aiwolos: Ch 896; Kelainos: Ch 896; Xouthos: Ch 900. For other names, see Ventris and Chadwick 1956, 427. See *LGPN*, s.v. Aiolos, Kelainos, and Xouthos.

15. Translation based on Ventris and Chadwick 1956, no. 83. For an alternative interpretation of the ox/BOS sign as hides, see J. Melena (pers. comm.), cited in Palaima 1989, 91. For discussion of C 902 and the suggestion that ideogram *170 represents bull testicles, signifying that these are uncastrated animals meant for sacrifice, see Landenius Enegren 2004.

16. *We-ka-ta* and *da-mo*: C 59 (five instances of six oxen and one of fifty); one pair: Ch 896.

17. B. Hallager 2001.

18. Godart 1972.

19. Ruipérez 1956, followed by Vernant 1982, 34; but see also Taillardat 1960. Chadwick 1976 is more circumspect and refers to the *ko-re-te* as a governor.

20. Palmer 1965, 314.

21. Copper ingots shaped like oxhides appear a number of Cretan sites in LM1. For references, see Rehak and Younger 1998, 123 and appendix 1.

22. Henkelman 2005, 165.

23. Schoep 1999.

24. Halstead 1996, 24.

25. Hallager and Hallager 1995.

26. S. Marinatos 1974.

27. For a mistaken reading of the scene as peaceful, see Warren 1979.

28. Hom. *Od.* 9.42–63; 9.459–70; 10.95–101;12.362–77

29. S. P. Morris 1989.

30. Loughlin 2004a.

31. E. Davis 1974.

32. Loughlin 2004b.

33. Burke 2005.

34. Evans 1921–1935, 310; Immerwahr 1990, 174–76.

35. Svoboda 1956; Tuchelt 1962.

36. Bull's-head rhyton image available at http://www.dilos.com/dilosimages/image/crete/museum_bull.jpg. The lyre shape of the horns and the ring of white around the nostrils have suggested to some that the animal may be identified as an example of the Greek steppe breed. See http://www.slowfoodfoundation.com/sf_premio/PREMIO/vincitori2002/pagine_en/Grecia_02.html (accessed November 21, 2006). For other similar breeds, all of the grey steppe/podolian group, see http://www.tiho-hannover.de/einricht/zucht/eaap/groups/c6_1.htm, as well as http://www.ansi.okstate.edu/breeds/cattle/greeksteppe/index.htm. Both characteristics, however, are common among a number of related breeds, such as the Italian maremmana and Hungarian grey.

37. Rehak and Younger 1998, 115.

38. N. Marinatos 1986.

39. Long 1974.

40. Rehak 1995.

41. Younger 1976, 136. For earlier studies, see Evans 1921; Sakellariou 1958, 85–89.

42. Press 1969.

43. MacGillivray 2000.

44. Beckmann 2005.

45. See White 1954 for approaches prior to the translation of Linear B. On theriomorphism, see Rousioti 2001. On context and materiality, see Alberti 2001.

46. For the various depictions of bull leaping, see Younger 1976: assistants (fig.23.III.16); two animals (fig. 6.II.10); seated animal (fig. 21.III.12); attacking altar (fig. 5.II.6).

47. Cordes and Ibrahim 1996; Pearson and Haney 1999; MacGillivray 2000; Sharpes 2006, 78. Most studies of modern bull sports do no more than repeat handbook information on the Cretan phenomenon, although Pointer (1985, 12) is notably worse, asserting that steer wrestling was on the program of the ancient Olympics! See also T. J. Mitchell 1986.

48. On the *course landaise*, see McCormick 2000, 210. Video clips can be viewed on the website of the Fédération Française de la Course Landaise at http://www.courselandaise.org/. For the similarity between the *course landaise* and Minoan bull leaping, see Ward 1970.

49. Loughlin 2004a.

50. German 2005. See also Damiani-Indelicato 1988.

51. Rehak 1995, 448.

52. Bietak et al. 2007.

53. On bull leaping outside of Crete, see Bietak 1996, especially pl. IV; Collon 1987, no. 708; 1994; 2003. Also see Brody 2002; Guillaume and Blockman 2004; Bietak et al. 2007.

54. Soles 1995; Rehak and Younger 1998, 129; German 2005.

55. N. Marinatos 1987; Gesell 1987.

56. Preziosi and Hitchcock 1999, 101.

57. Betts 1967; E. Hallager 1996.

58. A lively thread on the discussion list Aegeanet conducted in late February 2009 illustrates the continuing fascination with this topic and will lead the reader to many parallels from Hittite and other Bronze Age cultures. See http://people.ku.edu/~jyounger/aegeanet.html.

59. Harland 2007.

60. See esp. the Ta series from Pylos, but also K 434, K 740 and K 875. Take, e.g., K 872: *ke-ra-a* bull's head (2 bull's-head rhytons), one bull's-head rhyton decorated with *ne-pa*$_2$*-sa-pi*; Pylos Ta 711: "*Pu*$_2$*?-ke-qi-ri* made inspection, on the occasion when the king appointed Sigewas to be *damokoros*: one ewer of the queen's set, bull's head design (*qo-u-ka-ra*), decorated with seashells; one ewer of the queen's set, bull's-head design (*qo-u-ka-ra*), decorated with a running spiral."

61. Borgna 2004.

62. On the level of Mycenaean influence on Minoan culture in this period, see E. Hallager 1987.

63. See Wright 2004a and the other essays in the same volume (Wright 2004b). For an earlier, influential essay on banqueting, see Goody 1982. For a study of a similar feasting system focused on provincial centers hosting inclusive feasts in the context of late Mayan society, see LeCount 2001.

64. Palaima 2000. For the Pylos *megaron* fresco showing paired banqueters, see McCallum 1987.

65. For the identification of the tables, thrones, and stools, see Vandenabeele and Olivier 1979, 161–76. For the identity of the throne holders, see Palaima 2004, 115; also see Palmer 1958. The sixteen *koreteres* are attested on Jn 829, which is discussed by Bennet 1998.

66. Isaakidou et al. 2002; Stocker and Davis 2004.

67. Dabney et al. 2004.

68. Sacconi 2001.

69. Hamilakis and Konsolaki 2004.

70. Wright 2004a, 51.

71. Lindgren 1973. See also Godart et al. 1975; Rougement 2004.

72. KN Ch 896. See also KN Ch 897, 898, 899, 900, and 1015.

73. Palaima 2004, 106.

74. Hom. *Od.* 14.110–20; 20.234–44.

75. Sloan and Duncan 1978, 64; Shelmerdine 1981, 323.

76. Castleden (2005, 108) gives the figure of 234 hides without citation. The Ma series, as tabulated by Ventris and Chadwick (1956, 291) records 204 hides.

77. Wyatt 1962; Shelmerdine 1973.

78. Palmer 1958, 96. See also Finley 1957 and Dickinson 1994, 85, on categories of Mycenaean land.

79. Watrous 1984, 129. For an alternative reading of the cup as depicting adolescent initiation, see Koehl 1986.

80. Piteros et al. 1990.

81. Tartaron 2001, 22–24.

82. Ventris and Chadwick 1956, 190 and 206–7; Palmer 1958.

83. Killen 1994, 1998; Speciale 1999.

84. I follow the figures and readings in Palaima 2004, 103–4 and 122. See also Killen 1992, 376. The title *o-pi-te--ke-e-u* could refer to military equipment, but

Killen connects it with a more general term for equipment, in this case a reference to the banquet paraphernalia. See also Nikoloudis 2001.

85. Halstead 2002, 152–59 and 163–65.
86. Killen 1999. On Collectors, see Bennet 1992; Rougemont 2001.
87. Wijngaarden 2001, 192.
88. Kopaka and Chaniotakis 2003.
89. I. Morris 1997.
90. Nowicki 2002.
91. For discussion of the chief of Lefkandi as a Big Man, see Whitley 1991a.
92. Bintliff 1982, 107.
93. McDonald et al. 1983, 323.
94. Homer, *Od.* 19.460.
95. Hesiod, *Op.* 437–38.
96. Snodgrass 1971, 379–80. See also Muhly 1980b; Calligas 1988; and Thomas and Conant 1999.
97. Snodgrass 1987, 170–210.
98. Halstead 1996; see also Jameson 1977.
99. Cherry 1988.
100. Klippel and Snyder 1991.
101. Whitley 2001, 85–86.
102. Kindleberger 1951.
103. Chang and Koster 1986.
104. Chang 1993, 699. For the view that the break between the Bronze and Iron Ages did not dramatically affect food production, see Foxhall 1995.
105. Hertzfeld 1985.
106. On Iron Age networks on Crete and the importance of feasting as a vehicle for both elite self-definition and regional alliance building, see Haggis et al. 2004.
107. For the stock-raising elite, see Hanson 1995.

Chapter 4
Epic Consumption

1. On Lefkandi, see Popham et al. 1982a, 1982b. On the concept of the hero, see I. Morris 1997; Whitley 2002.
2. Kuhn 1853; Volk 2002.
3. On connections between cattle practices such as raiding and the formulation of a heroic code between the tenth and eighth centuries, see Jackson 1993. Jackson notes, "The great man of Lefkandi may simply have been one who stole from the poor to make himself rich" (p. 71). He also notes that Achilles, the supreme hero, is credited with the sacking of twenty-three cities (*Il.* 6.424; 20.91–92). See also Tandy 1997, 76–77 and 110. For oaths sworn over the entrails of a sacrificial animal, see Faraone 1993.
4. For prizes, see Homer, *Il.* 23.260, 700–710; for compensation, see Antonius Liberalis, *Met.* 38.5.

5. For the fragments of Euripides' *The Cretans*, see *NFE* 45–58 and Reckford 1974, 319–22. See also Apollodoros, *Bibl.* 3.1.3–4. For Pasiphae as the Divine Cow, see Burnell 1947. On Roman versions, see Armstrong 2006.

6. Andrews 1969. Astronomical calculations of the heliacal rising are attested as early as 2300 BC in Mesopotamian science; see BM cuneiform tablet 86378.

7. Ovid, *Met.* 2.80.

8. On the heliacal rising of the Pleiades, see Penrose 1893 and, more generally, P. L. Brown 1976, 115.

9. Hunt 2005.

10. On the significance of Taurus on Crete and in the ancient Near East, see Parisinou 2005.

11. For Europa's story, see Apollodoros, *Bibl.* 3.1.1 and Frazer's commentary (Loeb edition) ad loc.

12. For the marriage of the sun and moon, see Cook 1903; Frazer 1923, 71.

13. For Pasiphae as the moon, see Pausanias 3.26.1; and for her genealogy, see Apollodoros, *Bibl.* 1.9.

14. Minos' poisoned sperm: Antonius Liberalis, *Met.* 41; Talos as bull: Apollodoros, *Bibl.* 1.9.26. For a recent treatment of the Minotaur and Talos in connection with the supposed sacrifice of children in Semitic cultures, see Rundin 2004, 427–33.

15. Kirk 1972, 74–85.

16. Apollodoros, *Bibl.* 3.17.1–21; Hyginus, *Fab.* 146. Attic versions include Aeschylus, *Cretan Women*, Sophocles, *Seers*, and Euripides, *Polyeidos*.

17. Jeanmaire 1939, 444–50; Willetts 1959; Callaghan 1978; Koehl 1986; and Muellner 1998.

18. For similar color combinations, see the *Suda*, s.v. Isis, where Zeus changes Io into a white, then black, then violet-colored cow to avoid detection by Hera. Similarly, in the *Táin*, as Derdriu watches a raven drinking the blood of a skinned calf from the snow she says: "I could desire a man who had those three colors there: hair like the raven, cheeks like blood, and his body like snow" (trans. Kinsella).

19. Compare the French "élèves" to refer to both students and young cattle.

20. For Io, see Hesiod, *Cat.* frag. 124; Aeschylus, *Supp.* 291–315 and 531–94; Aeschylus, *Pr.* 642–82, 700–740, 785–854; Lycophron, *Alex.* 1291–95; Diodorus Siculus 5.60.4; Apollodoros, *Bibl.* 2.1.3; Ovid, *Met.* 1.588–750; and Hyginus, *Fab.* 145.

21. Lycophron, *Alex.* 1291–95 (trans. West). On the significance of *tauroparthenos* (bull's maid), see Griffiths 1986.

22. Herodotos 2.38. At 2.41 Herodotos remarks that in Egypt images of Isis are in the form of a woman horned like a cow, "exactly as the Greeks picture Io." On Egyptian elements in the Io story, see Hicks 1962, 93–97.

23. On the equivalence between Io's name and the word in the Bohairic dilaect of the Delta for moon, which would have sounded somewhat like "iooh," see Carpenter 1950. See also S. West 1984a, 1984b; Bernal 1987, 95. On Isis and Apis, see H. S. Smith 1972.

24. Veldhuis 1991, 9.

25. Friss Johansen and Whittle 1980, 408–10; M. L. West 1997, 557–64.

26. Bachvarova 2001.

27. Bachvarova 2001, 59–60.
28. Anhalt 1997.
29. *Od.* 17.189–94. Note that there is no mention of sacrificing these animals, merely slaughtering them. The suitors' gluttony is also impious.
30. Bergquist 1988, 30.
31. Kron 1988.
32. Hamilakis and Konsolaki 2004, 147.
33. On the power of sacrifice to restore social order, see Girard 1977.
34. Homer, *Iliad*, trans. S. Lombardo (Indianapolis, 1997).
35. Donlan 1989.
36. Odysseus' challenge recalls another episode involving his skill with a plow: when he feigned madness, plowing with an ox and an ass, to avoid joining Agamemnon's ranks. See Hyginus, *Fab.* 95; Servius, ad *Aen.* 2.81; Apollodoros, *Bibl.* 3.7.
37. On Homer's audience, see Doherty 1995, 73; Dalby 1995.
38. For a catalogue of his herds, see Johnson 1941.
39. On courtship and consumption, see *Od.* 1.266; 2.50–65; 4.335–45. For the twinning of woman and cow, see Brenneman 1989. On the suitors' devouring Odysseus' livestock, see van Wees 1992, 285–92.
40. For dowry practices in Homer, see Finley 1955; Snodgrass 1974. For a discussion of attitudes toward gift exchange in marriage contracts, see Lyons 2003.
41. Hesiod, *Cat.* frag. 7 (= *PBerol.* 7497)
42. Pausanias 4.36.3.
43. Hesiod, *Sc.* 10–12.
44. Westbrook 2002. Two passages address the peculiar situation directly: *Od.* 1.276–78 and *Od.* 2.52–54. In both passages Homer speaks of gifts that would accompany Ikarios' decision to give his daughter away. Neither here nor anywhere else does he raise the subject of the suitors' having to pay a bride price. It is also possible that in other versions Odysseus returned on the very day Telemachos came into his majority, freeing him to inherit and Penelope to remarry (in which case only her dowry would go to the successful suitor). See Danek 1998.
45. See Evans-Pritchard 1946; Fleisher 1999.
46. S. Reece 1993, 64–67. For the transformation of *xenia* into *philia*, see Konstan 1997, 35–37. Nestor is presented as a more ambiguous figure in Dickson 1995. For a discussion of reciprocity in Homeric society with a number of parallels from Nuer society, see Beidelman 1989.
47. Nestor's sacrifices while returning: Tenedos (*Od.* 3.175) and Geraistos (*Od.* 3.199). Proteus' advice to Menelaus: *Od.* 4.500–509.
48. Guests reduced to tears: *Od.* 4.117–24; Telemachos pleads to go to bed: *Od.* 4.314–15 (although Telemachos has by now been drugged by Helen); Telemachos asks to be allowed to leave: *Od.* 4.626–30; Menelaus' gifts (horses): *Od.* 4.633–43.
49. J. Ferguson 1985. See also Shipton 1989 (regarding the Luo).
50. Hutchinson 1992.

51. For a similar strengthening of "traditional" patterns of stock raising during times of change among the Tswana in response to the introduction of modern economic practices, see Comaroff and Comaroff 1990.
52. Finley 1954, 106; G. P. Rose 1969.
53. On Homer's attitude to trade, see Malkin 1998, 88–89; Schaps 2004, 74–77.
54. On supplication, see Giordano 1999. On the contradictions in the Cyclops, who drink milk rather than wine and who raise animals while the earth produces sustenance untended, see Pucci 1993. For a recent summary of interpretations of the Cyclops, see Hernández 2000.
55. Dougherty 2001, 98–127.
56. Seaford 2004. A few examples occur: at *Od.* 1.430 Laertes speaks of acquiring Eurykleia for goods worth twenty head of cattle. Mentes exchanges iron for bronze (*Od.* 1.200), and Euneus exchanges wine for various goods provided by the Greeks (*Il.* 7.467).
57. Buchan 2004.
58. Adkins 1972, 6–7.
59. The killing of the suitors is also foreshadowed by the description of the sacrifice to Apollo at *Odyssey* 20.276–78 and 21.258–59. See Aubriot 2003.
60. The suitors' impiety is also highlighted by the insertion of a sacrifice into the midst of their deliberations. As they contemplate killing Telemachos and busy themselves slaughtering sheep, goats, pigs, and a heifer, the poet cuts away to the town, where the long-haired Akhaians are preparing a hecatomb for Apollo (Homer, *Od.* 20.300–306). For insightful treatments of Homeric feasting with an emphasis on the social significance of the feast in the definition of hierarchies, see Saïd 1979; Rundin 1996.
61. On Zeus' justice, see Dimock 1989, 330; Segal 1992.

Chapter 5
Heroes and Gods

1. Fleisher 2000, 32.
2. Simoons 1968, 176.
3. Odysseus on raiding: *Od.* 23.370.
4. Katreus: Apollodoros, *Bibl.* 1.2.2; Herakles and the cattle of Geryon: *Bibl.* 2.5.10; Kastor: Proclus, *Kypria* frag. 1; Amphitryon: Hesiod *Sc.* 80; the Kikones: *Od.* 9.50; Autolykos: Apollodoros, *Bibl.* 2.6.2; Alkyoneus: Pindar, *I.* 6.31. For more examples, see Pausanias 4.36.4.
5. On reciprocity in general, see Sahlins 1968 and 1972. For reciprocity in Greek culture, see Gill et al. 1998 and, in Homer specifically, Donlan 1982. Sahlins' examples of "negative reciprocity" include haggling, cheating, and theft, actions in which the social distance between the participants is great.
6. Fleisher 2000.
7. Hesiod, *PBerol.* 10568. See Leaf 1910.
8. Lincoln 1976. The distance between Homer and the Indo-European cattle-raiding myth is demonstrated by his treatment of Cyclops. According to Thieme

(1951), Cyclops means originally "cattle thief." There is no sign of this in Homer's gigantic shepherd.

9. *Od.* 11.292–97; Pausanias 4.36.3 and 10.

10. On versions of the *Odyssey* from Crete and Thresprotia, see Danek 1998.

11. Pausanias 4.36.4.

12. The chief literary texts are Hesiod, *Th.* 287–94, Stesichoros, *Geryoneis* (*POxy.* 2617) and Apollodoros, *Bibl.* 2.106–9. For Stesichoros, see Page 1973. For the popularity of the story in vase painting, see M. Robinson 1969 and Brize 1980. Other depictions of Herakles' western adventures, including his acquisition of Helios' cup, in which he travels to the land of Geryon, are discussed by Pinney and Ridgway 1981. The episode also figured on two famous artworks, the throne of Amyklai and the chest of Kypselos (Pausanias 3.18.13; 5.18.7).

13. Bremmer 2002.

14. Burkert 1979, 84–97. Also see H. J. Rose 1954; Endsjø 2000.

15. On Christian versions of the story, see M. Davies 1988.

16. On the pairing of Kerberos and Orthos, see Schweitzer 1922, 87. There is no explicit mention of Cacus in Stesichoros, but the trope of the hero having to refight his first battle is, in fact, a common feature of myth. See M. Davies 1988, 286.

17. Virgil, *The Aeneid*, trans. T. C. Williams (Boston, 1908).

18. Arrian 2.16.

19. Aelian, *NA* 12.11, [Skylax] 26, Arrian, 2.16. On the reputation of Epiros as prime cattle country see Hesiod, *Cat.* frag. 57 and Aristotle, *HA* 8.7. The Epirote town Bouketa took its name from Themis' having gone there on the back of a bull to escape Deukalion's flood: *Suda*, s.v. *bouketa*.

20. Lippold 1952.

21. Lincoln 1976; Arabagian 1984.

22. Lincoln 1976, 58.

23. Lincoln 1976, 64. For the debate over Aryan migrations, see Parpola and Carpolan (for) and Lamberg-Karlovsky (against) in Bryant and Patton 2005; for a neutral stance, see Koryakova 1988.

24. Lincoln 1976, 63.

25. Walcot 1979. See also Bader 1980.

26. A. T. Lucas 1989, 4.

27. Pindar frag. 169; Plato, *Grg.* 484b. See Lloyd-Jones 1972, 55–56.

28. Lincoln argues that the subject of the verb "drove" is only implied and that Herakles' name occurs in the dative, leaving the identity of the verb's subject ambiguous. This is incorrect. Herakles' name occurs in an adjectival form in the nominative (*Heraklêeiê*), modifying the noun *biê* (strength). The sentence makes perfect sense: it is the "Heraklean strength" that vanquished Geryon *and* drove the cattle back to Greece. On Geryon breeding cattle, see Pausanias 1.35.6 (Lydia) and Arrian 2.16 (Ambrakia).

29. Page 1973, 150.

30. Bremmer 1999, 186. For adolescent male initiation, see Garland 1990, 170–73 and Johnston 2002.

31. Graf 1979. For a bull cult associated with ephebes outside Athens, see Strabo 14.1.44.

32. Laurence Frank, quoted by G. Martin, "The Lion, Once King of Vast African Savanna, Suffers Alarming Decline in Population," http://www.sfgate.com/cgi-bin/article.cgi?f=/c/a/2005/10/06/MNG1UF392D1.DTL (accessed June 10, 2005). For a tentative identification of the cult of Artemis Tauropolos with male initiation, see Lloyd-Jones 1983, 97.

33. Plutarch, *Thes.* 30.

34. On the practices at Brauron, see Kahil 1977; Sourvinou-Inwood 1988, 1990; Cole 1984; Dowden 1989; N. Marinatos 2002; Faraone 2003.

35. On contact with non-Greeks, see Dougherty 1993a; Malkin 1998; and Anello 2005.

36. Nafissi 2001.

37. Anello 2006.

38. Harari 2004, 164. For foreigners in Athenian culture, see Lissarrague 2002.

39. D'Agostino 1996. For an earlier reading emphasizing the image of the western lands as a threat, see Sbordone 1941. On Herakles' travels, see also Huttner 1997, 11–13.

40. Strabo, *Geography* 5.3.3. Strabo's account is that Evander entertained Herakles on his return from the west and made a sacrifice to him upon learning that he was destined to become a god.

41. Janni 1984 and 1998. On the derivation of Greek *pontos* ("deep sea") from Sanskrit *panthah* ("difficult passage"), see Montiglio 2005, 8.

42. Hesiod, *Theog.* 213–16; Mimnermus 11, 11a West; Hesiod, *Sc.* 313–17; Homer, *Il.* 18.607–8.

43. Braun 2004, 298. See also Nesselrath 2005.

44. Schol. ad Pind. *O.* 3.44–45. See Bosch-Gimpera 1944.

45. Pind. *N.* 4.69.

46. On the syncretism of Herakles and Melqart, see Appian, *Hisp.* 1.2; Diod. Sic. 5.20.2; Arrian, *An.* 2.16.4. On the cult at Gades, see Berchem 1967; Bonnet 1988, 203–41. For Herakles as *archegetes* and addressed as "Lord Melqart, Baal of Tyre," see Donner and Röllig 1962–64, no. 47.

47. Herodotos 2.44.

48. J. P. Brown 1968.

49. For Alebion and Derkynis, see Apollodoros, *Bibl.* 2.5.10; Pomponius Mela 2.5.39. For the rocks used to fight the Ligyes, see Hyginus, *Fab.* 2.6 and Strabo 4.1.7, preserving a fragment from Aeschylus, *Prometheus Unbound.*

50. Strabo 4.6.3. The epithet Monoiokos ("Living Alone") may derive from a Phoenician epithet of Melqart, Menouakh.

51. Dionysius of Halicarnassus, *Roman Antiquities*, trans. E. Cary (Cambridge, Mass., 1968).

52. Aulus Gellius, *NA* 11.1.1. Also see Hesychios, s.v. *italos.*

53. Aristotle, *Politics*, trans. H. Rackham (Cambridge, Mass., 1932).

54. Bonetto 2004.

55. Papadopoulos 2002, 37.

56. Virgil, *Aen.* 8.230–56 ; Propertius 4.9.1-20. The Cacus may have an eastern origin. See Gardner 1892–93.

57. Propertius, *Elegies*, trans. G. P. Goold (Cambridge, Mass., 1990).

58. Grandazzi 1997, 76. See Giovannini 1985, 380–81.
59. Apollodoros, *Bibl.* 2.5.10; Pausanias 4.36.4.
60. Apollodoros, *Bibl.* 1.135; Hyginus, *Fab.* 14.
61. On the Elymoi, see Thucydides 6.2 and Casson 1913.
62. Herodotos 5.42–48. See Visser 1982; Munson 2006.
63. For myths involving Herakles and the Neleidai in establishing relations between Athens, Metapontum, and Segesta, see Nafissi 1997.
64. Lincoln 1981, 19–22. The moral of the story naturalizes the enmity of the two: "And to this day the Dinka has always lived by robbery and the Nuer by war."
65. Herodotos 1.1–3.
66. Dougherty 1993a, 1993b.
67. Herodotos 4.8–10.
68. For an overview of Harrison's circle, see Ackerman 1991.
69. Fyske 2005.
70. Kirk 1972, 74.
71. Burkert 1985, 64–65, but see also Czachesz 2003, 92. For earlier comments on theriomorphism, see Farnell 1912, 3.
72. Goldhill 1991, 90.
73. Schnapp-Gourbeillon 1981; Dietrich 1983.
74. See the kathartic law from Cyrene, in *RÉG* 106 (1993): 27–28: "Pregnant women before birth will go down into the Nymphaion to Artemis and give to the She-Bear feet, hands, and skin."
75. Carpenter 1950, 181.
76. Homer, *Il.* 20.220–25.
77. Jakobson 1971. For other approaches to transformation in Greek myth, see Forbes Irving 1990; W. Hansen 2000.
78. Farnell 1896, ix.
79. Hogan and Schenker 2001. See also Parisi Presicce 1989.
80. Heath 1999, 17. On the anxiety over the boundaries between human and divine, also see Murnaghan 1992, 242.
81. On castration, cuckoldry, and the ambiguity of horns (possessed both by bull and ox), see Graber and Richter 1987.
82. In Eusebius, *Preparation for the Gospel*, trans. E. H. Gifford (Oxford, 1905).
83. For Akragas see Polybios, *Hist.* 9.27; for Olbia see Hurst 1903, 24 and 38.
84. Eissfeldt 1962; Hillmann 1990, 272.
85. Lowing bronze cattle: Schol. ad Pind. *O.* 7.87. See also Apollodoros, *Bibl.* 3.2.2. For dedications of cattle to Zeus, see *IG* XII 1.31 as well as Pugliese Caratelli 1950, 76. For Zeus Atabyrios, see Torr 1885, 75; Frazer's note (in his Loeb edition) to Apollodoros, *Bibl.* 3.2.2; Bresson 2000, 38.
86. For the Bouphonia, see Porphyry, *Abst.* 2.10–30. See Durand 1986.
87. Philodamos, *Paian to Dionysos* 2; Euripides, *Ba.* 90, 918; Athenaios 11.476; Euripides *IA* 275; Orphic Hymn 45 (to Dionysos).
88. Taurophagon: *Suda*, s.v. *Taurophagôn* (= Sophocles frag. 668); Arkadian grease wrestling: Pausanias 8.19.2.

89. For the name of the *boukoleion* and the *hieros gamos*, see [Xenophon], *Ath. Pol.* 3.5 and Guarducci 1962; for the marble relief see S. Miller 1979. For the location of the *boukoleion* near Lysikrates Square in the Plaka, see Schmalz 2006.

90. On dancing see Lucian, *Salt.* 79. For the *boukoloi* as priests of Dionysos, see C. Alexander 1932; Cumont 1933. For Kratinos' *Boukoloi*, see KA 130–31; Rosen 2003; Delneri 2006, 45.

91. For Poseidon as bull, see Hesiod, *Sc.* 104. For Poseidon's bull cults, see Philippson 1944. For Poseidon creating an outlet for the Peneios to the sea, see Herodotos 7.129.

92. Roman bull cult: Suetonius, *Claud.* 21. For earlier coinage, see Head 1911, 290–91.

93. Rothwell 2007, 47.

94. Poseidon as earthshaker: Hesiod, *Th.* 15. On dominion over the sea, Hesiod, *Th.* 881–85. Apollodoros, *Bibl.* 1.2.1. On Poseidon's connection to horses and earthquakes, see Mackay 1946.

95. Euripides, *Hipp.*; Schol. ad Homer, *Od.* 11.321; Apollodoros, *Bibl.* E1.9. The debate regarding Taraxippos is summarized by Detienne 1971, 169.

96. Poseidon Hippios and Demeter: Pausanias 8.25.5–8; Periklymenos, son of Neleus: Apollodoros, *Bibl.* 1.9.9, 2.7.3; Schol. ad Apollonius Rhodius 1.156. Mestra, daughter of Erysichthyon: Ovid, *Met.* 8.738–884.

97. Homer, *Od.* 1.24; 3.5–6; 3.8; 3.177–78; 11.130–31; 13.180–81; 13.184; 23.277–78. The only other mention of a bull is in the simile of 21.48, when Penelope opens the door of the storeroom in which Odysseus' bow is kept. Emily Anhalt (1997) argues convincingly that the simile is meant to evoke the other bull sacrifices made to Poseidon.

98. PY Un 6, 718.

99. For coins of Sybaris and Poseidonia, see Papadopoulos 2002. For the claim that the Poseidon of Magna Graecia was theriomorphic, see Elderkin 1942.

100. Pinney 1994.

101. Burkert 1985, 131.

102. Kardara 1960, 350–53.

103. For animal sacrifice at the Heraion, see Tuchelt 1992. For pigs and goats to Demeter, see Jarman 1973. For the bone deposit associated with the altar of Aphrodite Ourania in the Agora, see Foster 1984 and, more recently, D. S. Reece 1989.

104. Io: Apollodoros, *Bibl.* 2.1.3; Women of Kos: Ovid, *Met.* 7.350–403; daughters of Proitos: Bacchylides 11; Hera's gadfly: Apollodoros, *Bibl.* 2.5.10.

105. Hofmann 1949, s.v. Heros; Guarducci 1962.

106. Cook 1906, 369.

107. For Hera rousing the Achaians, see *Il.* 5.729. See O'Brien 1990. For the equivalence of *boôpis* with *megalophonos* see Hesychios, s.v. *boôpis*

108. Beck 1986.

109. Pausanias 3.4.

110. On the Great Daidala, see most recently Clark 1998, 24–25.

111. Tn 316. On metonymic connections of the epithet, see Reiser 1940, 214; McCartney 1951.

112. For Hera cults in Magna Graecia, the Peloponnese, and Samos, see La Genière 1997.

113. Jarosch 1994; *Anth. Gr.* 6.245.

114. Pausanias 5.13.8. For a summary of the evidence for cattle dedications and sacrifices to Samian Hera, see Kyrieleis 1981 and 1993.

115. Jeffery 1961, 253, 410, pl. 50.

116. *Etym. Mag.* s.v. *zeuxidia.*

117. Thucydides 2.2.

118. For Io as priestess, see Apollodoros, *Bibl.* 2.1.3. For the identification of Io with Hera, see Harrison 1893, 77. For myths involving sexual restrictions on priestesses, see Pausanias 9.27.6 (Herakles at Thespiai) and 6.19.3 (Artemis Triklaria).

119. Hesychios, s.v. *agôn chalkeios*; Schol. ad Pindar, *O.* 7.152 and *N.* 10.22–23.

120. Herodotus, *Histories*, trans. A. D. Godley (Cambridge, Mass., 1982).

121. Other versions of the story: Pausanias 2.20.3; Plutarch, *Solon* 27.5; *Mor.* 108e; Lucian, *Charon* 10; Hyginus, *Fab.* 254. The last is the most detailed, naming the mother (Kydippe) and making her a priestess of Juno. An excellent overview, with a judicious summary of the debate over the identification of the Kleobis and Biton *kouroi*, is to be found in Chiasson 2005.

122. Regenbogen 1965, 383–89.

123. Seaford 1988, 124. For other approaches see Sansone 1991, who sees them as models of perfect sacrificial victims, and O'Brien 1993, who sees in the story a parable of Hera's power as a nature goddess to bring the seasons to fulfillment.

124. Kurke 1999, 147.

125. The closeness of the young athlete to the animal is evoked in another story, told by Pausanias (2.19.5), according to which Biton won renown for carrying an ox on his shoulders to a sacrifice at the sanctuary of Nemean Zeus.

126. For "cows" (cakes) substituting for sacrificial animals, see Eustathios, Comm. ad Hom. *Il.* 18.575, Hesychios, s.v. *mazeinos bous* and *Suda*, s.v. *popana*

127. Lebessi 2002.

128. Lebessi 1992. Contra, see Schachter 1986, 97. See also Roesch 1985; Berlioz 2004.

Chapter 6
Gods, Cattle, and Space

1. *The Homeric Hymns*, trans. A. N. Athanassakis (Baltimore, 1976).

2. Herodotos 2.53. See Earp 1959.

3. Homer, *Il.* 24.58. See Murnaghan 1992.

4. See Polignac 1984, 49–60; and Morgan 1990, 12, who speaks of an Argive "propagandist attempt to legitimize a land claim."

5. For Helike, see Pausanias 7.24.5–6, Strabo 8.7.2 and Katsonopoulou 2002. Robertson 1984a offers a thorough survey of evidence linking Poseidon's cult to the winter solstice and shows that the cult was widespread. For Onchestos, see

Strabo 9.2.33; P. W. Wallace 1979, 134; Schachter 1986, 207–12; and Habicht 1985, 35–36. For the Kalaurian League, see Kelly 1966.

6. Homer, *Il.* 15.184–200.

7. For Amphiaraos, see Petrakos 1968; for Trophonios, see Cicero, *ND* 3.19 and Bonnechere 2003.

8. For Athena Alea, see Pausanias 8.47.1; Norman 1984, 1986; Østby 1986, 1994; Voyatzis 1990. *IPArk* 2.5, 11, 16, 24, dealing with the sanctuary's land, names only Alea. For Apollo Maleatas in Lakonia, see *IG* V 1.213; Jeffery 1961, 200 nn. 37 and 38; Phaklaris 1990; Shipley 2000 and 2004, 593; at Epidauros, see *IG* IV^2 1.128; Pausanias 2.27.8; Lambrinoudakis 1981; and Theodorou-Mavrommatidi 2004.

9. Sineux 2007.

10. Catling 1976–77, 36.

11. For Menidhi, see Lolling 1880 and Hägg 1987, 93–99.

12. Wärnlöf 2000.

13. Pausanias 10.4.7; Stewart 2004, 81.

14. Harrison 1890.

15. On the threats to Zeus' order, see Maitland 1999.

16. Homer, *Il.* 1.400. For the opposition between Zeus and the group consisting of Hera, Poseidon, and Athena, see Heiden 2002.

17. Sextus Empiricus, *M.* 9.193.

18. Hesiod, *Op.* 11–25. The literature on this passage and on *eris* is extensive. For a recent overview, see Thalmann 2004.

19. Clement of Alexandria, *Stromateis* 5.109.2 and 7.22.

20. On ritualized social competition, see Polignac 1994. For sanctuaries and panhellenic identity, see Morgan 1990, 137–46, and 1993.

21. On sanctuaries and mediation, see Polignac 1984. For the influence of Polignac, see Alcock and Osborne 1994.

22. Penglase 1994, 155, critiquing the thesis of Clay 1989.

23. Snodgrass 1971, 275. See also Dietrich 1986, 1–2.

24. C. Roy 1984, 163–73. This approach tends to oversimplify the situation with regard to cult, but does elucidate Hesiodic cosmology well. See also Holmberg 1997.

25. For Homer's treatment of the region around Olympia, see Wade-Gery 1948, 115–18. For Pytho, see *Il.* 2.520; 9.404–5; and *Od.* 8.72–82. For the view that the references to Pytho are late, see Defradas 1954.

26. Olympia at least could boast the pillar of Oinomaos. See Pausanias 5.20. Strabo reports that Pheidias claimed to have been inspired by Homer's description of Zeus for the colossal statue.

27. Pausanias 10.7.2. On performance, see Thucydides 3.104; Pavese 1998; Clay 1989; and Furley 1995. For an overview of the genre, see Parker 1991; Richardson 2003; and Rayor 2004. For topoi common to epic and the hymns, see Sowa 1984.

28. The expression *autar egôn humeôn te kai allês mnêsomai aoidês* ("Now I shall recall you and another song too"), or something similar, concludes Hymns 2, 3, 4, 5, 6, 9, 10, 19, 25, 27, 28, 29, 30, and 33 and is affixed as often to the longer hymns as to the shorter ones.

29. Depew 1998; Rutherford 1988.

30. On the kletic function of the *Hymns*, see García 2002. The god's story thus serves as an example of a historiola, a narration that seeks to reactivate the magic described in the story. See Frankfurter 1995.

31. McLeod 1961. On Euripides' use of hieratic language, see Furley 1995.

32. Käppel 1992, 207–84.

33. Herodotos 4.35.1.

34. For Kyllene, see Pausanias 8.17.1–2. The Tanagran claim that Hermes was born on Mount Kerykios in Boiotia is recorded without comment by Pausanias (9.20.3).

35. On this episode, see Kurke 2005, 89–98.

36. De Polignac 1994.

37. Sanchez 2001.

38. Hall 1995.

39. On the *Olympionikai* and the *prostasia* of Elis, see Mann 2001.

40. See Elderkin 1941, 116: "Poseidon as a bull seems to have been displaced by Erechtheus as an ox in Athenian worship."

41. *IG* II^2 1146.

42. Philippson 1944, 25–44; Heichelheim 1947, 69–70.

43. Schumacher 1993, 83. On the elemental associations of the horse and the contrasting *metis* of Athena, see Detienne 1971.

44. By the time of Pausanias a town had arisen close to the sanctuary, but it may be as late as Hellenistic. See Pausanias 9.26.3 and Bintliff and Snodgrass 1985, 140.

45. *Homeric Hymn XXII to Poseidon* 3; for Aigai on Euboia, see Strabo 9.2.13: "The temple is situated on a high mountain."

46. Pausanias 2.1.6.

47. Montiglio 2005, 62.

48. Bulloch 1984, 218.

49. For Kadmos, see Hellanikos *FGrH* 4 frag. 51 (= Schol. ad Homer, *Il.* 2.494), frag. 25a (Schol. [Tzetz.] Lyk. 29), and Apollodoros, *Bibl.* 3.4.1. See also Edwards 1979, 68; M. L. West 1997, 448; Kühr 2006, 83–133. For Ilos, see Hellanikos *FGrH* 4 frag. 25 (= Schol. ad Lyk. 29) and Apollodoros, *Bibl.* 3.12.3. For other examples of *bouplanoktisis*, see *Suda*, s.v. *bouketa*; Krappe 1942; Burnell 1947, 377–81. According to Strabo 5.4.12, the Sabines were also led to their future homeland by a bull sent by Mars.

50. The *Homeric Hymn II to Demeter* is traditionally regarded as supplying the *aition* for the Eleusinian mysteries. See, for example, Alderink 1982 and H. P. Foley 1994. Contra, see Clinton 1986, 43–49, who connects the poem to the Thesmophoria. For objections and a restating of the older view, see Parker 1991.

51. Homer, *Il.* 13.10–32.

52. For the geography of Io's wanderings, see Myres 1946, 2–4. See also L. G. Mitchell 2001; Kühr 2006, 94–95.

53. Davison 1991.

54. On the construction of space through cultural relations, see Lefebvre 1991.

55. On sanctuaries and the development of a panhellenic culture, see Cole 2004.

56. Polybios 12.4d.5–8, reporting the account of Timaios of Tauromenion.

57. Hera pretending to go to the edges of the world: Homer, *Il.* 14.300–310; Zeus and Olympians feasting with the Aithiopians: Homer, *Il.* 1.418–20; Poseidon visiting the Aithiopians: Homer, *Od.* 5.282. For Apollo among the Hyperboreans, see Aelian, *NA* 11.1. For the Hyperborean maidens at Delphi, see Herodotos 4.35.1. A similar emphasis on Delphi as the center of the world is reflected in Plutarch's accounts (*de Defectu Orac.* 409e–f) of eagles and grammarians meeting at Delphi, having come from opposite ends of the world. For Artemis on the edges of territory, see Cole 1998.

58. In his explanatory note on these lines Athansassakis remarks, "These lines read like a pilgrim's guide to Apollo's shrines" (1986, 81). On Apollo's travels, see also Detienne 1998, 28–31.

59. Hall 1997, 87–88.

60. Giovannini 1969.

61. Verrall 1894, 28. See also Hogan and Schenker 2001.

62. Versions of the Koronis story are told by Pindar, *P.* 3, and Apollodoros, *Bibl.* 3.118. For the Phlegyans as enemies of Apollo, see Pausanias 9.36.2 and McInerney 1999, 195.

63. Pausanias 8.4.2; 8.27.9; 8.37.1.

64. Robertson 1984a.

65. Dieuchidas of Megara, in Athenaios 6.82, 262e–263b.

66. Philostratos, *Im.* 2.19.

67. Hogan and Schenker 2001. See Parthenios, *Love Stories* 15; Pausanias 8.20.2.

68. Parke and Boardman 1957. For political interpretations of the episode, see Shapiro 1984, 271–74.

69. A persistent question concerning the *Homeric Hymn III to Apollo* is the issue of whether it is a single composition or two separate poems. Single composition: A. Miller 1986. Separate compositions: Janko 1982. One diagnostic feature, the ephelcystic *nu*, suggests that the two poems to Apollo were composed in different settings. In *Hymn. Hom. Ap. Del.* the occurrence is 61 percent; in *Hymn. Hom. Ap. P.*, 13.7 percent.

70. For Pyrrhos, see Fontenrose 1960. Pausanias claims Delphi was hostile to the hero, but this hard to reconcile with Pindar, *N.* 7 and Pherekydes. For the Ainianes, see Heliodoros, *Aithiopika*, and Farnell 1921, 315.

71. Antonius Liberalis, *Met.* 23. Apollo serving Admetos: Kallimachos, *Hymn to Apollo*; Diodorus Siculus 4.71.3; Hyginus, *Fab.* 49; Valerius Flaccus, *Argon.* 1.445; Apollo also serves Laomedon: Homer, *Il.* 21.440–50. Like many other elements in Apollo's story, it is echoed in Hermes' biography. According to the *Homeric Hymn to Pan*, Hermes entered the service of Dryops as a shepherd to win the hand of his daughter.

72. Admetos yoking the chariot was depicted on the spectacular throne of Apollo at Amyklai; see Pausanias 3.19.2. For the postponement of Admetos' death, see Euripides' *Alcestis*.

73. Apollo serving Admetos: Diodorus Siculus 4.71.3; Hyginus, *Fab.* 49; Valerius Flaccus, *Argon.* 1.445. Apollo serving Laomedon: Homer, *Il.* 21.440–50.

74. For these and other instances of the association between Iolkos and herding, see Matthews 1977, 192.

75. Hesiod frag. 296.

76. *IG* XII.9 1273/4. For discussion, see Cairns 1991.

77. Sourvinou-Inwood 1979. On the conflict between the cults of Apollo at Delphi and Eretria, see Bruneau 1976.

78. For Mykalessos, see Pausanias 9.19.4. For Teumessos, see Pausanias 9.19.1 and M. L. West 2002, 127.

79. *Hymn. Hom. Merc.* 555. The association of herding with youth is also apparent in the *Homeric Hymn to Aphrodite* (55–76), when the goddess first sees Anchises, "whose beauty was divine," tending the cattle alone on Mount Ida.

80. Herodotos 5.59. See also Pausanias 9.10.5.

81. The description of the broken chariot at lines 230–39 has produced a good deal of speculation concerning the ritual practiced at Onchestos. For a survey of views of varying implausibility, see Teffeteller 2001.

82. Homer, *Il.* 2.494–511.

83. A. Miller 1986.

84. Fontenrose 1969, 129.

85. Other episodes include Phorbas on Rhodes (snakes): Diodorus Siculus 5.58.5; and Byzas at Byzantion (a bull): Hesychios, *FGrH* IIIB 390 f1.11. See Buxton 1994, 90.

86. For suppression of female deities, see Clay 1994. For the argument that the evidence for these earlier cults is part of the Delphic tradition created for Apollo, see Sourvinou-Inwood 1991 and Chappell 2006. For dubious claims of Minoan cult prior to Apollo, see Bates 1925.

87. On the importance of such wordplay in epic and hymns (though not this one), see S. Reece 1997.

88. On impressing place into memory, see Yates 1966.

89. Reisman 1977, 24. UNESCO has recently classified the Yaaral and Degal festivals in the Niger Delta as a masterpiece of the Oral and Intangible Heritage of Humanity. See http://www.unesco.org/culture/ich/index.php?cp=ML&topic=mp#56.

90. Image and descriptions of the Diafarabé cattle crossing and the festival held to mark this are available at http://www.oxfam.org.uk/coolplanet/ontheline/schools/magicmali/diafara2. For a recent (December 2007) photo essay on the crossing, see http://www.africanews.com/site/list_messages/13930.

91. Detienne 1986, 49.

92. On the various meanings of *theoria* and the elisions between them, see Rutherford 2000b.

93. Frame 2006.

94. Wagner-Hasel 2000.

95. On the composition of the Amphictyony, see Hall 2002, 135–39.

96. On Hermes as a cattle rustler, see Johnston 2002; Haft 1996. For the relationship between the Greek Hermes the cattle thief and Iranian and Indian versions of the same story, see Sick 1996.

97. Jameson 1988, 89.

98. Hermes' "need for validation" is also emphasized by Greene 2005. That Hermes kills but does not eat the cattle has puzzled some commentators. It is possible that his abstinence serves to distinguish him from a local figure from the central Peloponnese, Bouphagos ("Cattle-Devourer"), the son of Iapetos, who was killed for laying hands on Artemis. See Pausanias 8.24–28.

99. Larson (2005) has recently pointed out that many themes such as the use of fire, the consumption of meat, and the origins of sacrifice are prominent in the tale of Lugalbanda in the Mountain Cave, a Neo-Sumerian text that shows remarkable affinities to the *Homeric Hymn to Hermes.*

100. N. O. Brown 1947, 90–101. Brown's influence can be seen in Grottanelli 1983.

101. Silver 1992, 264–80.

102. Pausanias 10.7.1. On the inviolability of sacred land and sanctuaries, see McInerney 2006.

103. The topos of exchanging cattle and lyre recurs in the famous story recorded in the Mnesiepes inscription from Paros, which tells how Archilochus was taking his cow to market, only to have it disappear. He then finds a lyre on the ground before him, signifying the god's approval of the poet. See Kontoleon 1955 and Tarditi 1956.

104. On the Thriaia, see Larson 1995 and McInerney 1997. On the themes of conflict and the threat posed by Hermes, see Harrell 1991. For a reading of the hymn that emphasizes competition between Apollo and Hermes as a model for athletic competition, see Allan 2004.

105. Vilatte 1988.

106. For Athena and Poseidon in dispute over Athens, see Pausanias 1.24.3, 1.26.5; for competition over Troizen, 2.30.5; Poseidon loses Corinth to Helios: 2.1.6. For the view that the pedimental statues of the Hekatompedon celebrate Athena's victory over the Poseidon substitute, Triton (the statue often referred to as Bluebeard), see T. P. Howe 1954.

107. On the exchange between Poseidon and Apollo of Kalauria and Delphi, see Ephoros apud Strabo 8.6.14, Pausanias 2.33.2, Kallimachos frag. 221 (= Schol. ad Aeschylus, *Eu.* 21).

Chapter 7
Sacred Economics

1. Aristotle, *Pol.* 1.1256a.

2. On *nomos* and its derivation, see Laroche 1949.

3. Linders 1992, 11.

4. Couelle 2005.

5. Diodorus Siculus 16.23; Androtion, *FGrH* III B 324 F 30; Philochoros, *FGrH* III B 328 F 155 (= Didymus, *Comm. in Dem.* 13.40–14.49); Thucydides 1.139.

6. For Delian terminology, see Kent 1948, 257.

7. Ban on tanning: *IG* XIV 352.1.69–71; *IG* I^3 257. Manure on sacred land: *IG* II^2 1126.21; *IG* II^2 2493; *IG* XII 7.62; *SIG*^{3} 986; *LSCG* 67.27–28; *IG* XI

2.146.76–77; Polybios 12.4d.5–8. On manure in premodern agriculture, see Columella, *Rust.* 2.14 and Slicher van Bath 1963, 3–25.

8. Frei 1988, 22; Piejko 1990, 149–150.

9. Attica-Boiotia: Thucydides 5.42; Ilion-Skamandrioi: *I.Ilion* 63 with restorations by F. Piejko, but also see *SEG* XLI 1055.

10. Sophocles, *OT* 1133–39.

11. Thucydides 1.139.

12. *IG* II² 204.25–30. See J. K. Davies 2001, 122. For commentary, see S. D. Lambert 2005, 132–35. For translation, see *GHI* 58. The *prostoios* may have been intended as a building to house visitors to Eleusis, similar to the *pastades* built at Krisa, the landing place for visitors coming to Delphi by sea. See *IG* II² 1126.22. This may correspond to a tetragonal stoa excavated close to the shore. See Lerat 1948 and Luce 1990.

13. "Outside the boundaries" is restored. For the view that the land under discussion was inside the boundaries, see Scafuro 2003.

14. Daverio-Rocchi 1988a, 108.

15. Skinner 1964, 1965; Rozman 1976.

16. Tuchelt 1992, 74.

17. *IG* I³ 78. See Clinton 1988; Cavanaugh 1996.

18. *IG* I³ 6C.40–45.

19. For estimates of meat yield at sacrifice, see Ruschenbusch 1982.

20. Aischines 3.108–13. The area calculated includes all land between sea level and 80 meters above sea level, bounded by Chrysso (NE), Sernikaki (NW), the slopes of Mount Ghiona (W), Xerovouni (E), and the Gulf of Corinth (S). See the topographical map of the vicinity of Delphi in Bommelaer 1991, fig. 1. On the Sacred Plain, see Kahrstedt 1953; Daverio-Rocchi 1988a and 1988b; Deslondes 1999; Rousset 2002, 183–204, and Howe 2003.

21. Demosthenes 18.151; *IG* II² 1126 (*CID* I 10).

22. *Pausanias: Guide to Greece,* trans. P. Levi (Harmondsworth, 1971).

23. Aischines 3.107, Diodorus Siculus 9.16; Thessalos, *Presb. Log.* (= Pomtow 1918, 318–20). On the date of composition, see W. D. Smith 1990, 7. Kirrha and Krisa are frequently confused in the ancient (and modern) sources. For the two sites, see McInerney 1999, 309–12.

24. Strabo 9.3.4. A similar account is given by the scholiast to Pindar, *Carm. hyp. a* 34.1.

25. The bibliography of the First Sacred War is as long as the war is obscure. For a confident midcentury reconstruction, see Forrest 1956. For subsequent (extreme) skepticism, see Robertson 1978. Contra see Tausend 1986. Most recently, see J. K. Davies 1994 and Howe 2003. For the retrojection of the Third and Fourth Sacred Wars onto the earlier wars, see Sanchez 2001, 58–80.

26. Luce 1991.

27. *Syll.*³ 604. The same men also conducted a twelve-animal sacrifice to Athena. See Alfaro Giner 1995, 21.

28. Içten and Engelmann 1995, no. 2: "the holy *horos* (boundary) of the land of Artemis in Hippoboton." For Nicias' endowment, see Plutarch, *Nicias* 3.16 and Kent 1948, 256.

29. Kent 1948 256.

30. Londey 1990.
31. Decrees of the *hieromnemones*, preserved in a dossier of documents supporting Delphi's claims in land disputes referred to the Roman senate: *Syll.*[3] 826 G IV.20–27; *Syll*[3] 293.19–20.
32. On the sanctuary of Zeus Temenites, see *IG* XII 7.62 and Isager 1992a, 15.
33. Plutarch, *de Pyth. Orac.* 9.
34. *Syll*[3] 407.
35. Diomedon's bequest: *LSCG* 177; Panathenaia stipulation: *IG* I[3] 34.42 and *IG* I[3] 46.15–16 (regulations for the Athenian colony at Brea).
36. Oinoanda regulations: *SEG* XXXVIII 1462.
37. Hatzopoulos and Loukopoulou 1989:17–29, 41–45. See Robert 1930.
38. On the expression *bous agelaia*, see Chandezon 2003, 99–100. Xenophon, *Eq.* 5.8 uses the adjective *agelaia* to refer to a brood mare, and it is in this sense that Paramonos' cow is described as a *bous agelaia*.
39. Regulations from Kos: *SEG* XXXVIII 812.
40. *IPArk* 2. For discussion and bibliography, see Chandezon 2003, 33–40.
41. Guarducci 1952. On leasing sacred land, see Behrend 1970 and Walbank 1991, 152.
42. *IMyl.* 801.3–8.
43. Kent 1948. More recently, see Reger 1994, who, however, has little to say regarding cattle husbandry.
44. According to Reger (1994, 220–30), the *hieropoioi* used the *hiera syngraphe* in order to proceed against defaulters and seized both property and livestock, which were then turned over to Apollo.
45. Isager 1992b. Seasonal payments are also attested in a lease from the Piraeus which refers to grazing lands (*ennomion*). See *IG* II[2] 2498.11–15.
46. Kent 1948, 278.
47. Records covering a fifty-year period that mention such items as a dead goose would certainly have preserved some trace of cattle sales had there been any.
48. For other instances of Delian officials legislating against speculation, see Reger 1994, 173; Descat 2000.
49. Osborne 1985 and 1988. For terracing and farming activities, see Brunet 1990.
50. *SEG* IX 11–44. See Chamoux 1988.
51. *SEG* XXXVIII 1898.
52. The stone reads *achra-*, which Dittenberger restores as *achrados* (wild pear). Other possibilities include *achrantos* (immaculate) or *achraes* (unusable).
53. Felsch et al. 1987; Yorke 1896.
54. The valley in question is located at 38°34′11.18″N by 22°51′18.27″E. Satellite imagery, available on Google Earth, reveals a fertile valley clearly visible running east–west for approximately 5.5 miles and north–south for approximately 1 mile.
55. For the festival of Artemis Elaphebolos, see Ellinger 1987.
56. The plethron is a common measurement of length, varying in different parts of the Greek world from 29.5 meters to over 33 meters. For convenience I have rounded to 30 meters in all calculations below. For its use as measurement

of area, corresponding to 10,000 square feet (Greek) or 900 square meters, see LSJ, s.v. *plethron* II.

57. Blümel 1995, 46–47.

58. Dionysos at Herakleia: Tab. Herac. (= *RIJG* XII); Hypodektes at Athens: *IG* II[2] 2501.

59. Manganaro 1989, 203–5.

60. Finley (1973) equated sacred land and its revenues with deme funds, denying that temples had any relevance in the matter. For extensive critiques of Finley, see Isager 1992a and Horster 2004, 9–15.

61. On calls for the redistribution of land in the fourth century and later, see De Ste. Croix 1981, 215 and 608 n.55.

62. Bull of Corcyra: Pausanias 10.9.3–4; Cyrene accounts: *SEG* XL 1596.

63. *SEG* XXXVIII 1170: Eumenes II exempting the sanctuary of Apollo at Tralles: "I recognize claim and remit to you the tenth owing to the royal fisc of the revenues that come in from the sacred land."

64. Pausanias 10.35.4. On the cattle of Augeas, see Apollodoros, *Bibl.* 2.5.5.

65. See, most recently, Dillon 2002, 273–74. For bibliography prior to 1977, see Rougemont in *CID* 1.1.9.

66. For full epigraphical commentary, see Rougemont's edition. The reading "at Phanateus" is secure. For the names Panopeus and Phanoteus [*sic*] see Strabo 9.3.14. Prott-Ziehen's reading is confirmed by the recently discovered copy of the same regulation at Panopeus (John McK. Camp, pers. comm.).

67. Homolle 1895, 5. See McInerney 1999, 106.

68. Helly 1992.

69. *IG* II[2] 2493 (Rhamnous) and *IG* XII 7.62 (Amorgos). For discussion of the Rhamnous lease, see Jameson 1982; for Amorgos, see *GHI* 59.

70. For these calculations, I follow Kenneth Harl's (low) estimation of the *medimnos* at 25 kilograms. The *hemihektos* is one-twelfth of a *medimnos*. See http://www.tulane.edu/~august/H310/handouts/Coinage.htm.

71. On the desirability of cow manure, see the *Suda*, s.v. *bolitos*, glossing Aristophanes, *Ach.* 1025–26: " 'Those two kept [me] in all that cow dung.' That is, in all luxury and all good things."

72. Manure-production figures are taken from the US Environmental Protection Agency Livestock Manure Management manual available at www.epa.gov/methane/reports/05-manure.pdf.

73. *GHI* 87.

74. Tab. Herac. (= *RIJG* XII). For recent treatment, see Ampolo 1992, 25–28.

75. Gallant 1991, 86.

76. Both the general undesirability of woodland and its suitability for providing fodder for grazing animals were recognized by Cato, *Agr.* 1.7.

77. Kritzas 1992.

78. Aristotle, *Pol.* 1267b and 1330a.

79. Dionysos and Thespiae: *IG* XII 1786.

80. French 1956.

81. On soil and pedology, see Bintliff 1977; Wagstaff 1981, 1987.

82. For grain laws, see Stroud 1998; for reliance on donors, see Shear 1978; for food shortages, see Garnsey 1988.

83. M. H. Hansen 2006b, but for a more nuanced understanding of what an urbanized countryside means, see Bintliff 2006.

84. Jameson 2002. Earlier studies tended to identify *eschatiai* with a specific geographical zone, either near hills or by the seaside. See D. M. Lewis 1973 (hills) and Walbank 1983, 117 (hills and seaside). Jameson insists that the term refers not to a specific location but describes the land's undeveloped qualities. For the *rationes centesimarum*, see most recently S. D. Lambert 1997.

85. Hanson 1995, 47–90.

86. Chevitarese 2005.

87. Aristotle, *Pol.* 1252b. On plow oxen see Gallant 1991, 124. The yoke of oxen serves as a byword for country life, a notion exploited by Aristophanes in *Acharnians*. See Compton-Engle 1999. For rituals associated with plowing, see Jameson 1951.

88. Thucydides 2.14.

89. Aristotle, *Pol.* 1305a. For discussion, see Lavelle 2005, 34.

90. Mitten and Morgan 1999, 314–20.

91. Sites comparable to Isthmia mentioned by Mitten and Morgan include Aphaia, Olympia, Kommos, the sanctuaries of Hera at Perachora and Samos, and of Zeus at Olympia and on Mount Apesas, the Bronze Age cult spot of Apollo Maleatas at Epidauros, as well as assorted shrines in the Athenian Agora and ash altars on Mount Arachnaion in the Argolid and Mount Oros on Aigina.

92. Cosmopoulos 2001, 74–75.

93. Sacrifice and incubation at the Amphiareion: Pausanias 1.34.5.

94. The expression *tôn de dêmoriôn* is unattested elsewhere in Greek inscriptions and is not found in LSJ. Etymology suggests a compound formed from demos ("people") and meiromai ("receive a share"). The Athenian equivalent is *kreanomia*.

95. The connection of the cult to the deme is suggested by the reference to "any demesman" (*tis demotes*) as opposed to an "outsider" (*xenos*) in lines 9–10.

96. See Gallant 1991, 70–72. Despite the interesting results obtained from the study of zinc/strontium ratios it is important to note that the data set consists of twenty-three skeletons from a five hundred–year period.

97. For fish consumption, see Davidson 1998; Gallant 1991, 120; and Bekker-Nielsen 2002.

Chapter 8
Cities and Cattle Business

1. Robertson 1969.

2. Neils 1996, 13. On agricultural conditions in Attica, see Sallares 1991, 309–13, who estimates that 40 percent of the region was under cultivation.

3. For Aphrodite, see D. S. Reece 1989. For variation in sacrificial animals in general, see Hägg 1998.

4. Rosivach 1994, 78.

5. Robert 1960. Contra, Langdon 1987.

6. For bibliography and commentary, see S. D. Lambert 2005, 144–47. Most recently, see Rhodes 2009, 11–12.

7. Pritchett and Pippin 1956, 255–58.

8. *IG* II^2 1635; *GHI* 28.35–36 (incorrectly translates the number of cows purchased as 154; the commentary, p. 145, gives the correct figure, 109.)

9. Jameson 1988, 91; Van Straten 1987, 159–70; Gallant 1991, 33.

10. Walbank (1994, 237) assumes that the sacrifice paid for by the 41 mnai from the Nea was a hecatomb, and that since the sacrifice involved one hundred head of cattle, it is possible to deduce the average price of the cattle at 41 drachmas a head. No hecatomb is specified, however, in *IG* II^2 334, and in *IG* II^2 1496 two separate sacrifices are attributed to the Panathenaia, only one of which is called a hecatomb. There is no reason a priori to assume that the Nea cattle were destined for a hecatomb.

11. *LSCG* 151.B.5–6. See *GHI* 62 for full bibliography.

12. See *Syll.*^{3} 398.24 and the discussion in Parker and Obbink 2000, 445.

13. Regulations of the *orgeones* of Echelos: see Meritt 1942, 283, (55.14–22).

14. For the Koresian *kreanomia*, see *IG* XII 5.647 and Ruschenbusch 1982. See also *Iscr. di Cos* ED 16 and 25 for *kreanomiai* on Kos. For a *kreanomia* associated with the cult of Asklepios, see *IG* II^2 47. For estimates of the Athenian population, based on the census of Demetrios of Phaleron, see Sealey 1993, 19–23. For higher figures, see M. H. Hansen 2006b.

15. Blümel 1995, 36.9–13.

16. *Syll.*^{3} 604.

17. *IG* II^2 847.11–26.

18. Pseudo-Xenophon, *Constitution of the Athenians*, trans. G. W. Bowersock (Cambridge, Mass., 1968).

19. Hurwit 2004, 207–10 and CD image 115 and 116.

20. For the role of the *boônai*, see Demosthenes 21.172 and schol. ad loc. For discussion, see Rosivach 1994. For *boônai* on Delos, see Jacquemin 1991.

21. For allied tribute of a "cow," see *IG* I^3 34 and *IG* I^3 46. For a description of the Panathenaia and a discussion of the regulations, see Parke 1977, 32–50. The month took its name for an earlier festival, the Hekatombaia, in honor of Apollo. This festival too matched sacrifice on an enormous scale with the calendar of animal husbandry.

22. Weber 1921, 742.

23. Poitevin et al. 1977, 33.

24. See Hodkinson 1992 and T. Howe 2008, 27–48.

25. Theognis 346–47 and 1197–1201. See Wees 2000, 56.

26. Demosthenes 19.265. T. Howe (2008, 45–46) mistakenly identifies Euthykrates as an Athenian. See, however, J. Engels' entry in *Der Neue Pauly*, s.v. Euthykrates.

27. Hesiod, *Op.* 405–6 and 435–40. Hesiod also offers advice on gelding and breaking oxen at 790–800.

28. Aristophanes, *Av.* 585, trans. O'Neill.

29. Chandezon 2003, no. 23 (Abdera, ca. 350 BC) and no. 43 (Knossos, third century BC).

30. On plows, see Pullen 1992. The same type of harness is widely depicted in Egyptian wall paintings and wooden figurines; see BM EA 52947 and the Tomb of Nakht, Luxor.

31. Pausanias 9.11.7–12.1, describing the Theban cult of Apollo Spodios in which, unusually, a draft ox was sacrificed to the god.

32. Thucydides 2.14.1.

33. Thucydides 7.27.

34. Aristotle, *Pol.* 2.1267b.

35. Isaios 11.41.

36. Pritchett 1953, 272 (stele VI.68–70) and Pritchett and Pippin 1956, 255–258, and Chandezon 2003, 17–21.

37. Xenophon, *Mem.* 2.7.6. T. Howe (2008, 60) claims, "The income derived from selling the animals was so great that Nausikydes was able to support his family and even undertake expensive liturgies." In fact, what Aristotle says is that the income from the mill was so great that Nausikydes could support his household and his animals. Aristotle presents them as an expense, not a source of revenue.

38. Xenophon, *Poroi* 5.3.

39. *IG* II^2 351.18–19. See Wilson 2009, 25; Rhodes 2009, 13. See also *IG* II^2 1656–57 for a Boiotian contractor, Demosthenes, who supplied yoke teams to transport stones for the rebuilding of the Long Walls. For yoke teams from Attica employed in transporting stone during the summer when they were not needed for farmwork, see Osborne 1987, 14–15; Burford 1960, 16–18; and Salmon 2001.

40. Polybios 4.38.4–5.

41. Aristotle, *HA* 3.21; Pausanias 10.13.1–3.

42. Akraiphia: *SEG* XXIX 439; Kopai: *SEG* XXII 432; Orchomenos: *IG* VII 3171. See Hodkinson 1988, 62; Chandezon 2003, 44–45.

43. Chandezon 2003, 40–49.

44. On Mount Tauros, see Pausanias 6.13.8. For revenue from cattle sales at Tauromenion, see *IGSI* VI 85 and Manganaro 1988, 188.

45. Plutarch, *de Pyth. orac.* 28.

46. Reese et al. 1987, 264.

47. Jameson 1988; Robert and Robert 1970, no. 511.

48. *IG* I^3 244.11–23. See also Isenberg 1975.

49. Berthiaume 1982, especially chap. 4.

50. Winand 1990.

51. Plato, *R.* 332d, *Grg.* 465b. For Aristophanes and the *mageiros*, see Fischer 1993; Compton-Engle 1999; and Ham 2004.

52. Theophrastus, *The Characters,* trans. R. C. Jebb (New York, 1979).

53. *LSCG* 151 and *GHI* 62.

54. For the terms used for different part of the sacrificial animal and the various ways of treating them (burning, eating, distribution), see Puttkammer 1912 and Meuli 1946.

55. Aristophanes, *Peace* 1048–64; for descriptions of perquisites, see, *inter alia, IG* I^2 190; *IG* II^2 1359; *IG* II^2 1360; *IG* I^3 35, *IG* I^3 250; *IG* I^3 255; *LSCG* 151; *LSCG* 157; *SEG* XXXV 113 (= *NGSL* 3). Among the best non-Athenian

examples are *Iscr. di Cos* ED 3A and ED 55. See Sokolowski's commentary in *LSCG* 156. For discussion of regulations regarding perquisites, see *NGSL* 42–43.

56. For *splanchna*, see Le Guen-Pollet 1991. For *theoxenia*, see Jameson 1994.

57. For discussion of *heuston*, see Prott-Ziehen 24; for *deisia kreon* see *LSCG* 28. The limited vocabulary employed in Greek butchery, as well as national variations in modern practice, make it difficult to offer accurate translations for Greek terminology. LSJ defines *kôlê* as the thighbone, meaning that it encompasses the bottom round (US cut), trumeau (French cut), or thick flank (British and Australian cut). The *pleuron ischiou* should include part of the hip joint and therefore approximates the top round (US cut), culotte (French cut), and silverside or topside (British and Australian cut).

58. Sokolowski 1954.

59. On perquisites as a *geras* and double portions, see Le Guen-Pollet 1991.

60. *LSAM* 7.

61. *IG* V 2.432.35–46.

62. Hide for the demarch: *IG* I^3 244.13–14. *Dermatikon* accounts: *IG* II^2 1496. See Harpokration, s.v. *dermatikon*. See Ferguson 1948, 134; Rosivach 1994.

63. Thompson 1942, 219; Mikalson 1998, 26–27. The title of a speech given by Lykourgos, *On Priestly Perquisites*, suggests that his reforms were also concerned with specifying how much of the sacrificial victim went to the other participants, the attending priest, and the sponsors of sacrifice. On the Athenian market in hides, see D. M. Robinson 1943, 193 n. 8.

64. Demosthenes 34.10.

65. *IG* II^2 1496 gives a total of 5,099 drachmas for the last six months of the archonship of Ktesikles (334/3). This figure is listed as*kephalaion*, or total. Immediately above it we have the astronomical total of 42 talents, 2,910 drachmas.

66. For an example prior to 400 BC, see *LSAM* 44 from Miletos. For Hellenistic examples, see *IG* XII 6, 170 (= *GNSL* 19) and *IG* XII 1197 from Samos; from Kos *Iscr. di Cos* ED 2, 3, 32, 62, 85, 89, 109, 144, 145, 177, 178, 180, 182, 216, 237, 238, 261 (?); from Theangela *SEG* XXIX 1088; from Erythrai *SEG* XXXVII 921. For discussion, see *NGSL* 48–53. Earlier bibliography is given in Parker and Obbink 2000, 21 n. 16.

67. *LSCG* 87.15–17. The denomination of the coins is not stated.

68. Parker and Obbink 2000, 2001.

69. For leg and hide as regular perquisites, see Puttkammer 1912, 7–8; and for a parallel from Kos, see *Iscr. di Cos* ED 3.8.

70. A half share of revenues for the priest is also stipulated in *Iscr. di Cos* ED 89.20.

71. *Iscr. di Cos* ED 2.A.14–15.

72. *Iscr. di Cos* ED 62.B.6–7 (=*LSCG* 161) Prices paid for priesthoods varied to an extraordinary degree. At Chalcedon in the third century, Aristonika purchased a priesthood for more than 300 drachmas (*LSAM* 4), while in the same city Matris son of Menios purchased the priesthood of Asklepios for over 5,000 (*LSAM* 5).

73. Parker and Obbink 2000, Text 1, lines 1–2. The clause is repeated at line 32.

74. *LSAM* 9.
75. *IG* V 1.1144.23–29.
76. *IG* II[2] 1343.
77. Pausanias 10.32.9; *IGLSyria* VII 4028.D.37–40. For discussion, see Dignas 2002, 157. Cattle and slaves are also grouped together in a law from Abdera regarding the rights of the buyer if the animal or slave falls sick. It is not known whether the regulations were formulated by a sanctuary or the city. For text and discussion, see Chandezon 2003, no. 23.
78. Diogenes Laertius, *vit. Pythag.* 6; Herakleides Pont. frag. 88 Wehrli. For skepticism, see Burkert 1960. On the *panegyris* see Nibley 1951 and for the *panegyris* as metaphor, see Rutherford 2000b; Nightingale 2004, 54–60 and 72–93.
79. Iamblichus, *On the Pythagorean Life*, trans. G. Clark (Liverpool, 1989).
80. Ligt 1993, 36–37.
81. *I.Ilion* 3; *LSCG* 67; *LSCG* Suppl. 45; *SEG* XXXVIII 1462; Ligt 1993, 42–45.
82. *Syll.*[3] 736.99–106
83. Aeschylus, *A.* 844–45; Thucydides 1.25.4.
84. Herodotos 2.58: "The Egyptians were the first among men to conduct festivals (*panegyreis*), processions, and supplications." See also 2.59, 2.63, and 6.111.
85. Hellanikos, *FGrH* 4 frag. 165; Plato, *Hp. Mi.* 363c; Isokrates, *Archid.* 95.
86. Demosthenes 10.50. For discussion, see Parker 2005b, 164. *IG*[3] 104 and Demosthenes 23.37–44 refer to the exclusion of killers from "the frontier market, games and Amphiktyonic sacrifices." Perhaps all three elements were once part of the same institution? See Stroud 1968, 53–54.
87. In *The Complete Greek Drama* 2, trans. E. O'Neill Jr. (New York, 1938).
88. For discussion of this passage, see Ligt 1993, 50; Kahrstedt 1953, 12. A sacred law from Morgantina may contain stipulations regarding similar sacrifices. See Manganaro 1999, 57–60.
89. *IPArk* 2.8–10. See Georgoudi 1974, 178.
90. On control of the hinterland by major sanctuaries in Hellenistic Asia Minor, see Briant 1996, 466–68.
91. Regulations from Kos authorize the *Hieropoioi* to purchase two cows, as beautiful as possible, with gilded horns. See *LSCG* 159.7–8.
92. Blümel 1995, 35–39; 1997, 153–55. For cattle breeders, see also Ritti 1995, 65–83.
93. Zimmermann 2000, 451.

Chapter 9
Sacred Law

1. A. T. Lucas 1989, 3.
2. Xenophon, *HG* 6.4.29.
3. The seizures of Delphi and Olympia were regional affairs. The only instance of internal discord is the expulsion of a faction of prominent Delphians led by Astykrates, exiled to Athens in 363 BC. See Buckler 1989, 9–13. Yet even this episode took place in the shadow of the Arkadian seizure of Olympia and on

the eve of the Third Sacred War, even as the prospect of Phokian interference at Delphi was growing more pressing. For events at this time, see McInerney 1999, 205–10.

4. Perikles' law: [Xenophon], *Ath. Pol.* 26.4; Plutarch, *Per.* 37.3. Inheritance at Gortyn: *ICret.* IV 72.4 l.40–5 l. 13.

5. Plutarch, *de Pyth. orac.* 28.

6. Pomtow 1883; Vokotopoulou 1973, 61 M42 and M65. Also see Plato, *Phdr.* 244b; Demosthenes 21.53; and Pausanias 10.12.10.

7. *SEG* XXXVII 945.

8. Demosthenes 59.73.

9. J. K. Davies 2004, 18–39.

10. Lysias 30.2 (Nichomachos' father was a public slave); Demosthenes 49.52 (Neaira and Stephanos giving Phano in marriage to Phrastor under the pretense she was Athenian). The procedure for introducing a son is also referred to at Demosthenes 49.61.

11. On bastards, see Patterson 1990. See *IG* XII Suppl. 303 for regulations for introducing bastards into a phratry. See also Isaios 6.57 and Ogden 1996, 78.

12. *IG* I^3 104.16–19. See Stroud 1968; Gagarin 1981; and Tulin 1996.

13. Carawan 1998.

14. See for example, Lykourgos, *in Leocrat.* 19; Isaios 9.18; Aischines 1.45; and Demosthenes 19.176; as well as MacDowell 1963, 90–100; and Blickman 1986, 193–208, especially app. 1. Also see Plescia 1970 and Thür 1996. On sacrifice, terror, and the implicit warning against perjury, see Dietrich 1988, 35.

15. Cole 1996, 231. On triple sacrifice, see Demosthenes 23.68 (bull, boar, ram); Schol. ad Aristophanes *Pl.* 820 (boar, goat, ram); and Istros *FGrH* 334 F51 (bull, goat, boar).

16. Demosthenes 47.69. On the importance of *hieromnemones*, see R. Thomas 1996.

17. [Aristotle], *Ath.Pol.* 57.3–4. See Boegehold 1995, 49–50.

18. Sealey 1983, 275–96.

19. Parker 1983.

20. Boegehold 1995, 47–48; and Robertson 2001.

21. Ruschenbusch 1960. On Demophon, see Pausanias 1.28.8.

22. Demosthenes 23.74. See Boegehold 1995, 48–49.

23. Pausanias 1.28.11.

24. Beloch 1913, 258–62. See also Sealey 1976, 99–106. On serpent worship on the Acropolis and the identification of Kekrops, Erechtheus, and Zeus Meilichios with snakes, see Kerényi 1952 and Burkert 1985, 201 and 229.

25. That the law itself was treated with special respect by the Athenians is demonstrated by the fact that the Athenians claimed (Andokides 6.2; Demosthenes 23.51) that the law of Drakon remained in effect even after Solon's archonship, and by the republication of the law by the *anagrapheis* in 409/8 BC.

26. Latte 1920. For more recent discussions of Latte's work, see Boer 1973 and Parker 2005a, 61–81.

27. Giovannini 2007, 42–47.

28. Demosthenes, *Against Meidias*, trans. A. T. Murray (Cambridge, Mass., 1939).

29. Xenophon, *HG* 3.4.11. See Bederman 2001.
30. Gagarin 1986; Hölkeskamp 1999; and Lanni 2006.
31. Robertson 1969.
32. Hesiod, *The Homeric Hymns and Homerica,* trans. H. G. Evelyn-White (Cambridge, Mass., 1920).
33. Gernet 1968. See also Meier 1988. For the categories of sacred and secular law, see Naiden 2006, 172.
34. Eder 1986; R. Thomas 1996.
35. Hölkeskamp 1999.
36. Wolff 1980, 557–79.
37. For the notion of the rule of law, see Sealey 1987.
38. Ruschenbusch 1966. For a recent treatment, see J. D. Lewis 2007.
39. On the fear of arbitrary power and abuse of authority in the Archaic states, see Humphreys 1988. The same issue is a leitmotif running through contemporary American constitutional debate; see Rozell 2002.
40. On *themis* and *nomos*, see Maine [1861] 2004, 3–4; Hölkeskamp 1992, 97–99. On the meaning of *nomos*, see Lloyd-Jones 1972, 55.
41. Lysias, *Against Nichomachos,* trans. W.R.M. Lamb (Cambridge, Mass., 1930).
42. Herakleitos, DK B114. On the gods' involvement in law, see E. M. Harris 2006, 51–52.
43. Plato, *R.* 427b–c.
44. In the introduction to *NGSL,* Lupu suggests topics that are characteristic of sacred laws: sacred space, officials, cult practice, and religious events.
45. Parker 2004, 57–70.
46. S. D. Lambert 2005, 125.
47. Humphreys 1988, 465.
48. Duchêne 1992. The translation is taken from D. M. Lewis 1993, 402–3, and incorporates the suggestions made by Graham 1998 and Henry 2002.
49. For a fuller articulation of the notion that a city is a community of humans bound by sacrifice to themselves and the gods, see Banno's appeal to the Romans in Appian, *Pun.* 84.
50. For the law, see Daux 1926; for the political reading, Davidson 1997.
51. Isaios 7, an inheritance dispute that involved Archedamos, the stepfather of Apollodoros, suing Apollodoros' uncle Eupolis, after which Apollodoros adopted Thrasyllos, his half-sister's son and Archedamos' grandson. When Apollodoros died, his cousin, the daughter of Eupolis, contested the adoption, and made a claim on the estate of Apollodoros.
52. For an analysis of the dispute between Achilles and Agamemnon, see Beidelman 1989.
53. Blümel 1997; *SEG* XLVII 1563. See also Jones 1999, 4.
54. For the full statement of his taxonomy, see Runciman 1989, 148–71, 182–265; for more specific application of the categories to Archaic Greece, see Runciman 1982. On secondary states, see Joffe 2002.
55. In *Greek Lyric: An Anthology in Translation,* trans. A. M. Miller (Indianapolis, 1996).

56. On Theognis' political stance, see Lane Fox 2000 and Wees 2000. On the antidemocratic tradition, see Roberts 1994. Similar eugenic sentiments are expressed by Plato (*R.* 459a–b). For the view that Plato, who is often characterized as virulently antidemocratic, was in fact only tepidly antidemocratic, see Monoson 2000.

57. *IG* II[2] 2498.12–13. For discussion of *ennomia* (pasturelands), see Chandezon 2003, 21–22.

58. *IG* II[2] 1196.

59. The various uses of the term *polis* have been exhaustively studied by M. H. Hansen, whose complete bibliography (up to 2000) can be found in Flensted-Jensen et al. 2000. A comprehensive summary of his views can be found in M. H. Hansen 1996.

60. Smertenko 1932. For the history of the Areopagos and the reforms of Ephialtes see R. W. Wallace 1989, 83–87, and Ostwald 1986 and 1993. For a judicious overview of the scholarship, see Samons 1999.

61. *IG* II[2] 337.

62. McCabe and Plunkett 1985, 261.29–35.

63. McCabe and Plunkett 1985, 2628–19.

64. Demosthenes 49.88.

65. Yunis 2005.

Chapter 10
Authority and Value

1. Snodgrass 2006. On bull dedications, see Shapiro 1988, 373–82. On bull sculptures, see Kawami 1986.

2. Strabo 8.3.25. See J. Roy 2000, 133–56.

3. Nafissi 2003.

4. Polybios 4.17–18. See also Sinn 1992.

5. Chaniotis 1988.

6. McInerney 2006.

7. Sartre 1979. For the Skorta plain, see Thucydides 5.42.

8. *IG* IX.2.521.15–18. See G. Lucas 1991.

9. Dio Chrysostom, *Euboian Discourse* (7). See Alcock 1993, 24–32; Brenck 2000; and Desideri 2000.

10. Pausanias 8.54.5.

11. For discussion, see Daverio-Rocchi 1988a, 134–42, as well as McInerney 1999, 76–80.

12. Rousset and Katzouros 1992, recording the decision of the *gaodikai* from Elis regarding the border between Panopeus and Stiris.

13. Daverio-Rocchi 1988a.

14. *IG* IV[2] 1.75.

15. For Miletos and Myus, see *Syll.*[3] 134. For other examples see Daverio-Rocchi 1988a, 92–129, and for earlier discussion, see Piccirilli 1973.

16. So fixed in Greek thought is the notion that a boundary is sacred that the term "sacred line" was even applied to the boundary between opposing sides on

a checkerboard. See Alcaeus frag. 56 with an explanation by the scholiast Eustathius, comm. ad Hom. *Od.* 1.107.

17. Runciman 1990.

18. On common peace treaties of the early fourth century, see Ryder 1965. On federalism and relations between *poleis*, see J.A.O. Larsen 1968; Rhodes 1993; and Ager 1996.

19. On *asylia*, see Rigsby 1996. The notion of the Hellenistic world as a network is developed by Ma 2003.

20. Chaniotis 2004, 188.

21. On *epinomia*, see Georgoudi 1974, 155–85.

22. Thucydides 5.53. The temple of Apollo at Asine is mentioned by Pausanias (2.36.5).

23. *IG* V 1.1429 and 1430. In the former it may be preferable to restore *epinomia* rather than *pronomia*. *IG* IX 2.3 makes reference to *epinomia* as an honor comparable to *proxenia* and *ektesis*.

24. *IG* IV^2 1.75B.19–21: "As regards the fruits and the pastures in the common land, neither side is to lodge an official complaint against the other concerning any incident that occurred before (the arbitration)."

25. *Hell. Oxy.* 21.5 (Chambers).

26. *I.Ilion* 63.

27. *FD* IV 4.352.3.2–7. The process regulated here is referred to by the terms *lôtis* and *lôtizein*, which are probably related to *lousis* (washing). See Bousquet 1965.

28. *ICret.* III 3.4 translated and discussed in Chaniotis 1999, 199.

29. *IG* XII 7.62.36–38.

30. *IPArk* 2.

31. For general treatments of herding and the economic activities conducted away from the center of the *polis*, see Skydsgaard 1988.

32. Aristotle, *Pol.* 1267b, 1330a.

33. For earlier discussions of the Lokrian law, see ML 24–25, but see subsequently Asheri 1965 and, for the most persuasive analysis, Cooper 1978. For early laws regularly being deposited in sanctuaries, see Hölkeskamp 1992, 100.

34. Strabo 13.1.42.

35. The literature on the introduction of coinage into Greece is enormous, but among the best recent treatments are Kroll and Waggoner 1984; Reden 1997; Kurke 1999; Kim 2001; Seaford 2004; and Schaps 2004.

36. Gordon 1958, 17, no. 5.32.

37. The episode of Laban's ewes was taken up later in the medieval and Renaissance discussions of the generative power of money (and alchemy). See, *inter alia*, Ibn Khaldun, *Maqaddimah* 3.277–79, and Shylock in *The Merchant of Venice* 1.3.74–86. For discussion of these and other texts, see Shell 1982, 48–55.

38. Herakleitos, frag. 90 DK. See Shell 1978, 49–62.

39. See Furtwängler 1986, on gold bars; Kim 2001, 15–18, on uncoined silver and weight systems.

40. On barter, exchange, and value, see Menger 1968–69.

41. Aristotle, *Pol.* 1257a.6–14. See Schaps 2004, 27–28, and Meissner 2004, 311–26.

42. See Giannisi 2004, 124–28.
43. Homer, *Il.* 11.244–45.
44. Homer, *Il.* 2.449.
45. Homer, *Il.* 6.219–36.
46. The observation was made by Laum (1924) and is discussed by Schaps (2004, 9).
47. Seaford 2004, 27. G. Davies (2002, 41) downplays the shift from object wealth to value when he writes, "One should not confuse the abstract concept of an ox as a unit of account or standard of value, which is its essential but not only monetary function, with its admittedly cumbersome physical form. Once that is realized (a position *quickly reached* by primitive man if not yet by all economists or anthropologists), the inclusion of cattle as money is easily accepted, in practice and logic" (emphasis added).
48. Masson and Sarianidi 1972, 125. For Greek branding, see *IG* VII 3171 and Jones 1987.
49. Nicolet-Pierre 2006, 17. See Homer, *Il.* 23.269 and 751.
50. Strøm 1992.
51. Strøm 1992, 50.
52. Laum 1924, 14–19.
53. *ICret* IV 1, *ICret* IV 8, *SEG* XXXV 991; see Reden 1997, 158.
54. Boessneck 1986.
55. On oxen and coinage, see Ridgeway 1892; for *pecus* and *pecunia*, see Sutherland 1974, 17, and Papadopoulos 2002.
56. Rehak and Younger 1998, 123.
57. Seltman 1924.
58. On the copper industry at Enkomi, see Muhly et al. 1980a and 1988. For the Ingot God, see Hadjisavvas 2003.
59. Comaroff and Comaroff 1990.
60. Ferguson 1985.
61. Strabo 8.6.16.
62. For this and other problems regarding Pheidon, notably chronological difficulties, see Schaps 2004, 101–2. For the (dubious) view that the Argive spits demonstrate a standardized weight and length see Courbin 1983.
63. Waldstein 1902, 61–62.
64. On coin as advertisements, see Papadopoulos 2002.
65. On fiduciarity, see Peacock 2006, 637–50.
66. Euripides, *IA*. 231–75.
67. Kurke 1999.
68. Tullock 1975, 491.
69. Seaford 2004, 67.
70. Plutarch, *Mor.* 709A.
71. For meat from hunting in the *syssition*, see Xenophon, *Lac.* 5.3 and Athenaios 139b–141e. For the regular contributions, see Dikaiarchos frag. 72 Wehrli and Plutarch, *Lyc.* 12, as well as Cartledge 2002, 146–50.
72. Plutarch, *Lyc.* 9.5. For Spartan iron spits, see Cartledge 2002, 149.
73. Athens: *IG* I^3 34, *IG* I^3 46; Boiotia: *IG* VII 351; Eretria: *IG* XII 9.189; Oinoanda *SEG* XXXVIII 1462.

74. On processions, see Chaniotis 2007 and Gödecken 1986. For further references, see Parker 1997, 91.

75. For signs of increasing care in labeling dining rooms and interpreting the evidence of ritual meals, see Bergquist 1998.

76. *LSCG* 80. Attalos also endowed the festival with 40 *metretai* of wine (approx. 360 gallons).

77. *LSAM* 9.

78. On sacred calendars, see Jameson 1965; Daux 1983; Van Straten 1987; Rosivach 1994; S. D. Lambert 2002; and Ekroth 2002, 150–69 and 343–55.

79. The hero Thorikos: Daux 1983; Athena Hellotis: *LSCG* 20 B.41; Zeus Polieus in the city: *LSCG* 18.3.15–18; Zeus Polieus in the deme: *LSCG* 18.3.61–63.

80. *IG* VII 1785. On the leasing of sanctuary land by individuals belonging to the deme in which the sanctuary was located, see N. Marinatos' remarks in the discussion appended to Lohmann 1992.

81. *Agora* XIX L 4a and *GHI* 37 for bibliography.

82. On *gene*, see Bourriot 1976; Roussel 1976; and Parker 1997, app. 2.

83. Pollux 8.107. On *orgeones*, see Ferguson 1944 and Leiwo 1997.

84. *IG* II^2 2499.

85. *LSCG* 12.6

86. Theophrastos, *On Piety* F 8 (ed. Pötscher).

87. In *Hellenistic Poetry: An Anthology*, trans. B. H. Fowler (Madison, 1990).

88. Theophrastos, *Char.* 9 (shamelessness).

89. Menander, *Dysk.* 394–425.

90. Theophrastos, *Char.* 15 (superstition).

91 *SEG* XXXVI 535.

92. Xenophon, *An.* 5.3.7–11. See Purvis 2003, 65–68.

93. Jameson et al. 1993, 63. See also Scullion 2000.

94. Graham 1995 interprets *hiara* as sacrificial victims.

95. Clinton 1996, 174.

96. Clinton 1996, 177–79. See also Johnston 1999, 47–49.

97. *LSCG* 119.2–3 and 6.

98. On regulations of the *orgeones* and the *demos* regarding Bendis, see *IG* 1^3 136, *IG* II^2 1324, 1361. On the coexistence of sacrificial groups, see Ustinova 1996. On Bendis in general, see Planeaux 2000.

99. On the repetition of patterns beneath the myriad varieties of religious festivities, see Motte 2003. See also Parker 2005b.

100. On the treatment of sacrificial animals, see Puttkammer 1912 and Le Guen-Pollet 1991.

101. Van Straten 1987, 164.

102. Pausanias 2.26.7.

103. *IG* XII Suppl. 414. See Picard 1923. For osteological evidence confirming the ban, see Des Courtils and Pariente 1991.

104. Cherry and Renfrew 1986. See also Van de Mieroop 2005 and Ma 2003. For criticisms, see Whitley 1991b, 40.

105. Cakes in the shape of deer were also sacrificed to Artemis: *Anecd. Bekk.* 1.249.7–9; Athenaios 646e. In Athens "seventh cows" were sacrificed to Pythian Apollo; see *LSCG* 25.

Chapter 11
Conclusions

1. Cohen and Drabkin 1958. On Archimedes and the cattle problem in the context of ancient traditions, see most recently Jaeger 2008, 200.
2. For the text, see I. Thomas 1993, 202. For a solution (of sorts) and summary of earlier work done on the problem, see Vardi 1998.
3. Finch and Western 1977. According to Timaios (*FGrH* 566 frag. 37), the cattle were herded on Cape Mylai. Nearby lies Taormina, named for Mount Tauros ("bull"). See Pausanias 6.13.8.
4. Aristotle, *HA* 8.7.
5. Pindar, *O.* 13.65–70.
6. For Euboia, see Aelian, *NA* 12.36. For Hera and Io, see Seaford 1988, 124.
7. Apollonius Rhodius 4.965.
8. On unblemished victims, see Schlaifer 1943, 52. Examples in Mosaic law and Islamic law also reflect a common concern that sacrificial animals should be unblemished (Leviticus 1.3; 21.19; Udhiyah, Saheeh, Saheeh al-Jaami', no. 886).
9. *Hymn. Hom. Ap.* 289; Herodotos 2.38. Porphyry (*Im.* frag. 10) reports that both the Mnevis and Apis bulls were entirely black.
10. Aelian, *NA* 17.45. Aelian's comparison with the bulls of Greece suggests a Greek rather than Roman source.
11. Eckl 2000.
12. Androtion, *FGrH* 324 55. See also Philochoros, *FGrH* 328 169.
13. Rowse 2001.
14. Pollex 1999, 546.
15. Günther 1990.
16. For the argument against the burials relating to draft animals, see Pollex 1999.
17. Zich 1992.
18. Sobocinski and Makowiecki 1991.
19. For a detailed treatment of possible Minoan and Mycenaean contact with northern Europe, see Kristiansen and Larsson 2005.
20. The literature devoted to ascribing "Homeric" society to a particular century is vast. My approach follows the conclusions reached by I. Morris 1986. For an exploration of gluttony and its social significance, see Wilkins 1997, 2000, and Steiner 2002.
21. Athenaios 1.15; 1.23.
22. Hardie 1923, 166–67.
23. Codellas 1945, 241.
24. *Suda*, s.v. *ataurôpê.*
25. Coldstream 1976; Snodgrass 1982; Antonaccio 1993 and 1995; Whitley 1988.

26. Hall 1997 and 2002.

27. Schedios in the Catalogue of Ships: *Il.* 2.517; Schedios killed by Hector: *Il.* 15.515 and 17.306 (with different patronymics). For local heroes in Homer, see McInerney 1999, 193–207. On epichoric heroes and regional identity, see Calame 1987, 153–86.

28. Gehrke 1986, 18.

29. Nagy 1979, 7.

30. Pindar, *O.* 13.18–19. See D'Angour 1997, 346–50.

31. T. J. Mitchell 1986, 395.

32. Hartog 1988.

33. Athenaios 7.299f–300a.

34. Wilson 2009, 22–25.

Bibliography

Abusch, T. 2001. "The Development and Meaning of the Epic of Gilgamesh: An Interpretive Essay." *JAOS* 121: 614–22.

Achilli, A., et al. 2007. "Mitochondrial DNA Variation of Modern Tuscans Supports the Near Eastern Origin of Etruscans." *AmJHumGen* 80: 759–68.

Ackerman, R. 1991. *The Myth and Ritual School: J. G. Frazer and the Cambridge Ritualists.* New York and London.

Adams, C. J. 2000. *The Sexual Politics of Meat: A Feminist-Vegetarian Critical Theory.* 10th anniversary edition. New York.

Adebayo, A. G. 1991. "Of Man and Cattle: A Reconsideration of the Traditions of Origin of Pastoral Fulani of Nigeria." *History in Africa* 18: 1–21.

Adkins, A.W.H. 1972. "Homeric Gods and the Values of Homeric Society." *JHS* 92: 1–19.

Ager, S. 1996. *Interstate Arbitrations in the Greek World, 337–90 B.C.* Berkeley and Los Angeles.

Alberti, B. 2001. "Faience Goddesses and Ivory Bull-Leapers: The Aesthetics of Sexual Difference at Late Bronze Age Knossos." *World Archaeology* 33: 189–205.

Alcock, S. E. 1993. *Graecia Capta: The Landscapes of Roman Greece.* Cambridge.

Alcock, S., and R. Osborne, eds. 1994. *Placing the Gods: Sanctuaries and Sacred Space in Ancient Greece.* Oxford.

Alderink, L. J. 1982. "Mythical and Cosmological Structure in the *Homeric Hymn to Demeter.*" *Numen* 29: 1–16.

Alderson, L. 1992. "The Categorisation of Types and Breeds of Cattle in Europe." *ArchZootec* 41: 325–34.

Alexander, C. 1932. "A Bacchic Inscription of the Second Century A.D." *BMMA* 27: 240–42.

Alexander, R. L. 1993. "The Storm-God at Yazilikaya: Sources and Influences," in *Aspects of Art and Iconography: Anatolia and Its Neighbours: Studies in Honor of Nimet Özgüç*, ed. M. J. Mellink, E. Porada, and T. Özgüç, 1–13. Ankara.

Allan, A. 2004. "From *Agonistes* to *Agonios*: Hermes, Chaos and Conflict," in *Games and Festivals in Classical Antiquity: Proceedings of the Conference Held in Edinburgh 10–12 July 2000* (*BAR- IS* 1220), ed. S. Bell and G. Davies, 45–54. Oxford.

Allen, R. 1994. "Agriculture during the Industrial Revolution," in *The Economic History of Britain since 1700*, vol. 1, *1700–1860*, ed. R. Floud and D. McCloskey, 96–122. Cambridge.

Alfaro Giner, C. 1995. "La *polis* y la explotación ganadera en los santuarios griegos." *Polis* 7: 5–37.

Alston, R. 1999. "The Revolt of the *Boukoloi*: Geography, History and Myth," in *Organised Crime in Antiquity,* ed. K. Hopwood, 129–54. London.

Ampolo, C. 1992. "The Economics of the Sanctuaries in Southern Italy and Sicily," in Linders and Alroth 1992, 25–28.
Andrewes, A. 1952. "Sparta and Arcadia in the Early Fifth Century." *Phoenix* 6: 1–5.
Andrews, P.B.S. 1969. "The Myth of Europa and Minos." *G&R* 16: 60–66.
Anello, P. 2005. "Cittadini e barbari in Sicilia," in *Il cittadino, lo straniero, il barbaro, fra integrazione ed emarginazione nell'antichità. Atti del I Incontro Internazionale di Storia antica, Genova, 22–24 maggio 2003.* (Serta Antiqua et Mediaevalia 7), ed. M.G.A. Bertinelli and A. Donati, 143–76. Rome.
———. 2006. "Ciclopi e Lestrigoni," in *Ethne e Religioni nella Sicilia Antica. Atti del convegno (Palermo, 6–7 Dicembre 2000)* (Supplementi a Kokalos XVIII), ed. P. Anello, G. Martorana and R. Sammartano, 71–86. Rome.
Anhalt, E. K. 1997. "A Bull for Poseidon: The Bull's Bellow in *Odyssey* 21.46–50." *CQ* 47: 15–25.
Antonaccio, C. M. 1993. "The Archaeology of Ancestors," in *Cultural Poetics in Archaic Greece*, ed. C. Dougherty and L. Kurke, 46–70. Cambridge.
———. 1995. *An Archaeology of Ancestors: Tomb Cult and Hero Cult in Early Greece*. Lanham, Md.
Arabagian, R. K. 1984. "Cattle Raiding and Bride Stealing." *Religion* 14: 107–42.
Arbuckle, B. S. 2005. "Experimental Animal Domestication and Its Application to the Study of Animal Exploitation in Prehistory," in *The First Steps of Animal Domestication: New Archaeological Approaches*, ed. J. D. Vigne, D. Helmer, and J. Peters, 18–33. Oxford.
Armstrong, R. 2006. *Cretan Women: Pasiphae, Ariadne, and Phaedra in Latin Poetry.* Oxford.
Asheri, D. 1965. "Distribuzioni di terre e legislazione agraria nella Locride occidentale." *Journal of Juristic Papyrology* 15: 313–28.
Athanassakis, A., trans. 1976. *The Homeric Hymns*. Baltimore.
Aubriot, D. 2003. "Autour de quelques fêtes épiques et lyriques: Homère et Pindare," in *Dieux, fêtes, sacré dans la Grèce et la Rome antiques*, ed. A. Motte and C.-M. Ternes, 53–80. Turnhout.
Bachechi, L., and F. Martini. 2002. "Le sepulture," in *Grotta del Romito*, ed. F. Martini, 29–36. Florence.
Bachvarova, M. R. 2001. "Successful Birth, Unsuccessful Marriage: Aeschylus' *Suppliants* and Mesopotamian Birth Incantations." *NIN. Journal of Gender Studies* 2: 49–90.
Bader, F. 1980. "Rhapsodies homériques et irlandaises," in *Recherches sur les religions de l'Antiquité classique*, ed. R. Bloch, 61–74. Paris.
Baker, C.M.A., and C. Manwell. 1991. "Population Genetics, Molecular Markers and Gene Conservation of Bovine Breeds," in *Cattle Genetic Resources*, ed. C. G. Hickman, 221–304. Amsterdam and New York.
Bates, W. N. 1925. "The E of the Temple at Delphi." *AJA* 29: 239–46.
Baudrillard, J. 1994. "The Animals: Territory and Metamorphoses," in *Simulacra and Simulation*, trans. S. F. Glaser, 129–42. Ann Arbor.
Beck, W. 1986. "Choice and Context: Metrical Doublets for Hera." *AJP* 107: 480–88.

Beckmann, S. 2005. "The Mystery of Taurus—Knowing the Way." *Journal of the Washington Academy of Sciences* 91: 55–68.
Bederman, D. J. 2001. *International Law in Antiquity.* Cambridge.
Bégouën, R., and J. Clottes. 1987. "Les Trois-Frères after Breuil." *Antiquity* 61: 180–87.
Behrend, D. 1970. *Attische Pachturkunden, Eine Beitr. z. Beschreibung der μισθώσις nach d. greich. Inschriften.* Munich.
Beidelman, T. O. 1989. "Agonistic Exchange: Homeric Reciprocity and the Heritage of Simmel and Mauss." *Cultural Anthropology* 4: 227–59.
Bekker-Nielsen, T. 2002. "Fish in the Ancient Economy," in *Ancient History Matters: Studies Presented to Jens Erik Skydsgaard on His Seventieth Birthday,* ed. K. Ascani et al., 29–38. Rome.
Bell, S., and G. Davies, eds. 2004. *Games and Festivals in Classical Antiquity. Proceedings of the Conference held in Edinburgh 10–12 July 2000* (*BAR-IS* 1220). Oxford.
Beloch, K. J. 1913. *Griechische Geschichte* 12.2. Strassburg.
Bennet, J. 1985. "The Structure of the Linear B Administration at Knossos." *AJA* 89: 231–49.
———. 1992. "Collectors or Owners? An Examination of Their Possible Functions within the Palatial Economy of LM III Crete," in *Mykenaïka. Actes du IXe Colloque international sur les textes mycéniens et égéens* (*BCH* Suppl. 25), ed. J.-P. Olivier, 65–101. Paris.
———. 1998. "The Linear B Archives and the Kingdom of Nestor," in *Sandy Pylos: An Archaeological History from Nestor to Navarino,* ed. J. L. Davis, 111–33. Austin.
Benton, S. 1961. "Cattle Egrets and Bustards in Greek Art." *JHS* 81: 44–55.
Berchem, D. van. 1967. "Sanctuaires d'Hercule-Melqart; Contribution à l'étude de l'expansion phénicienne en Mediterranée." *Syria* 44: 73–109.
Bergquist, B. 1988. "The Archaeology of Sacrifice: Minoan-Mycenaean versus Greek," in Hägg et al. 1988, 21–34.
———. 1998. "Feasting of Worshippers or Temple and Sacrifice? The Case of Herakleion on Thasos," in *Ancient Greek Cult Practice from the Archaeological Evidence,* ed. R. Hägg, 57–72. Stockholm.
Berlioz, S. 2004. "Vie del sacro, vie della transumanza: il Kabeirion di Tebe nella prima età del ferro," in Frizell 2004, 31–37.
Bernal, M. 1987. *Black Athena: The Afroasiatic Roots of Classical Civilization,* vol. 1, *The Fabrication of Ancient Greece 1785–1985.* New Brunswick, N.J.
Berthiaume, G. 1982. *Les rôles du mágeiros: Étude sur la boucherie, la cuisine et le sacrifice dans la Grèce ancienne.* Leiden.
Betts, J. H. 1967. "New Light on Minoan Bureaucracy: A Reexamination of Some Cretan Seals." *Kadmos* 6: 15–40.
Bietak, M. 1996. *Avaris, the Capital of the Hyksos: Recent Excavations at Tell el-Dab'a.* London.
Bietak, M., N. Marinatos, and C. Palivou. 2007. *Taureador Scenes in Tell el-Dab'a (Avaris) and Knossos.* Vienna.
Bintliff, J. 1977. *Natural Environment and Human Settlement in Prehistoric Greece* (*BAR* suppl. ser. 28). Oxford.

Bintliff, J. 1982. "Settlement Patterns, Land Tenure and Social Structure: A Diachronic Model," in *Ranking, Resource and Exchange*, ed. C. Renfrew, 106–111. Cambridge.

———. 2006. "City-Country Relationships in the 'Normal Polis,' " in *City, Countryside, and the Spatial Organization of Value in Classical Antiquity*, ed. R. M. Rosen and I. Sluiter, 13–32. Leiden.

Bintliff, J. L., and A. M. Snodgrass. 1985. "The Cambridge/Bradford Boeotian Expedition: The First Four Years." *JFA* 12: 123–61.

Bittel, K. 1976. *Die Hethiter.* Munich.

Blenkinsopp, J. 2004. *Treasures Old and New: Essays in the Theology of the Pentateuch*. Cambridge.

Blickman, D. R. 1986. "The Myth of Ixion and Pollution for Homicide in Archaic Greece." *CJ* 81: 193–208.

Blümel, W. 1995. "Inschriften aus Karien I." *EA* 25: 35–64.

———. 1997. "Vertrag zwischen Latmos und Pidasa." *EA* 29: 135–42.

Boegehold, A. L. 1995. *The Lawcourts at Athens: Sites, Buildings, Equipment, Procedure, and Testimonia* (*The Athenian Agora* 28). Princeton.

Boer, W. den. 1973. "Aspects of Religion in Ancient Greece." *HSCP* 77: 1–21.

Boessneck, J. 1962. *Die deutschenAusgrabungen auf der Argissa-Magula in Thessalien I*. Bonn.

———. 1986. "Zoologische Ergebnisse an den Tierknochen- und Molluskenfunden," in *Haus und Stadt im Klassischen Griechenland*, ed. W. Hoepfner and E.-L. Schwander, 136–40. Munich.

Bogin, B. 1999. *Patterns of Human Growth*^{2}. Cambridge.

Bommelaer, J.-F., ed. 1991. *Guide de Delphes: Le site*. Paris.

Bonatz, D. 2007. "The Divine Image of the King: Religious Representation of Political Power in the Hittite Empire," in *Representations of Political Power: Case Histories from Times of Change and Dissolving Order in the Ancient Near East*, ed. M. Heinz and M. H. Feldman, 111–36. Winona Lake, Ind.

Bonefas, S. 1989. "The Musical Inscription from Epidauros." *Hesperia* 58: 51–62.

Bonetto, J. 2004. "Agricoltura e allevamento in Cisalpina: Alcuni spunti per una riflessione," in Frizell 2004, 57–66.

Bonnechere, P. 2003. *Trophonios de Lébadée: Cultes et mythes d'une cité béotienne au miroir de la mentalité antique*. Leiden.

Bonnet, C. 1988. *Melqart, Cultes et Mythes de l'Héraclès Tyrien en Méditeranée* (Studia Phoenicia 7). Namur.

Borgna, E. 2004. "Aegean Feasting: A Minoan Perspective," in Wright 2004b, 127–59.

Bosch-Gimpera, P. 1944. "The Phokaians in the Far West: An Historical Reconstruction." *CQ* 38: 53–59.

Bossinger, W. 2005. *Die Heilende Kraft des Singens*. Norderstedt.

Boulestin, B., and J. Gomez de Soto. 1995. *La Mort: Passé, présent, conditionnel*. La Roche-sur-Yon.

Bourdieu, P. 1967. "Postface," in *Gothic Architecture and Scholasticism*, by E. Panofsky, trans. L. Petit, 135–67. Cleveland.

———. 1980. "Le capital social: Notes provisoires." *Actes de la Recherche en Sciences Sociales* 31: 2–3.
Bourriot, F. 1976. *Recherches sur la nature du genos: Études d'histoire sociale athénienne-périodes archaique et classique.* Lille.
Bousquet, J. 1965. "Convention entre Myania et Hypnia." *BCH* 89: 665–79.
Boyazoglu, J., and J.-C. Flamant. 1990. "Mediterranean Systems of Animal Production," in *The World of Pastoralism, Herding Systems in Comparative Perspective*, ed. J. G. Galaty and D. L. Johnson, 353–93. London.
Boyce, M. 1987. "Priests, Cattle and Men." *BSOAS* 50: 508–26.
Braun, T. 2004. "Hecataeus' Knowledge of the Western Mediterranean," in *Greek Identity in the Western Mediterranean: Papers in Honour of Brian Shefton*, ed. K. Lomas, 287–348. Leiden.
Bremmer, J. M. 1992. "Symbols of Marginality from Early Pythagoreans to Late Antique Monks." *G&R* 39: 205–14.
———. 1999. "Transvestite Dionysos," in *Rites of Passage in Ancient Greece: Literature, Religion, Society*, ed. M. W. Padilla, 183–200. London.
———. 2002. "Odysseus versus the Cyclops," in *Myth and Symbol* I: *Symbolic Phenomena in Ancient Greek Culture.* (Papers of the Norwegian Institute at Athens 5), ed. S. des Bouvrie, 135–50. Bergen.
Brenk, F. E. 2000. "Dio on the Simple and Self-Sufficient Life," in *Dio Chrysostom: Politics, Letters and Philosophy*, ed. S. Swain, 261–78. Oxford.
Brenneman, W. L. Jr. 1989. "Serpents, Cows, and Ladies: Contrasting Symbolism in Irish and Indo-European Cattle-Raiding Myths." *History of Religions* 28: 340–54.
Bresson, A. 2000. *La cité marchande.* Bordeaux.
Briant, P. 1996. *Histoire de l'empire perse, de Cyrus à Alexandre.* Paris.
Brize, P. 1980. *Die "Geryoneis" des Stesichoros und die frühe griechische Kunst.* Würzburg.
Brody, A. 2002. "From the Hills of Adonis through the Pillars of Hercules: Recent Advances in the Archaeology of Canaan and Phoenicia." *NEA* 65: 69–80.
Brown, J. P. 1968. "Cosmological Myth and the Tuna of Gibraltar." *TAPA* 99: 37–62.
———. 2001. *Israel and Hellas: The Legacy of Iranian Imperialism and the Individual.* Vol. 3. Berlin and New York.
Brown, N. O. 1947. *Hermes the Thief: The Evolution of a Myth.* New York.
Brown, P. L. 1976. *Megaliths, Myths and Men: An Introduction to Astro-archaeology.* New York.
Bruford, M. W., et al. 2003. "DNA Markers Reveal the Complexity of Livestock Domestication." *Nature Reviews Genetics* 4.11: 900–910.
Bruneau, P. 1976. "Hypothèses sur les vers 220–221 de l'*H.H.A.*: Delphes et Eretria." *RÉG* 89: 15.
Brunet, M. 1990. "Contribution à l'histoire rurale de Délos aux époques classiques et hellénistique." *BCH* 114: 669–82.
Bryant, E. F., and L. L. Patton, eds. 2005. *The Indo-Aryan Controversy: Evidence and Inference in Indian History.* London.
Buchan, M. 2004. *The Limits of Heroism: Homer and the Ethics of Reading.* Ann Arbor.

Buckler, J. 1989. *Phillip II and the Sacred War* (Mnemosyne Suppl. 109). Leiden.
Budiansky, S. 1992. *Covenant of the Wild: Why Animals Choose Domestication.* New York.
Bulloch, A. 1984. "The Future of a Hellenistic Illusion: Some Observations on Callimachus and Religion." *MusHelv* 41: 209–30.
Burke, B. 2005. "Materialization of Mycenaean Ideology and the Ayia Triada Sarcophagus." *AJA* 109: 403–22.
Burford, A. 1960. "Heavy Transport in Classical Antiquity." *EconHR*[2] 13: 1–18.
Burkert, W. 1960. "Plato oder Pythagoras? Zum Ursprung des Wortes 'Philosophie.' " *Hermes* 88: 159–70.
———. 1972. *Homo Necans: Interpretationen altgriechischer Opferriten und Mythen.* Berlin.
———. 1979. *Structure and History in Greek Mythology and Ritual.* Los Angeles.
———. 1985. *Greek Religion: Archaic and Classical,* trans. J. Raffan. Cambridge.
———. 1987. "The Problem of Ritual Killings," in Hamerton-Kelly 1987, 149–76.
Burnell, F. S. 1947. "The Holy Cow." *Folklore* 58: 377–81.
Burnett, A. P. 2005. *Pindar's Songs for Young Athletes of Aigina.* Oxford.
Buxton, R. 1994. *Imaginary Greece: The Contexts of Mythology.* Cambridge.
Cairns, F. 1991. "The 'Laws of Eretria' (*IG* XII, 9 1273 and 1274): Epigraphic, Legal, Historical, and Political Aspects." *Phoenix* 45: 296–313.
Calame, C. 1987. "Spartan Genealogies: The Mythological Representation of a Spatial Organization," in J. Bremmer and F. Graf, ed. 1987. *Interpretations of Greek Mythology,* 153–86. Totowa.
Callaghan, P. 1978. "KRS 1976: Excavations at a Shrine of Glaukos, Knossos." *BSA* 78: 1–30.
Calligas, P. G. 1988. "Hero-Cult in Early Iron Age Greece," in Hägg et al. 1988, 229–34.
Camps, G. 1992. "Guerre ou paix? Origines des conflits intraspécifiques humains." *Travaux du Laboratoire d'Anthropologie et de Préhistoire des Pays de la Méditerranée Occidentale* 1: 9–15.
Carawan, E. 1998. *Rhetoric and the Law of Draco.* Oxford.
Carlson, L. W. 2001. *Cattle: An Informal History.* Chicago.
Carpenter, R. 1950. "Argeiphontes: A Suggestion." *AJA* 54: 177–83.
Cartledge, P. 2002. *Sparta and Lakonia: A Regional History from 1300 to 362 BC*[2]. Oxford.
Casson, S. 1913. "The Dispersal Legend." *CR* 27: 153–56.
Castleden, R. 2005. *The Mycenaeans.* London.
Catling, H. W. 1976–77. "Excavations at the Menelaion, Sparta, 1973–1976." *AR* 23: 36.
Cauvin, J. 1978. *Les premiers villages de Syrie-Palestine du ix*[e] *au vii*[e] *millénaire avant Jésus-Christ.* Lyons.
———. 1994. *Naissance des divinités, naissance de l'agriculture: La révolution des symboles au Néolithique.* Paris.
Cavanaugh, M. B. 1996. *Eleusis and Athens: Documents in Finance, Religion and Politics in the 5th Century B.C.* Atlanta.
Chadwick, J. 1976. *The Mycenaean World.* Cambridge.

Chaix, L., and J. Hansen. 2003. "The 'Bent Horned' Cattle from Kerma (Sudan): A Cultural Marker?," in *A Delta-Man in Yebu: Occasional Volume of the Egyptologists' Electronic Forum No. 1*, ed. A. K. Eyma and C. J. Bennett, 119–20. Boca Raton.

Chamoux, F. 1988. "Les comptes des démiurges à Cyrène," in *Comptes et inventaires dans la cité grecque*, ed. D. Knoepfler, 143–54. Geneva.

Chandezon, C. 2003. *L'élevage en Grèce (fin V*[e]*-fin 1*[er] s. a.C.): L'apport des sources épigraphiques: Ausonius (Scripta Antiqua 5). Bordeaux.

Chang, C. 1993. "Pastoral Transhumance in the Southern Balkans as a Social Ideology: Ethnoarchaeological Research in Northern Greece." *American Anthropologist* 95: 687–703.

Chang, C., and H. Koster. 1986. "Beyond Bones: Toward an Archaeology of Pastoralism," in *Advances in Archaeological Method and Theory* 9, ed. M. Schiffer, 97–148. New York.

Chaniotis, A. 1988. "Habgierige Götter, habgierige Städte: Heiligtumsbesitz und Gebietsanspruch in den kretischen Staatsverträgen." *Ktema* 13: 21–39.

———. 1999. "Milking the Mountains: Economic Activities on the Cretan Uplands in the Classical and Hellenistic Period," in *From Minoan Farmers to Roman Traders*, ed. A. Chaniotis, 181–220. Stuttgart.

———. 2004. "Justifying Territorial Claims in Classical and Hellenistic Greece: The Beginnings of International Law," in *The Law and the Courts in Ancient Greece*, ed. E. M. Harris and L. Rubinstein, 185–213. London.

———. 2007. "Theatre Rituals," in *The Greek Theatre and Festivals: Documentary Studies*, ed. P. Wilson, 48–66. Oxford.

Chappell, M. 2006. "Delphi and the *Homeric Hymn to Apollo*." *CQ* 56: 331–48.

Cherry, J. F. 1988. "Pastoralism and the Role of Animals in the Pre- and Protohistoric Economies of the Aegean." In Whittaker 1988, 177–95.

Cherry, J. F., and C. Renfrew, eds. 1986. *Peer Polity Interaction and Sociopolitical Change*. Cambridge.

Chevitarese, A. L. 2005. "Water and Olive Oil: An Analysis of Rural Scenes in Black and Red Figure Attic Vases and the Construction of the Athenian Empire," in *Global Archaeological Theory: Contextual Voices and Contemporary Thoughts*, ed. P.P.A. Funari, A. Zarankin, and E. Stovel, 297–308. New York.

Chiasson, C. C. 2005. "Myth, Ritual, and Authorial Control in Herodotus' Story of Cleobis and Biton (Hist. 1.31)." *AJP* 126: 41–64.

Clark, I. 1988. "The *Gamos* of Hera: Myth and Ritual," in *The Sacred and the Feminine in Ancient Greece*, ed. S. Blundell and M. Williamson, 13–26. New York.

Clay, J. S. 1989. *The Politics of Olympus: Form and Meaning in the Major Homeric Hymns*. Princeton.

———. 1994. "Tendenz and Olympian Propaganda in the *Homeric Hymn to Apollo*," in *Apollo: Origins and Influences*, ed. J. Solomon, 23–36. Tucson and London.

Clinton, K. 1986. "The Author of the Homeric Hymn to Demeter." *OpAth* 16: 43–49.

Clinton, K. 1988. "Sacrifice at the Eleusinian Mysteries," in Hägg et al. 1988, 69–80.

———. 1996. "Review Article: A New *lex sacra* from Selinus: Kindly Zeuses, Eumenides, Impure and Pure Tritopatores, and Elasteroi." *CP* 91: 159–79.

Clottes, J. 1998. "The 'Three Cs': Fresh Avenues towards European Paleolithic Art," in *The Archaeology of Rock-Art*, ed. C. Chippindale and P. S. C. Taçon, 112–29. Cambridge.

Codellas, P. S. 1945. "Modern Greek Folklore: The Smerdaki." *Journal of American Folklore* 58: 236–44.

Cohen, M. R., and I. E. Drabkin. 1958. *A Sourcebook in Greek Science*. Cambridge, Mass.

Coldstream, J. N. 1976. "Hero Cults in the Age of Homer." *JHS* 96: 8–17.

———. 2003. *Geometric Greece: 900–700* BC^2. London.

Cole, S. G. 1984. "The Social Function of Rituals of Maturation: The Koureion and the Arkteia." *ZPE* 55: 233–44.

———. 1996. "Oath Ritual and the Male Community at Athens," in *Demokratia: A Conversation on Democracies, Ancient and Modern*, ed. J. Ober and C. Hedrick, 227–48. Princeton.

———. 1998. "Domesticating Artemis," in *The Sacred and the Feminine in Ancient Greece*, ed. S. Blundell and M. Williamson, 27–44. London.

———. 2004. *Landscapes, Gender, and Ritual Space: The Ancient Greek Experience*. Berkeley and Los Angeles.

Collon, D. 1987. *First Impressions: Cylinder Seals in the Ancient Near East*. Chicago.

———. 1994. "Bull Leaping in Syria." *Âgypten und Levante* 4:81–85.

———. 2003. "Dance in Ancient Mesopotamia." *NEA* 66: 96–102.

Comaroff, J. L., and J. Comaroff. 1990. "Goodly Beasts, Beastly Goods: Cattle and Commodities in a South African Context." *American Ethnologist* 17: 195–216.

Compton-Engle, G. 1999. "From Country to City: The Persona of Dicaeopolis in Aristophanes' *Acharnians*." *CJ* 94: 359–73.

Cook, A. B. 1903. "Zeus, Jupiter, and the Oak." *CR* 17: 406–12.

———. 1906. "Who Was the Wife of Zeus?" *CR* 20: 365–78.

Cooper, A. B. 1978. "The Family Farm in Greece." *CJ* 73: 162–75.

Cordain, L. 1999. "Cereal Grains: Humanity's Double-Edged Sword," in *Evolutionary Aspects of Nutrition and Health: Diet, Exercise, Genetics and Chronic Disease*, ed. A. P. Simopoulos. 19–73. Basel.

Cordes, K. A., and H. M. Ibrahim. 1996. *Applications in Recreation and Leisure for Today and the Future*. St. Louis.

Cosmopoulos, M. 2001. *The Rural History of Ancient Greek City States: The Oropos Survey Project* (*BAR-IS* 1001). Oxford.

Couelle, C. 2005. "Improbable paysages, jardins et 'paradis' dans le monde grec ancien," in *Espaces et paysages: Représentations et inventions du paysage de l'Antiquité à nos jours*, ed. S. Meitinger, 15–32. Paris.

Coupez, A., and T. Kamanzi. 1970. *Littérature de cour au Rwanda*. Oxford.

Courbin, P. 1983. "Obéloi d'Argolide et d'ailleurs," in *The Greek Renaissance of the 8th Century BC: Tradition and Innovation. Proceedings of the Second International Symposium of the Swedish Institute in Athens, 1–5 June 1981*, ed. R. Hägg, 149–56. Stockholm.
Crandall, D. P. 2000. *The Place of Stunted Ironwood Trees: A Year in the Lives of the Cattle-Herding Himba of Namibia*. New York and London.
Crouwel, J. H. 1981. *Chariots and Other Means of Land Transport in Bronze Age Greece*. Amsterdam.
Cumont, F. 1933. "Religious Commentary on the Bacchi Inscription in the Metropolitan Museum of Art." *AJA* 37: 265–70.
———. 1937. "St. George and Mithra 'The Cattle-Thief.' " *JRS* 27: 63–71.
Cymbron, T., et al. 2005. "Microsatellite Diversity Suggests Different Histories for Mediterranean and Northern European Cattle Populations." *Proceedings of the Royal Society of London B: Biological Sciences* 272: 1837–43.
Czachesz, I. 2003. "The Eagle on the Tree: A Homeric Motif in Jewish and Christian Literature," in *Jerusalem, Alexandria, Rome: Studies in Ancient Cultural Interaction in Honour of A. Hilhorst*, ed. F. García Martínez and G. P. Luttikhuizen, 87–100. Leiden.
Dabney, M. K., P. Halstead, and P. Thomas. 2004. "Mycenaean Feasting on Tsoungiza at Ancient Nemea," in Wright, 77–96.
D'Agostino, B. 1996. "The Colonial Experience in Greek Mythology," in *The Western Greeks*, ed. G. Pugliese Carratelli, 214–19. Venice.
Dalby, A. 1995. "The *Iliad*, the *Odyssey* and Their Audiences." *CQ* 45: 269–79.
———. 1996. *Siren Feasts: A History of Food and Gastronomy in Greece*. London.
Damiani-Indelicato, S. 1988. "Were Cretan Girls Playing at Bull-Leaping?" *Cretan Studies* 1: 39–47.
Danek, G. 1998. *Epos und Zitat: Studien zu den Quellen der "Odyssee"* (*Wiener Studien* Beiheft 22). Vienna.
D'Angour, A. 1997. "How the Dithyramb Got Its Shape." *CQ* 47: 331–51.
Daux, G. 1926. "Nouvelles inscriptions de Thasos 1921–24." *BCH* 50: 213–49.
———. 1983. "Le calendrier de Thorikos au Musée J. Paul Getty." *AntCl* 52: 150–74.
Daverio-Rocchi, G. 1988a. *Frontiera e confini nella Grecia Antica*. Rome.
———. 1988b. "La hierà chòra di Apollo, la piana di Cirra e i confini di Delfi," in *Mélanges Pierre Lévêque I: Religion*, ed. M.-M. Mactoux and E. Geny, 117–25. Paris.
Davidson, J. N. 1997. "A Ban on Public Bars in Thasos?" CQ 47: 392–95.
———. 1998. *Fishcakes and Courtesans: The Consuming Passions of Classical Athens*. New York.
Davies, G. 2002. *A History of Money from Ancient Times to the Present Day*[3]. Cardiff.
Davies, J. K. 1994. "The Tradition about the First Sacred War," in *Greek Historiography*, ed. S. Hornblower, 193–212. Oxford.
———. 2001. "Temples, Credit, and the Circulation of Money," Meadows and Shipton 2001, 117–28. Oxford.

Davies, J. K. 2004. "Athenian Citizenship: The Descent Group and the Alternatives," in *Athenian Democracy*, ed. P. J. Rhodes, 18–39. Oxford.

Davies, M. 1988. "Stesichorus' *Geryoneis* and Its Folk-Tale Origins." *CQ* 38: 277–90.

Davis, E. 1974. "The Vapheio Cups: One Minoan and One Mycenean?" *The Art Bulletin* 56: 472–87.

Davis, S.M.J. 1995. *The Archaeology of Animals*. Abingdon.

Davis, S.M.J., and F. Valla. 1978. "Evidence for the Domestication of the Dog 12,000 Years Ago in Natufian, Israel." *Nature* 276: 608–10.

Davison, J. M. 1991. "Myth and the Periphery," in *Myth and the Polis*, ed. D. C. Pozzi and J. M. Wickersham, 49–63. Ithaca.

De Bary, W. T., et al., eds. 2001. *Sources of Japanese Tradition: From Earliest Times to 1600*^{2} 1. New York.

Defradas, J. 1954. *Les thèmes de la propagande Delphique*. Paris.

Delneri, F. 2006. *I culti misterici stranieri nei frammenti della comedia attica antica*. Bologna.

Depew, M. 1998. "Delian Hymns and Callimachean Allusion." *HSCP* 98: 155–82.

Derks, H. 1995. *De Koe van Troje: De mythe van de Griekse oudheid*. Hilversum.

De Ste. Croix, G.E.M. 1981. *The Class Struggle in the Ancient Greek World: From the Archaic Age to the Arab Conquests*. Ithaca, N.Y.

Descat, R. 2000. "L'etat et les marchés dans le monde grec," in *Mercati permanenti e mercati periodici nel mondo romano. Atti degli Incontri capresi di storia dell'economia antica (Capri 13–15 ottobre 1997)*, ed. E. Lo Cascio, 13–29. Bari.

Des Courtils, J., and A. Pariente. 1991. "Problèmes topographiques et religieux à l'Hérakleion de Thasos," in Étienne and Le Dinahet 1991, 67–73. Paris.

Desideri, P. 2000. "City and Country in Dio," in *Dio Chrysostom: Politics, Letters and Philosophy*, ed. S. Swain, 93–107. Oxford.

Deslondes, O. 1999. "L'exploitation du territoire de Delphes: Contribution de la géographie à l'étude des paysages antiques," in *La nature et ses représentations dans l'Antiquité. Actes du colloque des 24 et 25 octobre 1996, École normale supérieure de Fontenay-Saint-Cloud*, ed. C. Cusset, 22–28. Paris.

Detienne, M. 1971. "Athena and the Mastery of the Horse." *History of Religions* 11: 161–84.

———. 1986. "Apollo's Slaughterhouse." *Diacritics* 16: 46–53.

———. 1989. "Culinary Practices and the Spirit of Sacrifice," in Detienne and Vernant 1989, 1–20.

———. 1998. *Apollon: Le couteau à la main: Une approche expérimentale du polythéisme grec*. Paris.

Detienne, M., and J.-P. Vernant. 1989. *The Cuisine of Sacrifice*, trans. P. Wissing. Chicago.

Dickinson, O. 1994. *The Aegean Bronze Age*. Cambridge.

Dickson, K. 1995. *Nestor: Poetic Memory in Greek Epic*. New York and London.

Dietrich, B. C. 1983. "Divine Epiphanies in Homer." *Numen* 30: 53–79.

———. 1986. *Tradition in Greek Religion*. Berlin and New York.

———. 1988. "The Instrument of Sacrifice," in Hägg et al. 1988, 35–40.

Dignas, B. 2002. *The Economy of the Sacred in Hellenistic and Roman Asia Minor.* Oxford.
Dillon, M. 2002. *Girls and Women in Classical Greek Religion.* London.
Dimock, G. E. 1989. *The Unity of the "Odyssey."* Amherst, Mass.
Doherty, L. E. 1995. *Siren Songs: Gender, Audiences, and Narrators in the "Odyssey."* Ann Arbor, Mich..
Dombrowski, D. A. 1984. *The Philosophy of Vegetarianism.* Amherst, Mass.
Donlan, W. 1982. "Reciprocities in Homer." *CW* 75: 137–75.
———. 1989. "The Unequal Exchange between Glaucus and Diomedes in Light of the Homeric Gift-Economy." *Phoenix* 43: 1–15.
Donner, H., and W. Röllig. 1962–64. *Kanaanäische und Aramäische Inscriften.* Wiesbaden.
Donovan, J. 1990. "Animal Rights and Feminist Theory." *Signs* 15: 350–75.
Dougherty, C. 1993a. "It's Murder To Found a Colony," in *Cultural Poetics in Archaic Greece*, ed. C. Dougherty and L. Kurke, 178–198. Cambridge.
———. 1993b. *The Poetics of Colonization: From City to Text in Archaic Greece.* Oxford.
———. 2001. *The Raft of Odysseus: The Ethnographic Imagination of Homer's "Odyssey".* Oxford.
Dowden, K. 1989. *Death and the Maiden: Girls' Intitiation Rites in Greek Mythology.* London.
Downs, J. F. 1960. "Domestication: An Examination of the Changing Social Relations between Man and Animals." *Kroeber Anthropological Papers* 22: 24–52.
Duchêne, H. 1992. *La stèle du port: Fouilles du port 1: Recherches sur une nouvelle insciption Thasienne* (Études Thasiennes 14). Paris.
Ducos, P. 1978. *Tell-Mureybet, étude archéozoologique et problèmes d'écologie humaine.* Paris.
Durand, J.-L. 1986. *Sacrifice et labour en Grèce ancienne: Essai d'anthropologie religieuse.* Paris.
Earp, F. R. 1959. "The Gods of Homer." *G&R* n.s. 6: 42–45.
Eckl, A. E. 2000. "Language, Culture and Environment: The Conceptualization of Herero Cattle Terms," in *People, Cattle and Land: Transformations of a Pastoral Society in Southwestern Africa*, ed. M. Bollig and J.-B. Gewald, 401–31. Cologne.
Eder, W. 1986. "The Political Significance of the Codification of Law in Archaic Societies," in *Social Struggles in Archaic Rome: New Perspectives on the Conflict of the Orders*, ed. K. A. Raaflaub, 262–300. Berkeley and Los Angeles.
Edwards, R. B. 1979. *Kadmos the Phoenician.* Amsterdam.
Ehret, C. 2001. "Sudanic Civilization," in *Agricultural and Pastoral Societies in Ancient and Classical History*, ed. M. Adas, 224–74. Philadelphia.
Eissfeldt, O. 1962. *Kleine Schriften* 2. Tübingen.
Ekroth, G. 2002. *The Sacrificial Rituals of Greek Hero-Cults* (*Kernos* Suppl. 12). Liège.
Elderkin, G. W. 1941. "The Cults of the Erechtheion." *Hesperia* 10: 113–24.
———. 1942. "Review of *Humanistic Studies in Honor of John C. Metcalf.* University of Virginia Studies I. Charlottesville, 1941." *AJA* 46: 149.

Ellinger, P. 1987. "Hyampolis et le sanctuaire d'Artémis Elaphebolos dans l'histoire, la légende et l'espace de la Phocide." *AA* 42: 88–99.
Endsjø, D. O. 2000. "To Lock up Eleusis: A Question of Liminal Space." *Numen* 47: 351–86.
Étienne, R., and M.-T. Le Dinahet. 1991. *L'espace sacrificiel dans les civilisations méditerranéennes de l'antiquité: Actes du colloque tenu à la Maison de l'Orient, Lyon, 4–7 juin 1988*. Paris.
Evans, A. 1921. "On a Minoan Bronze Group of a Galloping Bull and Acrobatic Figure from Crete: With Glyptic Comparisons and a Note on the Oxford Relief Showing the Taurokathapsia." *JHS* 41: 247–59.
———. 1921–35. *Palace of Minos* 2. London.
Evans-Pritchard, E. E. 1934. "Imagery in Ngok Dinka Cattle-Names." *Bulletin of the School of Oriental Studies* 7: 623–28.
———. 1940. *The Nuer*. Oxford.
———. 1946. "Nuer Bridewealth." *Africa: Journal of the International African Institute* 16: 247–57.
———. 1953. "The Sacrificial Role of Cattle among the Nuer." *Africa: Journal of the International African Institute* 23: 181–98.
Eyma, A. K., and C. J. Bennett, eds. 2003. *A Delta-Man in Yebu: Occasional Volume of the Egyptologists' Electronic Forum No. 1*. Boca Raton.
Faraone, C. A. 1993. "Molten Wax, Spilt Wine and Mutilated Animals: Sympathetic Magic in Near Eastern and Early Greek Oath Ceremonies." *JHS* 113: 60–80.
———. 2003. "Playing the Bear and Fawn for Artemis: Female Initiation or Substitute Sacrifice?" in *Initiation in Ancient Greek Rituals and Narratives: New Critical Perspectives*, ed. D. B. Dodd and C. Faraone, 43–68. London and New York.
Farnell, L. R. 1896. *The Cults of the Greek States*. Oxford.
———. 1912. *The Higher Aspects of Greek Religion*. London.
———. 1921. *Greek Hero Cults and Ideas of Immortality: The Gifford Lectures 1920*. Oxford.
Felsch, R., et al. 1987. "Bericht über die Grabungen im Heiligtum der Artemis Elaphebolos und des Apollon von Hyampolis 1978–1982." *AA* 42: 1–100.
Ferguson, J. 1985. "The Bovine Mystique: Power, Property and Livestock in Rural Lesotho." *Man* n.s. 20: 647–74.
Ferguson, W. S. 1944. "The Attic Orgeones." *HTR* 37: 62–140.
———. 1948. "Demetrius Poliorcetes and the Hellenic League." *Hesperia* 17: 112–36.
Fialkowski, K. R. 1986. "A Mechanism for the Origin of the Human Brain: A Hypothesis." *Current Anthropology* 27: 288–90.
Finch, C. E., and C. B. Stanford. 2004. "Meat-Adaptive Genes and the Evolution of Slower Aging in Humans." *The Quarterly Review of Biology* 79: 3–50.
Finch, V. A., and D. Western. 1977. "Cattle Colours in Pastoral Herds: Natural Selection or Social Preference?" *Ecology* 58: 1384–92.
Finley, M. I. 1954. *The World of Odysseus*. New York.
———. 1955. "Marriage, Sale and Gift in the Homeric World." *RIDA* 2: 177–86.

———. 1957. "Homer and Mycenae: Property and Tenure." *Historia* 6: 133–59.

———. 1973. *Studies in Land and Credit in Ancient Athens 500–200 B.C.: The Horos-Inscriptions*. New York.

Fischer, N.R.E. 1993. "Multiple Personalities and Dionysiac Festivals: Dicaeopolis in Aristophanes' *Acharnians*." *G&R* 40: 31–47.

Fleisher, M. L. 1999. "Cattle Raiding and Household Demography among the Kuria of Tanzania." *Africa: Journal of the International African Institute* 69: 238–55.

———. 2000. *Kuria Cattle Raiders: Violence and Vigilantism on the Tanzania/Kenya Frontier*. Ann Arbor.

Flensted-Jensen, P., T. H. Nielsen and L. Rubinstein, eds. 2000. *Polis and Politics: Studies in Ancient Greek History*. Copenhagen.

Foley, H. P., ed. 1994. *The Homeric Hymn to Demeter: Translation, Commentary, and Interpretive Essays*. Princeton.

Foley, R. 2001. "The Evolutionary Consequences of Increased Carnivory in Hominids," in *Meat-Eating and Human Evolution*, ed. C. G. Stanford and H. T. Bunn, 305–31. Oxford.

Fontenrose, J. 1960. *The Cult and Myth of Pyrrhos at Delphi*. Berkeley and Los Angeles.

———. 1969. "The Spring Telphusa." *TAPA* 100: 119–31.

Forbes Irving, P.M.C. 1990. *Metamorphosis in Greek Myths*. Oxford.

Forrest, W. G. 1956. "The First Sacred War." *BCH* 80: 34–44.

Foster, G. V. 1984. "The Bones from the Altar West of the Painted Stoa." *Hesperia* 53: 73–82.

Foxhall, L. 1995. "Bronze to Iron: Agricultural Systems and Political Structures in Late Bronze Age and Early Iron Age Greece." *ABSA* 90: 239–50.

Foxhall, L., and A.D.E. Lewis, eds. 1996. *Greek Law in Its Political Setting: Justifications Not Justice*. Oxford.

Frame, D. 2006. "The Homerizon: Conceptual Interrogations in Homeric Studies, The Homeric Poems after Ionia: A Case in Point," http://chs.harvard.edu/publications.sec/classics.ssp. Center for Hellenic Studies, Washington, DC. September, 2006.

Frankfurter, D. 1995. "Narrating Power: The Theory and Practice of the Magical Historiola in Ritual Spells," in *Ancient Magic and Ritual Power* (*RGRW* 129), ed. M. Meyer and P. Mirecki, 457–76. Leiden.

Frazer, J. G. 1923. *The Golden Bough: A Study in Magic and Religion* 4.3: *The Dying God*. London.

Frei, P. 1988. "Phrygische Toponyme." *EA* 11: 22.

French, A. 1956. "The Economic Background to Solon's Reforms." *CQ* n.s. 6: 11–25.

Friss Johansen, H., and E. W. Whittle. 1980. *Aeschylus: "The Suppliants."* Copenhagen.

Frizell, B. S., ed. 2004. *PECUS: Man and Animal in Antiquity. Proceedings of the Conference at the Swedish Institute in Rome, Sept. 9–12, 2002*. Rome.

Furley, W. D. 1995. "Praise and Persuasion in Greek Hymns." *JHS* 115: 29–46.

Furtwängler, A. 1986. "Neue Beobachtungen zur frühesten Münzprägung." *SNR* 65: 153–65.

Fyske, S. 2005. "Jane Harrison and the Afterlife of Dead Languages." *Journal of Modern Literature* 28: 130–64.
Gagarin, M. 1981. *Drakon and Early Athenian Homicide Law.* New Haven.
———. 1986. *Early Greek Law.* Berkeley and Los Angeles.
Galinsky, G. K. 1972. *The Herakles Theme: The Adaptations of the Hero in Literature from Homer to the Twentieth Century.* Oxford.
Gallant, T. W. 1991. *Risk and Survival in Ancient Greece: Reconstructing the Rural Domestic Economy.* Palo Alto.
García, J. F. 2002. "Symbolic Action in the Homeric Hymns: The Theme of Recognition." *CA*: 5–39.
García Soler, M. J. 2001. *El arte de comer en la antigua Grecia.* Madrid.
Gardner, P. 1892–93. "Cacus on a Black Figured Vase." *JHS* 13: 70–76.
Garland, R. 1990. *The Greek Way of Life.* Ithaca.
Garnsey, P.D.A. 1988. *Famine and Food Supply in the Graeco-Roman World.* Cambridge.
Gautier, A. 1990. *La Domestication: Et l'homme crea ses animaux.* Paris.
Gehrke, H. J. 1986. *Jenseits von Athen und Sparta: Das dritte Griechenland und seine Staatenwerk.* Munich.
Gejvall, N.-G. 1969. *Lerna: A Preclassical Site in the Argolid* 1: *The Fauna.* 1969.
Gellner, E. 2000. "Trust, Cohesion, and the Social Order," in *Trust: Making and Breaking Cooperative Relations*, ed. D. Gambetta. Electronic edition, Department of Sociology, University of Oxford. http://www.sociology.ox.ac.uk/papers/gellner142-157.doc.
Georgoudi. S. 1974. "Quelques problèmes de la transhumance dans la Grèce ancienne." *REG* 87: 155–85.
Georgoudis, A., and A. Baltas. 1998. "The Development and Maintenance of Animal Recording Systems in Greece: A Case Study." *Animal Genetic Resources Information* 23: 27–39.
German, S. 2005. *Performance, Power and the Art of the Aegean Bronze Age* (*BAR-IS* 1347). Oxford.
Gernet, L. 1968. "Droit et prédroit en Grèce ancienne." *L'année sociologique* 3: 21–119. Reprinted in *Anthropologie de la Grèce antique* (Paris, 1968).
Gesell, G. C. 1987. "Minoan Palace and Public Cult," in Hägg and Marinatos 1987, 123–28. Stockholm.
Giannisi, P. 2004. "The Cows and the Poet in Ancient Greece," in Frizell 2004, 125–28.
Gill, C., N. Postlethwaite, and R. Seaford, eds. 1998. *Reciprocity in Ancient Greece.* Oxford.
Giordano, M. 1999. *La supplica: Rituale, istituzione sociale e tema epico in Omero.* Naples.
Giovannini, A. 1969. *Étude historique sur les origines du Catalogue des Vaisseaux.* Bern.
———. 1985. "Le sel et la fortune de Rome." *Athenaeum* 73: 380–81.
———. 2007. *Les relations entre États dans la Grèce antique (Historia* Einzelschriften 193). Stuttgart.
Girard, R. 1977. *Violence and the Sacred.* Baltimore.

Godart, L. 1972. "Les tablettes de la série Co de Cnossos." In *Acta Mycenaea2* (*Minos* 12), ed. M. S. Ruipérez, 418–24. Salamanca.
Godart, L., et al. 1975. "Nouveaux textes en linéaire B de Tirynthe," in *Tiryns* 8, ed. U. Jantzen, 37–51. Mainz.
Gödecken, K. B. 1986. "Procession Road of the Molpoi." *ZPE* 66: 217–53.
Goldhill, S. 1991. *The Poet's Voice: Essays on Poetics and Greek Literature.* Cambridge.
Goody, J. 1982. *Cooking, Cuisine, and Class: A Study in Comparative Sociology.* New York.
Gordon, E. I. 1958. "Sumerian Animal Proverbs and Fables: Collection Five." *Journal of Cuneiform Studies* 12: 1–21.
Gould, S. J. 1979. "A Biological Homage to Mickey Mouse." *Natural History* 88: 30–36.
Graber, R. B., and G. C. Richter. 1987. "The Capon Theory of the Cuckold's Horns: Confirmation or Conjecture?" *Journal of American Folklore* 100: 58–63.
Graf, F. 1979. "Apollon Delphinios." *MH* 36: 2–22.
Graham, A. J. 1995. Review of Jameson et al. 1993. *Phoenix* 49: 366–67.
———. 1998. "The Woman at the Window: Observations on the 'Stele from the Harbour' of Thasos." *JHS* 118: 22–40.
Grandazzi, A. 1997. *The Foundation of Rome: Myth and History.* Trans. J. M. Todd. Ithaca.
Green, M. W. 1984. "The Uruk Lament." *JAOS* 104: 253–79.
Greene, E. S. 2005. "Revising Illegitimacy: The Use of Epithets in the *Homeric Hymn to Hermes*." *CQ* 55: 343–49.
Greenfield, H. J., and K. D. Fowler. 2005. *The Secondary Products Revolution in Macedonia: The Zooarchaeological Remains from Megalo Nisi Galanis, a Late Neolithic-Early Bronze Age Site in Greek Macedonia* (*BAR-IS* 1414). Oxford.
Griffiths, J. G. 1986. "Lycophron on Io and Isis," *CQ* n.s. 36: 472–77.
Grottanelli, C. 1983. "Tricksters, Scape-goats, Champions, Saviors." *History of Religions* 23: 117–39.
Guarducci, M. 1952. "La legge dei tegeati intorno ai pascoli di Alea." *RivFil* 80: 49–68.
———. 1962. "Bryaktes: Un contributo allo studio dei 'banchetti eroici.' " *AJA* 66: 273–80.
Guilaine, J., and J. Zammit. 2001. *Le sentier de la guerre: Visages de la violence préhistorique.* Paris.
Guillaume, P., and N. Blockman. 2004. "By My God, I Bull Leap (Psalm 18:30// 2 Samuel 22:30)." *lectio difficilior* 2: 1–8.
Günther, K. 1990. "Neolithische Bildzeichen an einem ehemaligen Megalithgrab bei Warburg, Kr. Höxter (Westfalen)." *Germania* 68: 39–65.
Habicht, C. 1985. *Pausanias' Guide to Ancient Greece.* Berkeley and Los Angeles.
Hadjisavvas, S. 2003. "Cyprus Discovers the World," in *From Ishtar to Aphrodite: 3200 Years of Cypriot Hellenism*, ed. S. Hadjisavvas, 21–25. New York.
Haft, A. 1996. "The Mercurial Significance of Raiding: Baby Hermes and Animal Theft in Contemporary Crete." *Arion* 4: 27–48.

Hägg, R. 1987. "Gifts to the Heroes in Geometric and Archaic Greece," in Linders and Nordquist 1987, 93–99. Uppsala.

———. 1998. "Osteology and Greek Sacrificial Practice," in *Ancient Greek Cult Practice from the Archaeological Evidence*, ed. R. Hägg, 49–56. Stockholm.

Hägg, R., and N. Marinatos, eds. 1987. *Function of the Minoan Palaces*. Stockholm.

Hägg, R., N. Marinatos, and G. C. Nordquist. 1988. *Early Greek Cult Practice. Proceedings of the Fifth International Symposium at the Swedish Institute at Athens, 26–29 June, 1986*. Stockholm.

Haggis, D. C., M. S. Mook, C. M. Scarry, L. M. Snyder, and W. C. West. 2004. "Excavations at Azoria, 2002." *Hesperia* 73: 339–400.

Hahn, E. 1896. *Die Haustiere und ihre Beziehungen zur Wirtschaft des Menschen*. Leipzig.

Hall, J. M. 1995. "How Argive Was the 'Argive' Heraion? The Political and Cultic Geography of the Argive Plain, 900–400 B.C." *AJA* 99: 577–613.

———. 1997. *Ethnic Identity in Greek Antiquity*. Cambridge.

———. 2002. *Hellenicity: Between Ethnicity and Culture*. Chicago and London.

———. 2004. "Culture, Cultures and Acculturation," in *Griechische Archaik: Interne Entwicklungen—Externe Impulse*, ed. R. Rollinger and C. Ulf, 35–50. Innsbruck.

Hallager, B. 2001. "The Ram in Cultic Contexts?" in Laffineur and Hägg 2001, 315–322. Liège.

Hallager, E. 1987. "The Inscribed Stirrup Jars: Implications for Late Minoan IIIB Crete." *AJA* 91: 171–90.

———. 1996. *The Minoan Roundel and Other Sealed Documents in the Neopalatial Linear A Administration*. Liège.

Hallager, E. and B. P. Hallager. 1995. "The Knossian Bull—Political Propaganda in Neo-Palatial Crete?" in Laffineur and Niemeier, 547–56.

Halperin, D. J., and G. D. Newby. 1982. "Two Castrated Bulls: A Study in the Haggadah of Ka'b Al-Ahbar." *JAOS* 102: 631–38.

Halstead, P. 1981a. "Counting Sheep in Neolithic and Bronze Age Greece," in *Patterns of the Past: Studies in Honour of David Clarke*, ed. I. Hodder et al., 307–39. Cambridge.

———. 1981b. "From Determinism to Uncertainty: Social Storage and the Rise of the Minoan Palace," in *Economic Archaeology: Towards an Integration of Ecological and Social Approaches* (*BAR-IS* 96), ed. A. Sheridan and G. Bailey, 187–213. Oxford.

———. 1996. "Pastoralism or Household Herding? Problems of Scale and Specialization in Early Greek Animal Husbandry." *World Archaeology* 28: 20–42.

———. 2002. "Texts, Bones, and Herders: Approaches to Animal Husbandry in Late Bronze Age Greece," in *A-NA-QO-TA: Studies Presented to J. T. Killen* (*Minos* 33–34), ed. J. Bennet and J. Driessen, 149–89. Salamanca.

Halstead, P., and J. O'Shea, eds. 1989. *Bad Year Economics*. Cambridge.

Ham, G. 2004. "Dionysiac Festivals in Aristophanes' *Acharnians*," in Bell and Davies 2004, 55–64. Oxford.

Hamerton-Kelly, Robert, ed. 1987. *Violent Origins: Ritual Killing and Cultural Formation*. Palo Alto.

Hamilakis, Y., and E. Konsolaki. 2004. "Pigs for the Gods: Burnt Animal Sacrifices as Embodied Rituals at a Mycenaean Sanctuary." *OJA* 23: 135–51.

Hamilton, W. J. Jr. 1989. "Omnivorous Primate Diets and Human Overconsumption of Meat," in *Food and Evolution: Toward a Theory of Human Food Habits*, ed. M. Harris and E. B. Ross, 117–32. Philadelphia.

Hansen, M. H. 1996. "Πολλαχῶς πόλις λέγεται (Arist. Pol. 1276a 23): The Copenhagen Inventory of Poleis and the Lex Hafniensis de Civitate." *CPCActs* 3: 7–72.

———. 2006a. *The Shotgun Method: The Demography of the Ancient Greek City-State Culture*. London.

———. 2006b. *Studies in the Population of Aigina, Athens and Eretria* (Historisk-filosofiske Meddelelser 94). Copenhagen.

Hansen, W. 2000. "Foam-Born Aphrodite and the Mythology of Transformation." *AJPh* 121: 1–19.

Hanson, V. D. 1995. *The Other Greeks: The Family Farm and the Agrarian Roots of Western Civilization*. Berkeley and Los Angeles.

Harari, M. 2004. "A Short History of Pygmies in Greece and Italy," in *Greek Identity in the Western Mediterranean: Papers in Honour of Brian Shefton*, ed. K. Lomas, 163–90. Leiden.

Hardie, M. M. 1923. "The Evil Eye in Some Greek Villages of the Upper Haliakmon Valley in Western Macedonia." *Journal of the Royal Anthropological Institute of Great Britain and Ireland* 53: 160–72.

Harland, P. A. 2007. " 'These People Are . . . Men Eaters': Banquets of the Anti-Associations and Perceptions of Minority Cultural Groups," in *Identity and Interaction in the Ancient Mediterranean: Jews, Christians and Others. Essays in Honour of Stephen G. Wilson*, ed. Z. A. Crook and P. A. Harland, 56–75. Sheffield.

Harrell, S. E. 1991. "Apollo's Fraternal Threats: Language of Succession and Domination in the *Homeric Hymn to Hermes*." *GRBS* 32: 307–29.

Harris, E. M. 2006. *Democracy and the Rule of Law in Classical Athens: Essays on Law, Society, and Politics*. Cambridge.

Harris, M. 1985. *Good to Eat: Riddles of Food and Culture*. New York.

Harrison, J. E. 1890. Introduction to M. de G. Verrall, *Mythology and Monuments of Ancient Athens*. London.

———. 1893. "Primitive Hera-Worship Illustrated from the Excavations at Argos." *CR* 7: 74–78.

Hartog, F. 1988. *The Mirror of Herodotus: The Representation of the Other in the Writing of History*. Berkeley and Los Angeles.

Hatzopoulos, M. B., and L. D. Loukopoulou. 1989. *Morrylos: Cité de la Crestonie (*ΜΕΛΕΤΗΜΑΤΑ 7). Athens.

Hayden, B. 1985. "Old Europe: Sacred Matriarchy or Complementary Opposition?" in *Archaeology and Fertility Cult in the Ancient Mediterranean*, ed. A. Bonanno, 17–30. Amsterdam.

Head, B. V. 1911. *Historia Nummorum: A Manual of Greek Numismatics*. Oxford.

Heath, J. 1999. "Disentangling the Beast: Humans and Other Animals in Aeschylus' *Oresteia*." *JHS* 119: 17–47.

Heath, J. 2005. *The Talking Greeks: Speech, Animals, and the Other in Homer, Aeschylus, and Plato*. Cambridge.
Hecker, H. M. 1982. "Domestication Revisited: Its Implications for Faunal Analysis." *JFA* 9: 217–36.
Heichelheim, F. M. 1947. "Zeus Peloris." *HTR* 40: 69–70.
Heiden, B. 1997. "The Ordeals of Homeric Song." *Arethusa* 30: 221–40.
———. 2002. "Structures of Progression in the Plot of the *Iliad*." *Arethusa* 35: 237–54.
Helly, B. 1992. "Incursions chez les Dolopes," in *Topographie antique et géographie historique en Pays grec*, ed. I. Blum et al., 49–91. Paris.
Helmer, D. 1992. *La domestication des animaux par les hommes préhistoriques*. Paris.
Helmer, D., L. Gourichon, and D. Stordeur. 2004. "À l'aube de la domestication animale: Imaginaire et symbolisme animal dans les premières sociétés néolithiques du nord du proche-Orient." *Anthropozoologica* 39: 143–63.
Helms, M. 1988. *Ulysses' Sail: An Ethnographic Odyssey of Power, Knowledge, and Geographical Distance*. Princeton.
Henkelman, W.F.M. 2005. "Animal Sacrifice and 'External' Exchange in the Persepolis Fortification Tablets," in *Approaching the Babylonian Economy. Proceedings of the START Project Symposium Held in Vienna, 1–3 July 2004*, ed. H. D. Baker and M. Jursa, 137–65. Münster.
Henrichs, A. 1978. "Greek Maenadism from Olympias to Messalina." *HSCPh*: 121–60.
———. 2003. "*Hieroi Logoi* and *Hierai Bibloi*: The (Un)written Margins of the Sacred in Ancient Greece." *HSCPh* 101: 207–26.
Henry, A. 2002. "Hookers and Lookers: Prostitution and Soliciting in Late Archaic Thasos." *ABSA* 97: 217–22.
Hernández, P. N. 1985. "Back in the Cave of the Cyclops." *AJPh* 121: 345–66.
Hertzfeld, M. 1985. *The Poetics of Manhood: Contest and Identity in a Cretan Mountain Village*. Princeton.
Hesse, B. 1995. "Husbandry, Dietary Taboos and the Bones of the Ancient Near East: Zooarchaeology in the Post-Processual World," in *Methods in the Mediterranean: Historical and Archaeological Views on Texts and Archaeology*, ed. D. B. Small, 197–222. Leiden.
Hicks, R. I. 1962. "Egyptian Elements in Greek Mythology." *TAPA* 93: 90–108.
Hillmann, R. 1990. "Otto Eissfeldt und die Orientalische Religionsgeschicht." *Oriens* 32: 260–92.
Hiltebeitel, A. 1980. "Rama and Gilgamesh: The Sacrifices of the Water Buffalo and the Bull of Heaven." *History of Religions* 19: 187–223.
Hodder, I. 1990. *The Domestication of Europe*. Oxford.
Hodkinson, S. 1988. "Animal Husbandry in the Greek Polis," in Whittaker 1988, 35–74.
———. 1992. "Imperialist Democracy and Market-Oriented Pastoral Production in Classical Athens." *Anthropozoologica* 16: 53–61.
Hofmann, J. B. 1949. *Etymologisches Wörterbuch des Griechischen*. Munich.

Hogan, J. C., and D. J. Schenker. 2001. "Challenging Otherness: A Reassessment of Early Greek Attitudes toward the Divine," in *Ancient Journeys: A Festschrift in Honor of Eugene N. Lane*, ed. C. Callaway (http://www.stoa.org/lane/).
Hölkeskamp, K.-J. 1992. "Written Law in Archaic Greece." *PCPhS* 38: 87–117.
———. 1999. *Schiedsrichter, Gesetzgeber und Gesetzgebung im archaischen Griechenland* (*Historia* Einzelschriften 131). Stuttgart.
Holmberg, I. E. 1997. "The Sign of ΜΗΤΙΣ. *Arethusa* 30: 1–33.
Homolle, T. 1895. "Inscriptions de Delphes." *BCH* 19: 1–69.
Horster, M. 2004. *Landbesitz griechischer Heiligtümer in archaischer und klassischer Zeit*. Berlin.
Howe, T. 2003. "Pastoralism, the Delphic Amphiktyony and the First Sacred War: The Creation of Apollo's Sacred Pastures." *Historia* 52: 129–46.
———. 2008. *Pastoral Politics: Animals, Agriculture and Society in Ancient Greece* (Publications of the Association of Ancient Historians 9). Claremont, Calif.
Howe, T. P. 1954. "Zeus Herkeios: Thematic Unity in the Hekatompedon Sculptures." *AJA* 59: 287–301.
Hubert, H., and M. Mauss. 1899. "Essai sur la nature et la fonction du sacrifice." *L'Année Sociologique* 2: 29–138.
Humphreys, S. C. 1988. "The Discourse of Law in Archaic and Classical Greece." *Law and History Review* 6: 465–93.
Hunt, P. 2005. *Rembrandt and Ovid: The Abduction of Europa, 1632; Metamorphoses II.849-59 and the Myth Tradition* (http://traumwerk.stanford.edu/philolog/2005/12/).
Hurst, G. M. 1903. "The Cults of Olbia." *JHS* 23: 24–53.
Hurwit, J. M. 2004. *The Acropolis in the Age of Pericles*. Cambridge.
Hutchinson, S. 1992. "The Cattle of Money and the Cattle of Girls among the Nuer, 1930–1983." *American Ethnologist* 19: 294–316.
Huttner, U. 1997. *Die politische Rolle der Heraklesgestalt im griechischen Herrschertum* (*Historia* Einzelschriften 112). Stuttgart.
Içten, Ç., and H. Engelmann. 1995. "Inschriften aus Ephesos und Metropolis." *ZPE* 108: 88–94.
Immerwahr, S. 1990. *Aegean Painting in the Bronze Age*. University Park, Pa.
Infield, M. 2003. *The Names of Ankole Cows*. Kampala.
Ingold, T. 1974. "On Reindeer and Men." *Man* 9: 523–38.
Isaac, E. 1962. "On the Domestication of Cattle." *Science* 137: 195–204.
Isaakidou, V., P. Halstead, J. Davis, and S. Stocker. 2002. "Burnt Animal Sacrifice at the Mycenaean 'Palace of Nestor,' Pylos." *Antiquity* 76: 86–92.
Isager, S. 1992a. "Sacred and Profane Ownership of Land," in *Agriculture in Ancient Greece*, ed. B. Wells, 119–22. Stockholm.
———. 1992b. "Sacred Animals in Classical and Hellenistic Greece," in Linders and Alroth 1992, 15–19. Uppsala.
Isenberg, M. 1975. "The Sale of Sacrificial Victims." *CP* 70: 271–73.
Jackson, A. 1993. "War and Raids for Booty in the World of Odysseus," in *War and Society in the Greek World*, ed. J. Rich and G. Shipley, 64–76. London and New York.

Jacquemin, A. 1991. "Remarques sur le budget sacrificiel d'une cité: Délos indépendante," in Étienne and Le Dinahet 1991, 93–98. Paris.
Jaeger, M. 2008. *Archimedes and the Roman Imagination*. Ann Arbor.
Jakobson, R. 1971. *Selected Writings* 2: *Word and Language*. The Hague.
Jameson, M. H. 1951. "The Hero Echetlaeus." *TAPA* 82: 49–61.
———. 1965. "Notes on the Sacrificial Calendar from Erchia." *BCH* 89: 154–72.
———. 1977. "Agriculture and Slavery in Classical Athens." *CJ* 73: 122–45.
———. 1982. "The Leasing of Land at Rhamnous," in *Studies in Attic Epigraphy, History and Topography Presented to Eugene Vanderpool* (*Hesperia* Suppl. 19). Princeton.
———. 1988. "Sacrifice and Animal Husbandry in Classical Greece," in Whittaker 1988, 87–119.
———. 1994. "Theoxenia," in *Ancient Greek Cult Practice from the Epigraphical Evidence, Proceedings of the Second International Seminar on Ancient Greek Cult, Organized by the Swedish Institute at Athens, 22–24 November 1991*, ed. R. Hägg, 35–57. Stockholm.
———. 2002. "Attic *Eschatia*," in *Ancient History Matters: Studies Presented to Jens Erik Skydsgaard on His Seventieth Birthday*, ed. K. Ascani et al., 63–69. Rome.
Jameson, M. H., D. R. Jordan and R. D. Kotansky. 1993. *A Lex Sacra from Selinous*. Durham, N.C.
Janko, R. 1982. *Homer, Hesiod and the Hymns*. Cambridge.
Janni, P. 1984. *La mappa e il periplo: Cartografia antica e spazio odologico*. Rome.
———. 1998. "Cartographie et l'art nautique dans le monde ancien," in *Geographica Historica* (Ausonius Études 2), ed. P. Arnaud and P. Counillon. Bordeaux and Nice.
Jarman, M. R. 1973. "Preliminary Report on the Animal Bones," in *Knossos: The Sanctuary of Demeter*, ed. J. N. Coldstream. London.
Jarosch, V. 1994. *Samos* 18: *Samische Tonfiguren des 10. bis 7. Jahrhunderts v. Chr. aus dem Heraion von Samos*. Bonn.
Jeanmaire, H. 1939. *Couroi et Courètes*. Lille.
Jeffery, L. H. 1961. *Local Scripts of Archaic Greece*. Oxford.
Jensen, A. E. 1963. *Myth and Cult among Primitive Peoples*. Trans. M. T. Choldin and W. Weissleder. Chicago.
Joffe, A. H. 2002. "The Rise of Secondary States in the Iron Age Levant." *Journal of the Economic and Social History of the Orient* 45: 425–67.
Johnson, F. P. 1941. "Odysseus' Livestock." *CP* 36: 273–74.
Johnston, S. I. 1999. *Restless Dead: Encounters between the Living and the Dead in Ancient Greece*. Berkeley and Los Angeles.
———. 2002. "Myth, Festival, and Poet: The 'Homeric Hymn to Hermes' and Its Performative Context." *CP* 97: 109–32.
Jones, C. P. 1987. "Tattooing and Branding in Graeco-Roman Antiquity." *JRS* 77: 139–55.
———. 1999. "The Union of Latmos and Pidasa." *EA* 31: 1–7.
Junod, H. A. 1912. *The Life of a South African Tribe* 1. Neuchatel.

Kahil, L. 1977. "L'Artémis de Brauron: Rites et mystère." *AK* 20: 86–98.

Kahrstedt, U. 1953. "Delphoi und Das Heilige Land des Apollon," in *Studies Presented to David Moore Robinson on His Seventieth Birthday*, ed. G. E. Mylonas and D. Raymond, 2: 749–57. St. Louis.

Käppel, L. 1992. *Paian: Studien zur Geschichte einer Gattung*. Berlin and New York.

Karageorghis, V. 1956. "Two Mycenaean Bull-Craters in the G. G. Pierides Collection, Cyprus." *AJA* 60: 143–49.

———. 1991. *The Coroplastic Art of Ancient Cyprus* 1: *Chalcolithic–Late Cypriote I*. Nicosia.

———. 1993. *The Coroplastic Art of Ancient Cyprus* 3: *The Cypro-Archaic Period: Large and Medium Size Sculpture*. Nicosia.

Kardara, C. 1960. "Problems of Hera's Cult-Images." *AJA* 64: 343–58.

Katsonopoulou, D. 2002. "Helike and Her Territory in the Light of New Discoveries," in *Gli Achei e l'identita etnica degli Achei d'occidente (*Tekmeria 3), ed. E. Greco, 205–16. Paestum.

Kawami, T. S. 1986. "Greek Art and Persian Taste: Some Animal Sculpture from Persepolis." *AJA* 90: 259–67.

Keen, I. 2005. "Stanner on Aboriginal Religion," in *Aboriginal Religion in Australia*: *An Anthology of Recent Writings*, ed. M. Charlesworth et al., 61–78. London.

Kelly, T. 1966. "The Calaurian Amphictyony." *AJA* 70: 113–21.

Kent, J. H. 1948. "The Temple Estates of Delos, Rheneia, and Mykonos." *Hesperia* 17: 243–338.

Kerényi, K. 1952. *Die Jungfrau und Mutter der griechischen Religion: Eine Studie über Pallas Athene*. Zurich.

Khare, R. S. 1966. "A Case of Anomalous Values in Indian Civilization: Meat-Eating among the Kanya-Kubja Brahmans of Katyayan Gotra." *Journal of Asian Studies* 25: 229–40.

Kikawada, I. M., and A. Quinn. 1985. *Before Abraham Was: The Unity of Genesis 1–11*. Nashville, Tenn.

Killen, J. T. 1964. "The Wool Industry of Crete in the Late Bronze Age." *ABSA* 59: 1–15.

———. 1992. "Observations on the Thebes Sealings," in *Mykenaïka. Actes du IXe Colloque international sur les textes mycéniens et égéens* (*BCH* Suppl. 25), ed. J.-P. Olivier, 365–80. Paris.

———. 1994. "Thebes Sealings, Knossos Tablets and Mycenaean State Banquets." *BICS* 40: 67–84.

———. 1998. "The Pylos Ta Tablets Revisited," in "Recherches récentes en épigraphie créto-mycénienne," ed. F. Rougemont and J.-P. Olivier. *BCH* 122: 421–22.

———. 1999. "Mycenaean *o-pa*," in *Floreant Studia Mycenaea. Akten des X. Internationalen Mykenologischen Colloquiums, Salzburg*, ed. S. Deger-Jalkotzy, S. Hiller, and O. Panagl, 325–41. Vienna.

Kim, H. S. 2001. "Archaic Coinage as Evidence for the Use of Money," in Meadows and Shipton 2001, 7–22. Oxford.

Kindleberger, C. P. 1951. "Group Behaviour and International Trade." *Journal of Political Economy* 50: 30–46.
Kirk, G. S. 1970. *Myth: Its Meaning and Functions in Ancient and Other Cultures*. Berkeley and Los Angeles.
———. 1972. "Greek Mythology: Some New Perspectives." *JHS* 92: 74–85.
Klippel, W. E., and L. M. Snyder. 1991. "Dark-Age Fauna from Kavousi: The Vertebrates from the 1987 and 1988 Excavations." *Hesperia* 60: 179–86.
Knell, H. 1990. *Mythos und Polis: Bildprogramme griechischer Bauskulptur.* Darmstadt.
Koehl, R. B. 1986. "The Chieftain Cup and a Minoan Rite of Passage." *JHS* 106: 99–110.
Konstan, D. 1997. *Friendship in the Classical World*. Cambridge.
Kontoleon, N. (1952) 1955. "Νέαι ἐπιγραφαὶ περὶ τοῦ Ἀρχιλόχου ἐκ Πάρου." *AEph* (1952) [1955]: 32–95.
Kopaka, K., and N. Chaniotakis. 2003. "Just Taste Additive? Bronze Age Salt from Zakros, Crete." *OJA* 22: 55–66.
Koryakova, L. 1988. "Cultural Relations in North Central Eurasia," in *Archaeology and Language 2: Correlating Archaeological and Linguistic Hypotheses*, ed. R. Blench and M. Spriggs, 209–19. London.
Krantz, G. S. 1968. "Brain Size and Hunting Ability in Earliest Man." *Current Anthropology* 9: 450–51.
Krappe, A. H. 1942. "Animal Guides." *Journal of American Folklore* 55: 228–46.
Kristiansen, K., and T. B. Larsson. 2005. *The Rise of Bronze Age Society: Travels, Transmissions and Transformations*. Cambridge.
Kritzas, C. 1992. "Aspects de la vie politique et économique d'Argos au Ve avant J.-C.," in *Polydypsion Argos* (*BCH* Suppl. 22), ed. M. Piérart, 231–40. Paris.
Kroll, J. H., and N. M. Waggoner. 1984. "Dating the Earliest Coins of Athens, Corinth and Aegina." *AJA* 88: 325–40.
Kron, U. 1988. "Kultmahle im Heraion von Samos archaischer Zeit," in Hägg et al. 1988, 136–48.
Kruijt, A. C. 1906. *Het animisme in den indischen Archipel*. The Hague.
Kuhn, A. 1853. "Über die durch Nasale erweiterten Verablstämme." *ZVS* 2: 455–71.
Kühr, A. 2006. *Als Kadmos nach Boiotien kam: Polis und Ethnos im Spiegel thebanischer Gründungsmythen*. Stuttgart.
Kurke, L. 1999. *Coins, Bodies, Games, and Gold: The Politics of Meaning in Archaic Greece*. Princeton.
———. 2005. "Choral Lyric as "Ritualization": Poetic Sacrifice and Poetic *Ego* in Pindar's Sixth Paian." *ClAnt* 24: 81–130.
Kyrieleis, H. 1981. *Führer Durch das Heraion von Samos*. Athens.
———. 1993. "The Heraion at Samos," in Marinatos and Hägg 1993, 125–53.
Laffineur, R., and W.-D. Niemeier, eds. 1995. *Politeia: Society and State in the Aegean Bronze Age*. Liège.
La Genière, J. de, ed. 1997. *Héra: Images, espaces, cultes. Actes du colloque international du Centre de Recherches Archéologiques de l'Université de Lille III et de l'Association P. R. A. C., Lille, 29–30 Novembre 1993*. Naples.

Laks, A., and M. Rashed, eds. 2004. *Aristote et le mouvement des animaux: Dix études sur le De motu animalium*. Villeneuve d'Ascq.
Lambert, M. 1993. "Ancient Greek and Zulu Sacrificial Ritual: A Comparative Analysis." *Numen* 40: 293–318.
Lambert, S. D. 1997. *Rationes Centesimarum: Sales of Public Land in Lykourgan Athens* (ΑΡΧΑΙΑ ΕΛΛΑΣ 3). Amsterdam.
———. 2002. "The Sacrificial Calendar of Athens." *ABSA* 97: 353–99.
———. 2005. "Athenian State Regulations and Decrees, 352/1–322/1: II. Religious Regulations." *ZPE* 154: 125–59.
Lambrinoudakis, V. K. 1981. "Remains of the Mycenaean Period in the Sanctuary of Apollo Maleatas," in *Sanctuaries and Cults in the Aegean Bronze Age. Proceedings of the First International Symposium at the Swedish Institute in Athens, 12–13 May 1980,* ed. R. Hägg and N. Marinatos, 59–65. Stockholm.
Landenius Enegren, H. 2004. "Animals and Men at Knossos—the Linear B Evidence," in Frizell 2004, 12–19.
Lane Fox, R. 2000. "Theognis: An Alternative to Democracy," in *Alternatives to Athens: Varieties of Political Organization and Community in Ancient Greece*, ed. R. Brock and S. Hodkinson, 35–51. Oxford.
Langdon, M. 1987. "An Attic Decree Concerning Oropos." *Hesperia* 56: 47–58.
Lanni, A. 2006. *Law and Justice in the Courts of Classical Athens*. Cambridge.
Lapatin, K. 2001. *Chryselephantine Statuary in the Ancient Mediterranean World*. Oxford.
Laroche, E. 1949. *Historie de la racine nem- en grec ancien (nemo, nemesis, nomos, nomizo)*. Paris.
Larsen, C. S. 1995. "Biological Changes in Human Populations with Agriculture." *Annual Review of Anthropology* 24: 185–213.
Larsen, J.A.O. 1968. *Greek Federal States: Their Institutions and History*. Oxford.
Larson, J. 1995. "The Corycian Nymphs and the Bee Maidens of the Homeric *Hymn to Hermes*." *GRBS* 36: 341–57.
———. 2005. "Lugalbanda and Hermes." *CP* 100: 1–16.
Latte, K. 1920. *Heiliges Recht: Untersuchungen zur Geschichte der sakralen Rechtsformen in Griechenland*. Tübingen.
Laum, B. 1924. *Heiliges Geld*. Tübingen.
Lavelle, B. M. 2005. *Fame, Money, and Power: The Rise of Peisistratos and Democratic Tyranny at Athens*. Ann Arbor.
Leaf, W. 1910. "Hesiod and the Dominions of Aias," *CR* 24: 179–80.
Lebessi, A. 1992. "Τὰ μέταλλινα ζώδια τοῦ θηβαικοῦ Καβιρίου. Μία ἑρμηνευτικὴ προστασή." *AEph* 131: 1–17.
———. 2002. *Το Ιερό του Ερμή και Αφροδίτης στη Σύμη Βιάννου Τα Χάλκινα Ανθρωπόμορφα Ειδώλια*. Athens.
LeCount, L. J. 2001. "Like Water for Chocolate: Feasting and Political Ritual among the Late Classic Maya at Xunantunich, Belize." *American Anthropologist* 103: 935–53.
Lefebvre, H. 1991. *The Production of Space*. Trans. D. Nicholson-Smith. Oxford.
Le Guen-Pollet, B. 1991. "Espace sacrificiel et corps des bêtes immolées: Remarques sur le vocabulaire designant la part du prêtre dans la Grèce antique,

de l'époque classique à l'époque imperiale," in Étienne and Le Dinahet 1991, 13–23. Paris.

Leiwo, M. 1997. "Religion, or Other Reasons? Private Associations in Athens," in *Early Hellenistic Athens: Symptoms of Change. Papers and Monographs of the Finnish Institute at Athens* 6, ed. Jaakko Frösén, 103–17. Helsinki.

Lepetz, S. 1996. "Effets de la romanisation sur l'élévage dans les éstablissements ruraux du nord de la Gaule: L'exemple de l'augmentation de la stature des animaux domestiques," in *De la ferme indigène à la villa romaine; la romanisation des campagnes de la Gaulle. Actes du IIième colloque de l'association AGER tenu à Amiens (Somme) du 23 au 25 Septembre 1993*, ed. D. Bayard and J.-L. Collart, 317–24. Amiens.

Lerat, L. 1948. "Krisa." *RA* 31–32: 621–32.

Lévy-Bruhl, L. 1996. *L'âme primitive*. Paris.

Lévi-Strauss, C. 1963. *Totemism*. Trans. R. Needham. Boston.

Lewis, D. M. 1973. "The Athenian Rationes Centesimarum," in *Problèmes de la terre en Grèce*, ed. M. I. Finley, 187–212. Paris.

———. 1993. "Keeping Roads Clean in Thasos." *CR* 43: 402–3.

Lewis, J. D. 2007. *Early Greek Lawgivers*. London.

Lienhardt, G. 1961. *Divinity and Experience: The Religion of the Dinka*. Oxford.

Ligt, L. de. 1993. *Fairs and Markets in the Roman Empire: Economic and Social Aspects of Periodic Trade in a Pre-Industrial Society*. Amsterdam.

Lincoln, B. 1976. "The Indo-European Cattle-Raiding Myth." *History of Religions* 16: 42–65.

———. 1981. *Priests, Warriors and Cattle*. Berkeley and Los Angeles.

Linders, T. 1992. "Sacred Finances: Some Observations," in Linders and Alroth 1992, 9–12.

Linders, T., and B. Alroth, eds. 1992. *Economics of Cult in the Ancient Greek World. Proceedings of the Uppsala Symposium 1990* (Boreas: Uppsala Studies in Ancient Mediterranean and Near Eastern Civilizations 21). Uppsala.

Linders, T., and G. Nordquist, eds. 1987. *Gifts to the Gods. Proceedings of the Uppsala Symposium 1985* (Boreas: Uppsala Studies in Ancient Mediterranean and Near Eastern Civilizations 15). Uppsala.

Lindgren, M. 1973. *The People of Pylos: Prosopographical and Methodological Studies in the Pylos Archives* (Boreas: Uppsala Studies in Ancient Mediterranean and Near Eastern Civilizations 3). Uppsala.

Lippold, G. 1952. "Vasen und Münzen." *JDAI* 67: 78–92.

Lissarrague, F. 2002. "The Athenian Image of the Foreigner." In *Greeks and Barbarians*, ed. T. Harrison, 101–24. New York.

Lloyd-Jones, H. 1972. "Pindar *Fr.* 169." *HSCP* 76: 45–56.

———. 1983. "Artemis and Iphigenia." *JHS* 103: 87–102.

Lohmann, H. 1992. "Agriculture and Country Life in Classical Athens," in *Agriculture in Ancient Greece. Proceedings of the Seventh International Symposium at the Swedish Institute at Athens 16–17 May 1990*, ed. B. Wells, 29–57. Stockholm.

Lolling, H. G. 1880. *Das Kuppelgrab bei Menidi*. Athens.

Londey, P. 1990. "The Outbreak of the 4th Sacred War." *Chiron* 20: 239–60.

Long, C. 1974. *The Ayia Triadha Sarcophagus: A Study of Late Minoan and Mycenaean Funerary Practice and Beliefs*. Göteborg.

Lonsdale, S. H. 1979. "Attitudes towards Animals in Ancient Greece." *G&R* 26: 146–59.

Loraux, N. 1990. "Herakles: The Supermale and the Feminine," in *Before Sexuality: The Construction of Erotic Experience in the Ancient Greek World,* ed. D. M. Halperin, J. J. Winkler, and F. I. Zeitlin, 21–52. Princeton.

Lorenz, K. 1971. *Studies in Animal and Human Behaviour* 2. London.

Loughlin, E. 2004a. "The Calf in Bronze Age Cretan Art and Society," in Frizell 2004, 183–89.

———. 2004b. "Grasping the Bull by the Horns: Minoan Bull Sports," in Bell and Davies 2004, 1–8.

Lucas, A. T. 1989. *Cattle in Ancient Ireland*. Kilkenny.

Lucas, G. 1991. "Askyris, un cité dans le Bas-Olympe." *ZPE* 89: 135–44.

Luce, J. M. 1990. "Kirrha, Port de Delphes," in *Delphes: Oracles, Cultes et Jeux* (Les Dossiers d'Archéologie 151), ed. C. Rolley et al., 28-29.

———. 1991. "Delphes 3: L'aire du char des Rhodiens B: L'étude de l'environnement." *BCH* 115: 691–97.

Lyons, D. 2003. "Dangerous Gifts: Ideologies of Marriage and Exchange in Ancient Greece." *ClAnt* 22: 93–134.

Ma, J. 2003. "Peer Polity Interaction in the Hellenistic Age." *Past and Present* 180: 9–40.

MacDowell, D. M. 1963. *Athenian Homicide Law*. Manchester.

MacGillivray, J. A. 2000. "Labyrinths and Bull-Leapers." *Archaeology* 53: 53–56.

Mackay, L. A. 1946. "The Earthquake-Horse." *CP* 41: 150–54.

Maine, H. S. [1861] 2004. *Ancient Law*. Repr., London.

Maitland, J. 1999. "Poseidon, Walls, and Narrative Complexity in the Homeric *Iliad*." *CQ* n.s. 49: 1–13.

Malkin, I. 1998. *The Returns of Odysseus: Colonization and Ethnicity*. Berkeley and Los Angeles.

Manganaro, G. 1988. "Le tavola finanziarie di Tauromenion," In *Comptes et Inventaires dans la Cité grecque,* ed. D. Knoepfler, 155–90. Geneva.

———. 1989. "Testi e monumenti." *PP* 44: 189–216.

———. 1999. *Sikelika: Studi di Antichità e di Epigrafia della Sicilia Greca*. Pisa.

Mann, C. 2001. *Athlet und Polis im archaischen und frühklassischen Griechenland* (Hypomnemata 138). Göttingen.

Marciniak, A. 2005. *Placing Animals in the Neolithic: Social Zooarchaeology of Prehistoric Farming Communities*. London.

Marinatos, N. 1986. *Minoan Sacrificial Ritual: Cult Practice and Symbolism*. Göteborg.

———. 1987. "Public Festivals in the West Courts of the Palaces," in Hägg and Marinatos 1987, 135–43. Stockholm.

———. 2002. "The Arkteia and the Gradual Transformation of the Maiden into a Woman," in *Le orse di Brauron: Un rituale di iniziazione femminile nel santuario di Artemide*, ed. B. Gentile and F. Perusino, 29–42. Pisa.

Marinatos, N., and R. F. Hägg, eds. 1993. *Greek Sanctuaries: New Approaches*. London.
Marinatos, S. 1974. *Excavations at Thera* VI. Athens.
Mason, I. L. 1996. *A World Dictionary of Livestock Breeds, Types and Varieties*[4]. Wallingford.
Masson, M., and V. I. Sarianidi. 1972. *Central Asia: Turkmenia before the Achaemenids*. New York.
Matthews, V. J. 1977. "Naupaktia and Argonautika." *Phoenix* 31: 189–207.
McCabe, D. F., and M. A. Plunkett. 1985. *Teos Inscriptions*. Princeton.
McCallum, L. R. 1987. "Decorative Program in the Mycenaean Palace of Pylos: The Megaron Frescoes." Ph.D. diss., Univ. of Pennsylvania.
McCartney, E. 1951. "The Epithet 'Boopis.' " *CJ* 46: 348–50.
McCormick, J. 2000. *Bullfighting: Art, Technique and Spanish Society*. London.
McDonald, W. A., W.D.E. Coulson, and J. Rosser. 1983. *Excavations at Nichoria in Southwest Greece III: Dark Age and Byzantine Occcupation*. Minneapolis.
McInerney, J. 1997. "Parnassus, Delphi, and the Thyiades." *GRBS* 38: 263–83.
———. 1999. *The Folds of Parnassos*. Austin.
———. 2006. "On the Border: Sacred Land and the Margins of the Community," in *City, Countryside, and the Spatial Organization of Value in Classical Antiquity*, ed. R. M. Rosen and I. Sluiter, 33–60. Leiden.
McLeod, W. E. 1961. "Oral Bards at Delphi." *TAPA* 92: 317–25.
Meadow, R. H. 1989. "Osteological Evidence for the Process of Animal Domestication," in *The Walking Larder: Patterns of Domestication, Pastoralism and Predation*, ed. J. Clutton-Brock, 80–90. London.
Meadows, A., and K. Shipton, eds. 2001. *Money and Its Uses in the Ancient Greek World*. Oxford.
Meier, C. 1988. *Die politische Kunst der griechischen Tragödie*. Munich.
Meissner, B. 2004. "Über Modelle antiker Geldverkehrssysteme," in *Commerce and Monetary Systems in the Ancient World: Means of Transmission and Cultural Interaction* (Oriens et Occidens 6), ed. R. Rollinger and C. Ulf, 311–26. Stuttgart.
Menger, C. 1968–69. *Gesammelte Werke* 4: *Schriften über Geld und Währungspolitik*. Tübingen. Orig. published 1909.
Meritt, B. D. 1942. "Greek Inscriptions." *Hesperia* 11: 275–98.
Meuli, K. 1946. "Greichische Opferbäuche," in *Phyllobolia: Festschrift für Peter von der Mühll*, 185–288. Basel.
Michalowski, P. 2003. "The Libraries of Babel: Text, Authority, and Tradition in Ancient Mesopotamia," in *Cultural Repertoires: Structure, Function and Dynamics*, ed. G. J. Dorleijn and H. L. J. Vanstiphout, 105–29. Leuven.
Mikalson, J. D. 1998. *Religion in Hellenistic Athens*. Berkeley and Los Angeles.
Milgrom, J. 1981. "The Paradox of the Red Cow (Num. XIX)." *Vetus Testamentum* 31: 62–72.
Miller, A. 1986. *From Delos to Delphi: A Literary Study of the Homeric Hymn to Apollo* (Mnemosyne Suppl. 93). Leiden.
Miller, S. 1979. "Old Discoveries from Old Athens." *Hesperia* 39: 223–31.
Mirov, N. T. 1945. "Notes on the Domestication of Reindeer." *American Anthropologist* 47: 393–408.

Mitchell, L. G. 2001. "Euboean Io." *CQ* n.s. 51: 339–52.
Mitchell, T. J. 1986. "The Ritual Origin of Scholarly Myths." *Journal of American Folklore* 99: 394–414.
Mitten, D. and C. Morgan. 1999. "The Figurines." In *Isthmia VIII: The Late Bronze Age Settlement and Early Iron Age Sanctuary*, ed. C. Morgan. Princeton.
Monoson, S. S. 2000. *Plato's Democratic Entanglements: Athenian Politics and the Practice of Philosophy*. Princeton.
Montiglio, S. 2005. *Wandering in Ancient Greek Culture*. Chicago.
Morgan, C. 1990. *Athletes and Oracles: The Transformation of Olympia and Delphi in the Eighth Century BC*. London.
———. 1993. "The Origins of Panhellenism," in Marinatos and Hägg 1993, 18–44.
Morris, H. F., ed. 1964. *The Heroic Recitations of the Bahima of Ankole*. Oxford.
Morris, I. 1986. "The Use and Abuse of Homer." *ClAnt* 5: 81–138.
———. 1997. "Homer and the Iron Age," in *A New Companion to Homer*, ed. I. Morris and B. Powell, 535–59. Leiden.
Morris, S. P. 1989. "A Tale of Two Cities: The Miniature Frescoes from Thera and the Origins of Greek Poetry." *AJA* 93: 511–35.
Motte, A. 2003. "Fêtes chez les hommes, fêtes chez les Dieux," in *Dieux, fêtes, sacré dans la Grèce et la Rome antiques*, ed. A. Motte and C.-M. Ternes, 113–31. Turnhout.
Muellner, L. 1998. "Glaucus redivivus." *HSCPh* 98: 1–30.
Muhly, J. D., et al. 1980a. "The Oxhide Ingots from Enkomi and Mathiati and Late Bronze Age Copper Smelting in Cyprus." *RDAC* 1980: 84–95.
———. 1980b. Review of *Excavations at Nichoria in Southwest Greece* 1. AJA 84: 101–2.
———, et al. 1988. "Cyprus, Crete and Sardinia: Copper Ox-Hide Ingots and the Bronze Age Metals Trade." *RDAC* 1988: 281–98.
Mullin, M. H. 1999. "Mirrors and Windows: Sociocultural Studies of Human-Animal Relationships." *Annual Review of Anthropology* 28: 201–24.
Munson, R. V. 2006. "An Alternate World: Herodotus and Italy," in *The Cambridge Companion to Herodotus*, ed. C. Dewald and J. Marincola, 257–73. Cambridge.
Murnaghan, S. 1992. "Maternity and Mortality in Homeric Poetry." *ClAnt* 11: 242–64.
Mussi, M. 2001. *Earliest Italy: An Overview of the Italian Paleolithic and Mesolithic*. New York.
Myres, J. L. 1946. "The Wanderings of Io: Aeschylus, *Prometheus*, 707–869." *CR* 60: 2–4.
Nafissi, M. 1997. "Atene e Metaponto: Ancora sulla *Melanippe Desmotis* e i Neleidi." *Ostraka* 6: 337–57.
———. 2001. *Problemi della "Chora" Coloniale dall'Occidente al Mar Nero*. Taranto.
———. 2003. "Elei e Pisati: Geografia, storia e istituzioni politiche della regione di Olimpia." *Geographia Antiqua* 12: 23–55.
Nagy, G. 1979. *The Best of the Achaeans*. Baltimore.
Naiden, F. S. 2006. *Ancient Supplication*. Oxford.

Narotzky, S. 2007. "The Project in the Model." *Current Anthropology* 48: 403–24.
Neils, J., ed. 1996. *Worshipping Athena: Panathenaia and Parthenon.* Madison.
Nesselrath, H.-G. 2005. " 'Where the lord of the sea grants passage to sailors through the deep-blue mere no more': The Greeks and the Western Seas." *G& R* 52: 153–71.
Neve, P. 1993. *Hattusa: Stadt der Götter und Tempel.* Mainz.
Nibley, H. 1951. "The Hierocentric State." *The Western Political Quarterly* 4: 226–53.
Nicolet-Pierre, H. 2006. "Les talents d'Homère," in *Agoranomia: Studies in Money and Exchange Presented to John H. Kroll*, ed. P. G. van Alfen, 1–20. New York.
Nightingale, A. W. 2004. *Spectacles of Truth in Classical Greek Philosophy.* Cambridge.
Nikoloudis, S. 2001. "Animal Sacrifice in the Mycenaean World." *Journal of Prehistoric Religion* 15: 32–38.
Norman, N. J. 1984. "The Temple of Athena Alea at Tegea." *AJA* 88: 169–94.
———. 1986. "Asklepios and Hygieia and the Cult Statue at Tegea." *AJA* 90: 425–30.
Nowicki, K. 2002. "From Late Minoan IIIC Refuge Settlements to Geometric Acropoleis: Architecture and Social Organization of Dark Age Villages and Towns in Crete," in *Habitat et urbanisme dans le monde grec de la fin des palais mycéniens à la prise de Milet* (494 av.J.–C.). ed. J. M. Luce, 149–74. Toulouse.
O'Brien, J. V. 1990. "Homer's Savage Hera." *CJ* 86: 105–25.
———. 1993. *The Transformation of Hera: A Study of Ritual, Hero and the Goddess in the Iliad.* Lanham, Md.
Ogden, D. 1996. *Greek Bastardy in the Classical and Hellenistic Period.* Oxford.
Ohly, D. 1976. *Die Aegineten: Die Marmorskulpturen des Tempels der Aphaia auf Aegina: Ein Katalog der Glyptothek München* 1: *Die Ostgiebelgruppe.* Munich.
Oppong, Y.M.P.A. 2002. *Moving through and Passing on: Fulani Mobility, Survival, and Identity in Ghana.* New Brunswick.
Ornan, T. 2003. "Picture and Legend: The Case of Humbaba and the Bull of Heaven." *Eretz Israel* 27: 18–32.
Ortega y Gasset, J. 1972. *Meditations on Hunting*, trans. H. B. Wescott. New York.
Osborne, R. 1985. "Buildings and Residence on the Land in Classical and Hellenistic Greece: The Contribution of Epigraphy." *ABSA* 80: 119–28.
———. 1987. *Classical Landscape with Figures: The Ancient Greek City and Its Countryside.* London.
———. 1988. "Social and Economic Implications of the Leasing of Land and Property in Classical and Hellenistic Greece." *Chiron* 18: 279–323.
Østby, E. 1986. "The Archaic Temple of Athena Alea at Tegea." *OAth* 16: 75–102.
———. 1994. "Recent Excavations in the Sanctuary of Athena Alea at Tegea (1990–1993)," in *Archaeology in the Peloponnese: New Excavations and Research*, ed. K. A. Sheedy, 39–63. Oxford.

Ostwald, M. 1986. *From Popular Sovereignty to the Sovereignty of Law.* Berkeley and Los Angeles.
———. 1993. "The Areopagus in the 'Αθηναίων Πολιτεία," in *Aristote et Athènes*, ed. M. Piérart, 139–53. Paris.
Page, D. 1973. "Stesichorus: The *Geryoneïs*." *JHS* 93: 138–54.
Palaima, T. G. 1989. "Perspectives on the Pylos Oxen Tablets," in *Studia Mycenaea* 1988 (*Ziva Antika* Monograph 7), 85–124. Skopje.
———. 2000. "The Pylos Ta Series: From Michael Ventris to the New Millennium." *BICS* 44: 236–37.
———. 2004. "Sacrificial Feasting in the Linear B Documents," in Wright 2004b, 97–126.
Palmer, L. R. 1958. "The Mycenaean Tablets and Economic History." *The Economic History Review* 11: 87–96.
———. 1965. Review of *Mycenaean Studies: Proceedings of the Third International Colloquium for Mycenaean Studies*, ed. E. L. Bennet Jr. (Madison, 1964). *Language* 41: 312–29.
Papadopoulos, J. K. 2002. "Minting Identity: Coinage, Ideology and the Economics of Colonization in Akhaian Magna Graecia." *CArchJ* 12: 21–55.
Parisinou, E. 2005. "Brightness Personified: Light and Divine Image in Ancient Greece," in *Personification in the Greek World: From Antiquity to Byzantium*, ed. E. Stafford and J. Herrin, 29–44. Aldershot.
Parisi Presicce, C. 1989. "Animals and Divinities: Hera and the Cuckoo." *MAL* 32: 83–177.
Parke, H. W. 1977. *Festivals of the Athenians*. London, 1977.
Parke, H. W., and J. Boardman. 1957. "The Struggle for the Tripod and the 1st Sacred War." *JHS* 77: 276–82.
Parker, R. 1983. *Miasma: Pollution and Purification in Early Greek Religion.* Oxford.
———. 1991. "The 'Hymn to Demeter' and the 'Homeric Hymns.' " *G&R* 38: 1–17.
———. 1997. *Athenian Religion: A History*. Oxford.
———. 2004. "What Are Sacred Laws?" in *The Law and the Courts in Ancient Greece*, ed. E. M. Harris and L. Rubinstein, 57–70. London.
———. 2005a. "Law and Religion," in *The Cambridge Companion to Ancient Greek Law*, ed. M. Gagarin and D. J. Cohen, 61–81. Cambridge.
———. 2005b. *Polytheism and Society at Athens*. Oxford.
Parker, R., and D. Obbink. 2000. "Aus der Arbeit der 'Inscriptiones Graecae' VI: Sales of Priesthoods on Cos I." *Chiron* 30: 415–49.
———. 2001. "Aus der Arbeit der 'Inscriptiones Graecae' VI: Sales of Priesthoods on Cos II." *Chiron* 31: 229–52.
Patou-Mathis, M. 1997. "Analyses taphonomique et paleoethnographique du matériel osseux de Krapina (Croatie): Nouvelles données sur la faune et les restes humains." *Préhistoire européenne* 10: 63–90.
Patterson, C. B. 1990. "Those Athenian Bastards." *ClAnt* 9: 40–73.
Pavese, C. O. 1998. "The Rhapsodic Epic Poems as Oral and Independent Poems." *HSCPh* 98: 63–90.

Payne, S. 1985. "Zoo-Archaeology in Greece: A Reader's Guide," in *Contributions to Aegean Archaeology: Studies in Honor of William A. McDonald*, ed. N. C. Wilkie and W.D.E. Coulson, 211–44. Minneapolis.

Peacock, M. S. 2006. "The Origins of Money in Ancient Greece: The Political Economy of Coinage and Exchange." *Cambridge Journal of Economics* 30: 637–50.

Pearson, D. W., and C. A. Haney. 1999. "The Rodeo Cowboy: Cultural Icon, Athlete or Entrepreneur." *Journal of Sport and Social Issues* 23: 308–27.

Peires, J. B. 1989. *The Dead Will Arise: Nongquawuse and the Great Xhosa Cattle-Killing Movement of 1856–7*. Johannesburg.

Pellecchia, M., et al. 2007. "The Mystery of Etruscan Origins: Novel Clues from Bos Taurus Mitochondrial DNA." *Proc. R. Soc. B: Biological Sciences* 274: 1175–79.

Penglase, C. 1994. *Greek Myths and Mesopotamia: Parallels and Influence in the Homeric Hymns and Hesiod*. London.

Penrose, F. C. 1893. "On the Results of an Examination of the Orientations of a Number of Greek Temples." *Philosophical Transactions of the Royal Society of London* 184: 805–34.

Perlès, C. 2001. *The Early Neolithic in Greece: The First Farming Communities in Europe*. Cambridge.

Petrakos, B. 1968. *Ο Ωρωπος και το Ιερον του Αμφιαραου*. Athens.

Phaklaris, P. V. 1990. *Αρχαία Κυνουρία. ανθρωπινή δραστηριότητα και περίβαλ λον* (2nd ed.). Athens.

Philippson, P. 1944. *Thessalische Mythologie*. Zurich.

Picard, C. 1923. "Un rituel archaïque d'Héraklès trouvé à Thasos." *BCH* 47: 241–74.

Piccirilli, L. 1973. *Gli arbitrati interstatali greci*. Pisa.

Piejko, F. 1990. "To the Inscriptions of Labraunda." *OAth* 18: 133–56.

Pilafidis-Williams, K. 1998. *The Sanctuary of Aphaia on Aigina in the Bronze Age*. Munich.

Pinney, G. F. 1994. *Symposium on "Parthenon and Panathenaia." Princeton University, September 18, 1993*. Reported by Sarah Peirce, Fordham University, and Ann Steiner, Franklin and Marshall College. *Bryn Mawr Classical Review* 94.03.09.

Pinney, G. F., and B. S. Ridgway. 1981. "Herakles at the Ends of the Earth." *JHS* 101: 141–44.

Piteros, C., J.-P. Olivier, and J. L. Melena. 1990. "Les inscriptions en Linéaire B des nodules de Thèbes (1982): La fouille, les documents, les possibilités d'interprétation." *BCH* 114: 103–84.

Planeaux, C. 2000. "The Date of Bendis' Entry into Attica." *CJ* 96: 165–92.

Plescia, J. 1970. *The Oath and Perjury in Ancient Greece*. Tallahassee.

Pointer, L. 1985. *Rodeo Champions: Eight Memorable Moments of Riding, Wrestling and Roping*. Albuquerque.

Poitevin, J., M. Mollard, and E. Picon. 1977. *Beef Production in Southern Europe*. Paris.

Polignac, F. de. 1984. *La naissance de la cité grecque: Cultes, espace et société, VIIIe–VIIe siècles avant J.-C.* Paris.

———. 1994. *Cults, Territory and the Origins of the Greek City-State*. Trans. J. Lloyd. Chicago and London.
Pollan, M. 2006. "The Modern Hunter-Gatherer." *The New York Times Magazine*, March 26.
Pollex, A. 1999. "Comments on the Interpretation of the So-Called Cattle Burials of Neolithic Central Europe." *Antiquity* 73: 542–50.
Pomtow, H. 1883. "Die Orakelinschriften von Dodona." *Jahrbücher für Philologie* 29: 306–60.
Popham, M. R., E. Touloupa, and L. H. Sackett. 1982a. "Further Excavation of the Toumba Cemetery at Lefkandi, 1981." *ABSA* 77: 213–48.
———. 1982b. "The Hero of Lefkandi," *Antiquity* 56: 159–64.
Portefaix, L. 1982. "Concepts of Ecstasy in Euripides' 'Bacchanals' and their Interpretation," in *Religious Ecstasy: Papers Read at the Symposium on Religious Ecstasy Held at Abo, Finland, on the 26th–28th of August 1981*, ed. N. G. Holm, 201–10. Stockholm.
Prescendi, F. 2007. *Décrire et comprendre le sacrifice: Les réflexions des Romains sur leur propre religion à partir de la littérature antiquaire* (Potsdamer Altertumswissenschaftliche Beiträge 19). Stuttgart.
Press, L. 1969. "The Location of Minoan Displays: A Problem of Iconography and Reality." *Man* n.s. 4: 250–55.
Preziosi, D., and L. A. Hitchcock. 1999. *Aegean Art and Architecture*. Oxford.
Pritchett, W. K. 1953. "The Attic Stelai, Part I." *Hesperia* 22: 225–99.
Pritchett, W. K., and A. Pippin. 1956. "The Attic Stelai, Part II." *Hesperia* 25: 178–328.
Pucci, P. 1993. "L'io e l'altro nel racconto di Odisseo sui Ciclopi." *SIFC* 11: 26–46.
Pugliese Caratelli, G. 1950. "Epigrafi rodie inedite." *PP* 5: 76–80.
Pullen, D. J. 1992. "Ox and Plow in the Early Bronze Age Aegean." *AJA* 96: 45–54.
Purvis, A. L. 2003. *Singular Dedications: Founders and Innovators of Private Cults in Classical Greece*. London.
Puttkammer, F. 1912. *Quomodo Graeci victimarum carnes distribuerint*. Königsberg.
Rapp, A. 1872. "Die Mänade im griechischen Cultus, in der Kunst und Poesie." *RhM* 27: 1–22, 562–611.
Rayor, D. J. 2004. *Introduction* to *The Homeric Hymns*. Trans. D. Rayor. Berkeley and Los Angeles.
Reckford, K. 1974. "Phaedra and Pasiphae: The Pull Backward." *TAPA* 104: 307–28.
Reden, S. von. 1997. "Money, Law and Exchange: Coinage in the Greek Polis." *JHS* 117: 154–76.
Reece, S. 1993. *The Stranger's Welcome: Oral Theory and the Aesthetics of the Homeric Hospitality Scene*. Ann Arbor.
———. 1997. "A Figura Etymologica in the Homeric Hymn to Hermes." *CJ* 93: 29–39.
Reese, D. S. 1989. "Faunal Remains from the Altar of Aphrodite Ourania, Athens." *Hesperia* 58: 63–70.

Reese, D. S, et al. 1987. "A Bone Assemblage at Corinth of the Second Century after Christ." *Hesperia* 56: 255–74.
Regan, T. 1983. *The Case for Animal Rights*. Berkeley and Los Angeles.
Regenbogen, O. 1965. "Die Geschichte von Solon und Krösus," in *Herodot: Eine Auswahl aus der neueren Forschung*², ed. W. Marg, 375–403. Darmstadt.
Reger, G. 1994. *Regionalism and Change in the Economy of Independent Delos*. Oxford.
Rehak, P. 1995. "The Use and Destruction of Minoan Stone Bull's Head Rhyta," in Laffineur and Niemeier, 435–59.
Rehak, P., and J. G. Younger. 1998. "Review of Aegean Prehistory VII: Neopalatial, Final Palatial, and Postpalatial Crete." *AJA* 102: 91–173.
Reinhardt, B., K. Wehrberger, and G. Bosinski. 1994. *Der Löwenmensch: Tier und Mensch in der Kunst der Eiszeit*. Ulm.
Reinold, J. 2005. "Note sur le monde animal dans le funéraire néolithique du Soudan." *Revue de Paléobiologie* 10: 107–19.
Reiser, M. 1940. "Analysis of the Poetic Simile." *Journal of Philosophy* 37: 209–17.
Reisman, P. 1977. *Freedom in Fulani Social Life: An Introspective Ethnography*. Trans. M. Fuller. Chicago.
Rhodes, P. J. 1972. *The Athenian Boule*. Oxford.
———. 1993. "The Greek Poleis: Demes, Cities and Leagues," in *The Ancient Greek City-State: Symposium on the Occasion of the 250th Anniversary of the Royal Danish Academy of Sciences and Letters. July 1–4 1992*, ed. M. H. Hansen, 161–82. Copenhagen.
———. 2009. "State and Religion in Athenian Inscriptions." *G&R* 56: 1–13.
Rice, M. 1998. *The Power of the Bull*. London.
Richardson, N. 2003. Introduction to *The Homeric Hymns*. Trans. J. Cashford. Harmondsworth.
Ridgeway, W. 1892. *The Origin of Metallic Currency and Weight Standards*. Cambridge.
Rigby, P. 1969. *Cattle and Kinship among the Gogo*. Ithaca and London.
Rigsby, K. J. 1996. *Asylia: Territorial Inviolability in the Hellenistic World*. Berkeley and Los Angeles.
Ritti, T. 1995. "Associazioni di mestiere," in *Viaggi e Commerci nell'antichita*, ed. B. M. Giannattasio, 65–83. Genoa.
Robert, J., and L. Robert. 1970. "Lydie." *BE* 6: no. 511.
Robert, L. 1930. "Notes d'épigraphie hellénistique: Lettre royale à la ville d'Ilion." *BCH* 54: 348–51.
———. 1960. "Sur une loi d'Athènes relative aux Petites Panathénées," in *Hellenica* XI–XII, 189–203. Paris.
Roberts, J. T. 1994. *Athens on Trial: The Antidemocratic Tradition in Western Thought*. Princeton.
Robertson, N. 1969. "How To Behave at a Sacrifice: Hesiod, Erga 755–756." *CP* 64: 164–69.
———. 1978. "The Myth of the First Sacred War." *CQ* 28: 38–73.
———. 1984a. "Poseidon's Festival at the Winter Solstice." *CQ* 34: 1–16.

———. 1984b. "The Ritual Background of the Erysichthon Story." *AJP* 105: 369–408.
———. 2001. "Athena as Weather Goddess: The Aegis in Myth and Ritual," in *Athena in the Classical World*, ed. S. Deacy and A. Villing, 29–56. Leiden.
Robinson, D. M. 1943. "A New Arcadian Inscription." *CP* 38: 191–99.
Robinson, M. 1969. "*Geryoneis*: Stesichorus and the Vase Painters." *CQ* 19: 207–21.
Roesch, P. 1985. "Les taureaux de bronze du Kabirion de Thebes et l'ecriture archaïque beotienne," in *Proceedings of the Third International Conference on Boeotian Antiquities (Montreal-Quebec 1979)*, ed. M. Fossey and H. Giroux, 137–49. Amsterdam.
Rose, G. P. 1969. "The Unfriendly Phaeacians." *TAPA* 100: 387–406.
Rose, H. J. 1954. "Chthonian Cattle." *Numen* 1: 213–27.
Rosen, R. M. 2003. "Revisiting Sophocles' *Poimenes*: Tragedy or Satyr Play?" in *Shards from Kolonos: Studies in Sophoclean Fragments* (Rane Series, Studi 34), ed. A. H. Sommerstein, 373–86. Bari.
———. 2007. *Making Mockery: The Poetics of Ancient Satire*. Oxford.
Rosivach, V. J. 1994. *The System of Public Sacrifice in Fourth Century Athens*. Atlanta.
Rothwell, K. S. Jr. 2007. *Nature, Culture and the Origins of Greek Comedy*. Cambridge.
Rougemont, F. 2001. "Some Thoughts on the Identification of the 'Collectors' in the Linear B Tablets," in *Economy and Politics in the Mycenaean Palace States*, ed. S. Voutsaki and J. Killen, 129–38. Cambridge.
———. 2004. "The Administration of Mycenaean Sheep Rearing (Flocks, Shepherds, 'Collectors')," in Frizell 2004, 20–30.
Rousioti, D. 2001. "Did the Mycenaeans Believe in Theriomorphic Divinities?" in Laffineur and Hägg 2001, 305–14.
Roussel, D. 1976. *Tribu et Cité: Études sur les groupes sociaux dans les cités grecques aux époques archaïque et classique (*Annales littéraires de l'Université de Besançon 193). Paris.
Rousset, D. 2002. *Le territoire de Delphes at la terre d'Apollon*. Paris.
Rousset, D., and P. Katzouros. 1992. "Une délimitation de frontière en Phocide." *BCH* 116: 197–215.
Rowse, T. 2001. "Terra nullius," in *The Oxford Companion to Australian History*, ed. G. Davison, J. Hirst, and S. Macintyre, http://www.oxfordreference.com/views/ENTRY.html?subview=Main&entry=t127.e1449.
Roy, C. le. 1984. "Mémoire et tradition. Réflexions sur la continuité," in *Aux origines de l'Hellénisme, La Crète et la Grèce: Hommage à H. Van Effenterre*. Paris.
Roy, J. 2000. "The Frontier between Arkadia and Elis in Classical Antiquity," in *Polis and Politics: Studies in Ancient Greek History*, ed. P. Flensted-Jensen et al., 133–56. Copenhagen.
Rozell, M. J. 2002. *Executive Privilege: Presidential Power, Secrecy, and Accountability*[2]. Lawrence, Kans.
Rozman, G. 1976. *Urban Networks in Russia, 1750–1800, and Premodern Periodization*. Princeton.

Rubin, J. 2005. *Des vaches savantes aux quotas laitiers: Un siecle de mutations dans une société rurale à Sagy, village bressan*. Dijon.

Rudhardt, J. 1992. *Notions fondamentales de la pensée réligieuse et actes constitutifs du culte dans la Grèce classique*[2]. Paris.

Ruipérez, M. S. 1956. "*Ko-re-te-re* et *po-ko-re-te-re* à Pylos: Remarques sur l'organisation militaire mycénienne," in *Études mycéniennes. Actes du colloque international sur les textes mycéniens, Gif-sur-Yvette, 3–7 avril 1956,* ed. M. Lejeune, 105–19. Paris.

Runciman, W. G. 1982. "Origin of States: The Case of Archaic Greece." *Comparative Studies in Society and History* 24: 351–77.

———. 1989. *A Treatise on Social Theory* 2: *Substantive Social Theory.* Cambridge.

———. 1990. "Doomed to Extinction: The Polis as an Evolutionary Dead-End," in *The Greek City: From Homer to Alexander*, ed. O. Murray and S. Price, 347–67. Oxford.

Rundin, J. S. 1996. "A Politics of Eating: Feasting in Early Greek Society." *AJPh* 117: 179–215.

———. 2004. "Pozo Moro, Child Sacrifice, and the Greek Legendary Tradition." *JBL* 123: 425–47.

Runia, E. 2006. "Presence." *History and Theory* 45: 1–29.

Runnels, C. N. , C. Payne, N. V. Rifkind, C. White, N. P. Wolff, and S. A. LeBlanc. 2009. "Warfare in Neolithic Thessaly: A Case Study," *Hesperia* 78: 165–94.

Ruschenbusch, E. 1960. "ΦΟΝΟΣ: Zum Recht Drakons und seiner Bedeutung für das Werden des athenischen Staates." *Historia* 9: 129–54.

———. 1966. *ΣΟΛΩΝΟΣ ΝΟΜΟΙ: Die Fragmente des solonischen Gesetzeswerkes mit einer Text- und Überlieferungsgeschichte* (*Historia* Einzelschriften 9). Wiesbaden.

———. 1982. "Ein Bürgersliste von Koresia und Iulis auf Keos." *ZPE* 48:175–88.

Rutherford, I. 1988. "Pindar on the Birth of Apollo." *CQ* 38: 65–75.

———. 2000a. "The Genealogy of the *Boukoloi*: How Greek Literature Appropriated an Egyptian Narrative-Motif." *JHS* 120: 106–21.

———. 2000b. "Theoria and Darsan: Pilgrimage and Vision in Greece and India." *CQ* n.s. 50: 133–46.

Ryder, T.T.B. 1965. *Koine Eirene: General Peace and Local Independence in Ancient Greece*. London.

Sacconi, A. 2001. "Les repas sacrés dans les textes mycéniens," in Laffineur and Hägg 2001, 467–70.

Sahlins, M. D. 1968. *Tribesmen*. Englewood Cliffs, N.J..

———. 1972. *Stone Age Economics*. Chicago.

Saïd, S. 1979. "Les crimes des pretendants, la maison d'Ulysse et les festins de l'Odyssé." *Etudes de litterature ancienne* 1: 9–49.

Sakellariou, A. 1958. *Les cachets minoens*. Paris.

Sallaberger, W. 2008. *Das Gilgamesch-Epos*. Munich.

Sallares, R. 1991. *The Ecology of the Ancient Greek World*. Ithaca.

Salmon, J. 2001. "Temples the Measure of Men: Public Building in the Greek Economy," in *Economies beyond Agriculture in the Classical World*, ed. D. J. Mattingly and J. Salmon, 195–208. London.

Samons, Loren J. II. 1999. "Aeschylus, the Alkmaeonids and the Reform of the Areopagos." *CJ* 94: 221–33.
Sanchez, P. 2001. *L'Amphictionie des Pyles et de Delphes: Recherches sur son rôle historique, des origines au Iie siècle de notre ère* (*Historia* Einzelschriften 148). Stuttgart.
Sansone, D. 1991. "Kleobis and Biton in Delphi." *Nikephoros* 4: 121–32.
Sartre, M. 1979. "Aspects économiques et aspects religieux de la frontière dans les cités grecques." *Ktema* 4: 213–24.
Sbordone, F. 1941. "Il ciclo italico di Eracle." *Athenaeum* 19: 72–96.
Scafuro, A. C. 2003. "*IG* II^2 204: Boundary Setting and Legal Process in Classical Athens," in *Symposion 1999: Vorträge zur griechischen und hellenistischen Rechtsgeschichte (Pazo de Mariñán, La Coruña, 6–9 September 1999),* ed. G. Thür and F.J.F. Nieto, 123–43. Cologne.
Schachter, A. 1986. *Cults of Boiotia* 2: *Herakles to Poseidon*. London.
Schapera, I., ed. 1965. *Praise Poems of Tswana Chiefs*. Oxford.
Schaps, D. M. 2004. *The Invention of Coinage and the Monetization of Ancient Greece*. Ann Arbor.
Schlaifer, R. 1943. "The Cult of Athena Pallenis: (Athenaeus VI 234-235)." *HSCP* 54: 35–67.
Schmalz, G.C.R. 2006. "The Athenian Prytaneion Discovered?" *Hesperia* 75: 33–81.
Schnapp-Gourbeillon, A. 1981. *Lions, héros, masques: Les interpretations de l'animal chez Homère*. Paris.
Schoep, I. 1999. "Tablets and Territories? Reconstructing Late Minoan IB Political Geography through Undeciphered Documents." *AJA* 103: 201–21.
Schumacher, R.W.M. 1993. "Three Related Sanctuaries of Poseidon: Geraistos, Kalaureia and Tainaron," in Marinatos and Hägg 1993, 62–87.
Schwabe, C. W. 1994. "Animals in the Ancient World," in *Animals and Human Society: Changing Perpectives*, ed. A. Manning and J. Serpell, 36–58. London.
Schweitzer, B. 1922. *Herakles*. Tübingen.
Scullion, S. 2000. "Heroic and Chthonian Sacrifice: New Evidence from Selinous." *ZPE* 132 163–71.
Seaford, R. 1988. "The Eleventh Ode of Bacchylides: Hera, Artemis, and the Absence of Dionysos." *JHS* 108: 118–36.
———. 2004. *Money and the Early Greek Mind: Homer, Philosophy, Tragedy*. Cambridge.
Sealey, R. 1976. *A History of the Greek City States 700–338 B.C.* Berkeley and Los Angeles.
———. 1983. "The Athenian Courts for Homicide." *CP* 78: 275–96.
———. 1987. *The Athenian Republic: Democracy or the Rule of Law?* University Park, Penn., and London, 1987.
———. 1993. *Demosthenes and His Times: A Study in Defeat*. Oxford.
Segal, C. 1992. "Divine Justice in the *Odyssey*: Poseidon, Cyclops and Helios." *AJPh* 113: 489–518.
Seltman, C. 1924. *Athens, Its History and Coinage before the Persian Invasion*. Cambridge.

Shaner, D. E., S. Nagatomo, and Y. Yasuo. 1989. *Science and Comparative Philosophy: Introducing Yuasa Yasuo*. Leiden.
Shapiro, H. A. 1984. "Herakles, Kyknos and Delphi," in *Ancient Greek and Related Pottery: Proceedings of the International Vase Symposium*, ed. H.A.G. Brijder, 271–74. Amsterdam.
———. 1988. "The Marathonian Bull on the Athenian Akropolis." *AJA* 92: 373–82.
Sharpes, D. K. 2006. *Sacred Bull, Holy Cow*. New York.
Shaw, B. D. 1984. "Bandits in the Roman Empire." *P&P* 105: 3–52.
Shear, T. L. Jr. 1978. *Kallias of Sphettos and the Revolt of Athens in 286 BC (Hesperia* Suppl. 17). Princeton.
Shell, M. 1978. *The Economy of Literature*. Baltimore.
———. 1982. *Money, Language, and Thought: Literary and Philosophical Economies from the Medieval to the Modern Era*. Berkeley and Los Angeles.
Shelmerdine, C. W. 1973. "The Pylos Ma Tablets Reconsidered." *AJA* 77: 261–75.
———. 1981. "Nichoria in Context: A Major Town in the Pylos Kingdom." *AJA* 85: 319–25.
Sherratt, A. 1983. "The Secondary Exploitation of Animals in the Old World." *World Archaeology* 15: 90–104.
Shipley, G. 2000. "The Extent of Spartan Territory in the Late Classical and Hellenistic Periods." *ABSA* 95: 367–90.
———. 2004. "Lakedaimon," in *An Inventory of Archaic and Classical Poleis: An Investigation Conducted by the Copenhagen Polis Centre for the Danish National Research Foundation*, ed. M. H. Hansen and T. H. Nielsen, 569–98. Oxford.
Shipton, P. 1989. *Bitter Money: Cultural Economy and Some African Meanings of Forbidden Commodities*. Washington, D.C..
Sick, D. H. 1996. "Cattle, Sacrifice and the Sun: A Mythic Cycle in Greece, Iran, and India." Ph.D. diss., University of Minnesota.
Siimets, Ü. 2007. "The Sun, the Moon and Firmament on Chukchi Mythology and on the Relations of Celestial Bodies and Sacrifice." *Folklore* 32: 129–56.
Sikdar, S. 1982. "Tribalism vs. Colonialism: British Capitalistic Intervention and Transformation of Primitive Economy of Arunachal Pradesh in the Nineteenth Century." *Social Scientist* 10: 15–31.
Silver, M. 1992. *Taking Ancient Mythology Economically*. Leiden.
Simoons, F. J. 1968. *A Ceremonial Ox of India: The Mithan in Nature, Culture, and History, with Notes on the Domestication of Common Cattle*. Madison.
Sineux, P. 2007. *Amphiaraos: Guerrier, devin et guérisseur*. Paris.
Singer, P. 1975. *Animal Liberation*. New York.
Sinn, U. 1992. "The Sacred Herd of Artemis at Lusoi," in *The Iconography of Greek Cult in the Archaic and Classical Periods* (*Kernos* Suppl. 1), ed. R. Hägg, 177–88. Athens and Liège.
Skinner, G. W. 1964. "Marketing and Social Structure in Rural China, Part 1." *Journal of Asian Studies* 24: 3–43.
———. 1965. "Marketing and Social Structure in Rural China, Part 2." *Journal of Asian Studies* 24: 195–228.

Skydsgaard, J. E. 1988. "Transhumance in Ancient Greece," in Whittaker 1988, 35–74.
Slicher van Bath, B. H. 1963. *The Agrarian History of Western Europe, A.D. 500–1850*. Palo Alto.
Sloan, R. E., and M. A. Duncan. 1978. "Zooarchaeology of Nichoria," in *Excavations at Nichoria in Southwest Greece* 1: *Site, Environs, and Techniques*, ed. G. Rapp Jr. and S. E. Aschenbrenner, 60–77. Minneapolis.
Smertenko, C. E. 1932. "The Political Sympathies of Aeschylus." *JHS* 52: 233–35.
Smith, H. S. 1972. "Dates of the Obsequies of the Mothers of the Apis." *Revue d'Égyptologie* 24: 176–87.
Smith, J. Z. 1987. "The Domestication of Sacrifice," in Hamerton-Kelly 1987, 191–205.
Smith, M. S. 2001. *The Origins of Biblical Monotheism: Israel's Polytheistic Background and the Ugaritic Texts*. Oxford.
Smith, W. D., ed. 1990. *Hippokrates: Pseudepigraphic Writings*. Leiden.
Snodgrass, A. M. 1971. *The Dark Age of Greece*. Edinburgh.
———. 1974. "An Historical Homeric Society?" *JHS* 94: 114–25.
———. 1982. "Les origines du culte des héros dans la Grèce antique," in *La mort, les morts dans les sociétés anciennes*, ed. G. Gnoli and J.-P. Vernant, 89–105. Cambridge.
———. 1987. *An Archaeology of Greece: The Present State and Future Scope of a Discipline*. Berkeley and Los Angeles.
———. 2006. "The Economics of Dedication at Greek Sanctuaries." *Scienze dell'antichità* 3–4 (1989–90): 287–94. Reprinted in *Archaeology and the Emergence of Greece*. Ithaca, 2006.
Sobocinski, M., and D. Makowiecki. 1991. "Current State of Knowledge of Archaeozoological Materials in the Globular Amphorae Culture in Kuiavia," in *New Tendencies in Studies of Globular Amphorae Culture*, ed. A. Cofta-Broniewska, 143–53. Warsaw.
Sokolowski, F. 1954. "Fees and Taxes in the Greek Cults." *HTR* 47: 153–64.
Soles, J. 1995. "The Functions of a Cosmological Center: Knossos in Palatial Crete," in Laffineur and Niemeier, 404–14.
Sorabji, R. 1993. *Animal Minds and Human Morals: The Origins of the Western Debate*. London.
Sourvinou-Inwood, C. 1979. "The Myth of the First Temples at Delphi." *CQ* n.s. 29: 231–51.
———. 1988. *Studies in Girls' Transitions: Aspects of the Arkteia and Age Representation in Attic Iconography*. Athens.
———. 1990. "Ancient Rites and Modern Constructs: On the Brauronian Bears Again." *BICS* 37: 1–14.
———. 1991. "Myth as History: The Previous Owners of the Delphic Oracle," in *"Reading" Greek Culture: Texts and Images, Rituals and Myths*. Oxford, 217–43.
Sowa, C. 1984. A. *Traditional Themes and the Homeric Hymns*. Chicago.
Spear, T. T. 1993. "Being 'Maasai,' but not 'People of Cattle': Arusha Agricultural Maasai in the Nineteenth Century," in Spear and Waller 1993, 120–36.

Spear, T. and R. Waller, ed. 1993. *Being Maasai: Ethnicity and Identity in East Africa*. London.

Speciale, M. S. 1999. "La tavoletta PY Ta 716 e i sacrifici di animali," in *Επὶ πόντον πλαζόμενοι: Simposio italiano di studi egei dedicato a Luigi Bernabò Brea e Giovanni Pugliesi Caratelli*, ed. V. La Rosa, D. Palermo, and L. Vagnetti, 290–97. Rome.

Spenser, E. 1934. *A View of the Present State of Ireland*, ed. W. L. Renwick. London.

Stanford, C. B., and H. T. Bunn, eds. 2001. *Meat-Eating and Human Evolution*. Oxford.

Stanley, P. 1995. *Robert Bakewell and the Longhorn Breed of Cattle*. Ipswich.

Stanner, W.E.H. 1963. *On Aboriginal Religion*. Sydney.

Stehle, E. 1997. *Performance and Gender in Ancient Greece: Nondramatic Poetry in its Setting*. Princeton.

Steiner, D. 2002. "Indecorous Dining, Indecorous speech: Pindar's First *Olympian* and the Poetics of Consumption." *Arethusa* 35: 297–314.

Stewart, R. 2004. *The Places in Between*. London.

Stocker, S. R., and J. L. Davis. 2004. "Animal Sacrifice, Archives and Feasting at the Palace of Nestor," in Wright 2004b, 59–76.

Strøm, I. 1992. "Obeloi of Pre-or Proto-monetary Value in the Greek Sanctuaries." In Linders and Alroth 1992, 41–51.

Stroud, R. S. 1968. *Drakon's Law on Homicide*. Berkeley and Los Angeles.

———. 1998. *The Athenian Grain Tax Law of 374/3 BC* (*Hesperia* Suppl. 29). Princeton.

Sutherland, C.H.V. 1974. *Roman Coins*. New York.

Sutton, J.E.G. 1993. "Becoming Maasailand," in Spear and Waller 1993, 38–60.

Svoboda, B., and D. Concev. 1956. *Neue Denkmäler antiker Toreutik* (Monumenta Archaeologica: Acta Praehistorica et Historica Instituti Archaeologici Academiae Scientiarum Bohemoslovenicae 4). Prague.

Taillardat, J. 1960. "Notules mycéniennes: Mycénienne *ko-re-te* et homérique *Καλήτωρ*." *REG* 73: 1–5.

Tandy, D. W. 1997. *Warriors into Traders: The Power of the Market in Early Greece*. Berkeley and Los Angeles.

Tarditi, G. 1956. "La nuova epigrafe archilochea e la tradizione biografica del poeta." *PP* 11: 122–39.

Tartaron, T. F. 2001. "Glykys Limin: A Mycenaean Port of Trade in Southern Epirus?" in *Prehistory and History: Ethnicity, Class and Political Economy*, ed. D. W. Tandy, 1–40. Montreal.

Tausend, K. 1986. "Die Koalitionen im 1. Heiligen Krieg." *RSA* 16: 49–66.

Taylor, C.C.W., ed. 1997. *From the Beginning to Plato*. London.

Teffeteller, A. 2001. "The Chariot Rite at Onchestos: *Homeric Hymn to Apollo* 229–38." *JHS* 121: 159–66.

Thalmann, W. G. 2004. " 'The Most Divinely Approved and Political Discord': Thinking about Conflict in the Developing Polis." *ClAnt* 23: 359–99.

Theodorou-Mavrommatidi, A. 2004. "An Early Helladic Settlement in the Apollon Maleatas Site at Epidauros," in *Die ägäische Frühzeit: Forschungsbericht 1975–2002 2.2: Die Frühbronzezeit in Griechenland, mit Ausnahme von Kreta*

(Veröffentlichungen der Mykenischen Kommission 21), ed. E. Alram-Stern, 1167–82. Vienna.
Thieme, P. 1951. "Etymologische Vexierbilder." *Zeitschrift für vergleichende Sprachforschungen* 69: 177–78.
Thomas, C. G., and C. Conant. 1999. *Citadel to City-State: The Transformation of Greece, 1200–700 B.C.E.* Bloomington.
Thomas, I., ed. 1993. *Greek Mathematical Works* 2. London.
Thomas, R. 1996. "Written in Stone? Liberty, Equality, Orality and the Codification of Law," in Foxhall and Lewis 1996, 9–31.
Thompson, M. 1942. "Coins for the Eleusinia." *Hesperia* 11: 213–29.
Thür, G. 1996. "Oaths and Dispute Settlement in Ancient Greek Law," in Foxhall and Lewis 1996, 57–72.
Torr, C. 1885. *Rhodes in Ancient Times*. Cambridge.
Trinkhaus, E. 1985. "Cannibalism and Burial at Krapina." *Journal of Human Evolution* 14: 203–16.
Troy, C. S., et al. 2001. "Genetic Evidence for Near Eastern Origins of European Cattle." *Nature* 410: 1088–91.
Tuchelt, K. 1962. *Tiergefässe in Kopf-und Protomengestalt: Untersuchungen zur Formengeschichte tierförmiger Giessgefässe*. Berlin.
———. 1992. "Tieropfer in Didyma—Ein Nachtrag." *AA* 1992: 61–81.
Tulin, A. 1996. *Dike Phonou: The Right of Prosecution and Attic Homicide Procedure* (Beiträge zur Altertumskunde 76). Stuttgart and Leipzig.
Tullock, G. 1975. "Competing Monies." *Journal of Money, Credit and Banking* 7: 491–97.
Ustinova, Y. 1996. "Orgeones in Phratries: A Mechanism of Social Integration in Attica." *Kernos* 9: 227–42.
Van de Mieroop, M. 2005. "The Eastern Mediterranean in Early Antiquity." In *Rethinking the Mediterranean*, ed. W. V. Harris. 117–40. Oxford.
Vandenabeele, F., and J.-P. Olivier. 1979. *Les Idéogrammes archéologiques du Linéaire B (Ét. Cret.)*. Paris.
Van Straten, F. 1987. "Greek Sacrificial Representations: Livestock Prices and Religious Mentality." In Linders and Nordquist 1987, 159–70.
———. 1988. "The God's Portion in Greek Sacrificial Representations: Is the Tail Doing Nicely?" in Hägg et al. 1988, 51–68.
Vardi, I. 1998. "Archimedes' Cattle Problem." *The American Mathematical Monthly* 105: 305–19.
Veldhuis, N. 1991. *A Cow of Sîn: Library of Oriental Texts*. Groningen.
Ventris, M., and J. Chadwick. 1956. *Documents in Mycenaean Greek: 300 Selected Tablets from Knossos, Pylos and Mycenae*. Cambridge.
Vernant, J. 1981. "Théorie générale du sacrifice et mise à mort dans la *Thusia* grecque," in *Le sacrifice dans l'antiquité. Entretiens sur l'antiquité classique* 27. Geneva.
———. 1982. *The Origins of Greek Thought*. Ithaca.
Verrall, A. W. 1894. "The Hymn to Apollo: An Essay in the Homeric Question." *JHS* 14: 1–29.
Verrall, M. de G. 1890. *Mythology and Monuments of Ancient Athens*. London.

Vigne, J.-D., D. Helmer, and J. Peters, eds. 2005. *The First Steps of Animal Domestication: New Archaeological Approaches*. Oxford.

Vilatte, S. 1988. "Apollon-le dauphin et Poséidon l'Ébranleur: Structure familiale et souveraineté chez les Olympiens; à propos du sanctuaire de Delphes," in *Mélanges Pierre Lévêque* 1: *Religion*, ed. M.-M. Mactoux and E. Geny, 308–30. Paris.

Villa, P., J. Courtin, et al. 1986. "Un cas de cannibalisme au Néolithique: Boucherie et rejet de restes humains et animaux dans la grotte de Fontbregoua (Salernes, Var)." *Gallia Préhistoire* 29: 143–71.

Visser, M. 1982. "Worship Your Enemy: Aspects of the Cult of Heroes in Ancient Greece." *HTR* 75: 403–28.

Vokotopoulou, J. P. 1973. *Οδηγος Μουσειου Ιωαννινων*. Athens.

Volk, K. 2002. "ΚΛΕΟΣ ΑΦ ΘΙΤΟΝ Revisited." *CP* 97: 61–68.

Voyatzis, M. 1990. *The Early Sanctuary of Athena Alea at Tegea*. Göteborg.

Vulpe, N. 1994. "Irony and the Unity of the Gilgamesh Epic." *Journal of Near Eastern Studies* 53: 275–83.

Wade-Gery, W. T. 1948. "What Happened in Pylos." *AJA* 52: 115–18.

Wagner-Hasel, B. 2000. *Der Stoff der Gaben: Kultur und Politik des Schenkens und Tauschens im archaischen Griechenland*. Frankfurt.

Wagstaff, J. N. 1981. "Buried Assumptions: Some Problems in the Interpretation of the 'Younger Fill' Raised by Recent Data from Greece." *Journal of Archaeological Science* 8: 247–64.

———, ed. 1987. *Landscape and Culture: Geographical and Archaeological Perspectives*. Oxford.

Wahl, J., and H.-G. König. 1987. "Anthropologisch traumatologisch Untersuchung der menschlichen Skelettreste aus dem bandkeramischen Massengrab bei Talheim, Kreis Heilbronn." *Funderberichte aus Baden-Württemberg* 12: 65–193.

Walbank, M. B. 1983. "Leases of Sacred Properties in Attica, Part I." *Hesperia* 52: 100–135.

———. 1991. "Leases of Public Lands," in *Inscriptions: Horoi, Poletai, Leases of Public Lands* (*The Athenian Agora* XIX), by G. V. Lalonde, M. K. Langdon, and M. B. Walbank. Princeton.

———. 1994. "A *Lex Sacra* of the State and of the Deme of Kollytos." *Hesperia* 63: 233–39.

Walcot, P. 1979. "Cattle Raiding, Heroic Tradition, and Ritual: The Greek Evidence." *History of Religions* 18: 326–51.

Waldstein, C. 1902. *The Argive Heraeum* 1. Boston.

Wallace, P. W. 1979. *Strabo's Description of Boiotia: A Commentary*. Heidelberg.

Wallace, R. W. 1989. *The Areopagos Council to 307 BC*. Baltimore.

Walter, H. 1993. *Ägina: Die archäologische Geschichte einer griechischen Insel*. Munich.

Ward, A., ed. 1970. *The Search for Theseus*. New York.

Wärnlöf, C. 2000. "The Politics of Death: Demarcating Land through Ritual Performance," in *People, Cattle and Land: Transformations of a Pastoral Society in Southwestern Africa*, ed. M. Bollig and J.-B. Gewald, 449–69. Cologne.

Warren, P. 1979. "The Miniature Fresco from the West House at Akrotiri, Thera, and Its Aegean Setting." *JHS* 99: 115–29.
Watrous, L. B. 1984. "Ayia Triada: New Perspectives on the Minoan Villa." *AJA* 88: 123–34.
Wayne, R. K. 2001. "Consequences of Domestication: Morphological Diversity of the Dog," in *The Genetics of the Dog*, ed. A. Ruvinsky and J. Sampson, 43–60. Oxford.
Weber, M. 1921. "Die Stadt. Eine Soziologische Untersuchung." *Archiv von Sozialwissenschaft und Sizialpolitik* 47: 621–772.
Wees, H. van. 1992. *Status Warriors: War, Violence, and Society in Homer and History.* Amsterdam.
———. 2000. "Megara's Mafiosi: Timocracy and Violence in Theognis," in *Alternatives to Athens: Varieties of Political Organization and Community in Ancient Greece*, ed. R. Brock and S. Hodkinson, 52–67. Oxford.
Wendorf, F., ed. 1968. *The Prehistory of Nubia.* Dallas.
Wengrow, D. 2001. "Rethinking 'Cattle Cults' in Early Egypt: Toward a Prehistoric Perspective on the Narmer Palette." *CArchJ* 11: 91–104.
West, M. L. 1997. *The East Face of Helicon.* Oxford.
———. 2002. " 'Eumelos': A Corinthian Epic Cycle?" *JHS* 122: 109–33.
West, S. 1984a. "Io and the Dark Stranger." *CQ* n.s. 34: 292–302.
———. 1984b. "Lycophron on Isis." *JEA* 70: 151–54.
Westbrook, R. 2002. "Penelope's Dowry and Odysseus' Kingship," in *Women and Property in Ancient Near Eastern and Mediterranean Societies*, ed. D. Lyons and R. Westbrook, 1–18. Cambridge, Mass.
White, K. D. 1954. "The Sacred Grove: A Comparative Study of Some Parallel Aspects of Religious Ritual in Ancient Crete and the Near East." *G&R* 1: 112–27.
Whitley, A.J.M. 1988. "Early States and Hero Cults: A Re-appraisal." *JHS* 108: 173–82.
———. 1991a. "Social Diversity in Dark Age Greece." *ABSA* 86: 341–65.
———. 1991b. *Style and Society in Dark Age Greece: The Changing Face of a Pre-literate Society 1100-700 BC.* Cambridge.
———. 2001. *The Archaeology of Ancient Greece.* Cambridge.
———. 2002. "Objects with Attitude: Biographical Facts and Fallacies in the Study of Late Bronze Age and Early Iron Age Warrior Graves." *CArchJ* 12: 217–32.
Wijngaarden, G. J. van. 2001. *Use and Appreciation of Mycenaean Pottery in the Levant, Cyprus and Italy (1600–1200 BC).* Amsterdam.
Wilkins, J. 1997. "Comic Cuisine: Food and Eating in the Comic Polis," in *The City as Comedy: Society and Representation in Athenian Drama*, ed. G. W. Dobrov, 250–69. Chapel Hill.
———. 2000. *The Boastful Chef: The Discourse of Food in Ancient Greek Comedy.* Oxford.
Willetts, R. F. 1959. "The Myth of Glaucus and the Cycle of Birth and Death." *Klio* 37: 21–28.
Willis, R. G., and P. Curry. 2004. *Astrology, Science and Culture: Pulling Down the Moon.* Oxford.

Wilson, P. 2009. "Tragic Honours and Democracy: Neglected Evidence for the Politics of the Athenian Dionysia." *CQ* 59: 8–29.
Wilson, S. M., L. W. Ngige, and L. J. Trollinger. 2003. "Kamba and Maasai Paths to Marriage in Kenya," in *Mate Selection across Cultures*, ed. B. B. Ingoldsby and R. R. Hamon, 95–118. Thousand Oaks, Calif.
Winand, J. 1990. *Les Hiérothytes: Recherche institutionelle* (Académie royale de Belgique: Mémoires de la Classe des lettres 2e série 4). Brussels.
Winkler, J. 1980. "Lollianos and the Desparadoes." *JHS* 100: 151–81.
Wolff, H. J. 1980. "Vorgeschichte und Entstehung des Rechtsbegriff im frühen Griechentum," in *Entstehung und Wandel rechtlicher Traditionen: Historische Anthropologie* II, ed. W. Fikentscher et al., 557–79. Freiburg and Munich.
Wright, J. C. 2004a. "A Survey of Evidence for Feasting in Mycenaean Society," in Wright 2004b, 13–58.
———, ed. 2004b. *The Mycenaean Feast.* Princeton.
Wyatt, W. F. Jr. 1962. "The Ma Tablets from Pylos." *AJA* 66: 21–41.
Yates, F. A. 1966. *The Art of Memory.* Chicago.
Yorke, V. W. 1896. "Excavations at Abae and Hyampolis in Phocis." *JHS* 16: 291–312.
Younger, J. G. 1976. "Bronze Age Representations of Aegean Bull-Leaping." *AJA* 80: 125–37.
Yunis, H. 2005. "The Rhetoric of Law in 4th Century Athens," in *The Cambridge Companion to Ancient Greek Law*, ed. M. Gagarin and D. J. Cohen, 191–210. Cambridge.
Zervas, G. 1998. "Quantifying and Optimizing Grazing Regimes in Greek Mountain Systems." *Journal of Applied Ecology* 35: 983–86.
Zich, B. 1992. "Ausgrabungen auf dem stein- und bronzezeitlichen Grabhügelfeld von Flintbek, Kreis Rendsburg-Eckernförde: Ein Vorbericht." *Archäologische Nachrichten aus Schleswig-Holstein* 3: 6–21.
Zimmermann, K. 2000. "Späthellenistische Kultpraxis in einer karischen Kleinstadt: Eine neue *lex sacra* aus Bargylia." *Chiron* 30: 451.
Zumbrunnen, J. 2004. "Elite Domination and the Clever Citizen: Aristophanes' *Acharnians* and *Knights*." *Political Theory* 32: 656–77.

Index